Essays about Modern Algebra

From Classical Topics to New Problems

Franz Rothe

Introduction

The *front cover* shows the vertices of a regular 257-gon, connected in the ordering

$$r \mapsto \exp\left[\frac{2\pi i \cdot 3^r}{257}\right]$$

with $r = 0, 1, 2, \ldots 255$. See also formula (II.0.8). Such an ordering of the vertices is used for the construction of the Gaussian periods. On these sums or periods is based the derivation of a closed formula for the coordinates of the vertices, and hence in the end, is proved that the regular 257-gon is constructible with straightedge and compass, in Euclidean geometry.

The *back cover* shows formulas for the x-coordinate of the first vertex in the upper half plane of a regular 17-gon with circum radius 16. The first formula is the one given by Gauss. The second formula I have obtained by my simplified method for the regular 17-gon, which shall be explained in the section about the 17-gon starting on page 39. For the simplified method, one does not need to know anything about the Gaussian periods, instead one uses the trigonometric addition theorem and clever guesses.

Preface

This book has been growing over the years out of several resources. Firstly, the courses in number theory and modern algebra I had been teaching at UNC Charlotte. Secondly, my interest to invent mathematical problems for competitions, and individual work with gifted students. In the end, the shear size of material I had gathered made it convenient and necessary to take a break, and think about publication.

The material is much more concrete and computational than a course in modern algebra usually is. I do not try to deliver a complete course. The danger to sound to elaborate at the beginning is avoided. Instead the book is encouraging own creativity as soon as possible, on a level accessible to a good student, and not only adressing the specialists. Besides complete proofs, simple as well as advanced problems, finally computation on all levels are elaborated. Many generally known as well own problems about the cyclotomic and other special polynomials, formulas for π, are solved completely.

Secondly, the book concentrates on the part of modern algebra that is related to geometry, and especially geometric constructions. The construction of Fermat polygons is done via computations with mathematica, and a second computer language, finally numeric calculations for check of correctness. Luckily enough, with these means, I could completely work out the constructions of the regular 17, 257 and even the 65 537-gon.

I prove the basic theorems on totally real and totally positive algebraic numbers. These numbers come up in the geometric construction by Hilbert tools,—which are a bid more restrictive tools than the classical tools compass and straightedge. The theory of geometric constructions with different sets of tools,—starting from Hilbert tools, via the classical compass and straightedge, and finally the use of two-marked straightedge and their treatment with Galois theory,—are explained in all details. The Galois theory is treated in Artin's elegant approach via characters. Thus both characteristic zero and prime characteristic are covered. The solvability of equations is treated as far as it is related to geometric constructions, and tailored to prove the steps needed is that context, as simple as possible. Much less known is the relation between the lunes of Hippocrates and transcendental numbers to which the final section is dedicated.

Here is a paragraph by Hermann Weyl fitting nicely to the intention of the present book.

> Important though the general concepts and propositions may be with which the modern and industrious passion for axiomatizing and generalizing has presented us, in algebra perhaps more than everywhere else, nevertheless I am convinced that the special problems in all their complexity constitute the stock and core of mathematics, and that to master their difficulties requires on the whole the harder labor.

MSC2020-Mathematics Subject Classification

11-01 Introductory exposition (textbooks, tutorial papers, etc.) pertaining to number theory

11-04 Software, source code, etc. for problems pertaining to number theory

12-01 Introductory exposition (textbooks, tutorial papers, etc.) pertaining to field theory

12D10 Polynomials in real and complex fields: location of zeros (algebraic theorems)

12D15 Fields related with sums of squares (formally real fields, Pythagorean fields, etc.)

12E05 Polynomials in general fields (irreducibility, etc.)

12F10 Separable extensions, Galois theory

About the author: Franz Rothe graduated from high school in Karlsruhe and studied mathematics, physics and music there. Graduated with a diploma in mathematics from the E T H Zürich, a doctorate in Tübingen. He got his Habilitation and venia legendi from the university at Tübingen (1984) and the Ludwig Maximilian university of Munich (1988). He has published about 40 articles, and meanwhile three books about mathematics. Since 1990 a professorship at the University of North Carolina at Charlotte, USA. His is now retired because of health problems.

In addition, Rothe and pianist Thomas Turner have developed a repertoire of classical music for flute and piano, and have recorded and released three CDs. This collection also contains several of their own transcriptions.

Some recent information about activity in all fields, is gathered on the website
https://franzrothe.com

Contents

IX   The Lunes of Hippocrates and Transcendental Numbers   417

Part I

Some Number Theory

I.1 The Euclidean Algorithm

Definition 1 (Division with remainder). Given any two integers a and positive integer $b > 0$, the *quotient* q and *remainder* r are defined as the integers such that

$$(\text{I.1.1}) \qquad a = qb + r \quad \text{and} \quad 0 \le r < b$$

In that case, we write

$$(\text{I.1.2}) \qquad a : b = q \quad \text{rem } r$$

and

$$(\text{I.1.3}) \qquad a \equiv r \mod b$$

Definition 2 (Greatest common divisor). The *greatest common divisor* of two positive integers a and b is the greatest positive integer that is a divisor of both a and b. We denote the greatest common divisor of a and b by $\gcd(a, b)$.

Definition 3 (Least common multiple). The *least common multiple* of two positive integers a and b is the positive integer l such that

(a) l is a multiple of both a and b.

(b) If any other integer k is a multiple of both a and b, then k is a multiple of l.

We denote the least common multiple of a and b by $\text{lcm}(a, b)$.

Intuitively, the greatest common divisor is the <u>greatest common measure</u> for the lengths a and b. The least common multiple is the <u>least common period</u> of two simultaneous processes with individual periods a and b.

I.1.1 The Common Euclidean Algorithm

Let a, b be positive integers. The greatest common divisor $\gcd(a, b)$ can be calculated by successive divisions with remainders. The algorithm starts with dividing a by b. The last nonzero remainder is the greatest common divisor.

Example I.1.1. *Take $a = 42, b = 16$.*

$$
\begin{array}{rcl}
42 : 16 &=& 2 \text{ rem } 10 \\
16 : 10 &=& 1 \text{ rem } 6 \\
10 : 6 &=& 1 \text{ rem } 4 \\
6 : 4 &=& 1 \text{ rem } 2 \\
4 : 2 &=& 2 \text{ rem } 0
\end{array}
$$

The last nonzero remainder is 2, and hence $\gcd(42, 16) = 2$.

4

Reason. Let subscript i count the rows. One start with the given numbers as remainders $r_0 = a$ and $r_1 = b$. The i-th row of the scheme is

$$(\text{I.1.4}) \qquad r_{i-1} : r_i = q_i \quad \text{rem } r_{i+1}$$

where the successive remainders r_i and quotients q_i satisfy

$$(\text{ri}) \qquad r_{i+1} = r_{i-1} - q_i r_i \quad \text{and} \quad 0 \le r_{i+1} < r_i$$

The algorithm stops when a zero remainder $r_{m+1} = 0$ appears for the first time, say in row m. The last nonzero remainder is r_m. We check that r_m is the greatest common divisor. From the identity

$$(\text{I.1.5}) \qquad \gcd(a, b) = \gcd(a - qb, b) = \gcd(b, a - qb)$$

which holds for all integers q, we get inductively

$$\gcd(a, b) = \gcd(r_0, r_1)$$
$$\gcd(a, b) = \gcd(r_{i-1}, r_i) = \gcd(r_i, r_{i+1}) \quad \text{for all } i = 1, 2 \ldots m$$
$$\gcd(a, b) = \gcd(r_m, r_{m+1}) = r_m$$

Hence the last nonzero remainder r_m is the greatest common divisor. $\qquad\square$

Problem 1. *Find the greatest common divisor* $\gcd(4321, 1234)$.

I.1.2   The Extended Euclidean Algorithm

Definition 4 (Integer combination). Any number $sa + tb$ with (positive or negative) integers $s, t \in \mathbf{Z}$ is called an *integer combination* of a and b.

Again, a, b are positive integers. Beyond calculating their greatest common divisor, the extended Euclidean algorithm yields $\gcd(a, b)$ as an integer combination

$$(*) \qquad \gcd(a, b) = sa - tb$$

Actually, the convenient smallest solution s, t of equation (I.2.10) is calculated by the extended Euclidean algorithm. Furthermore, $(-1)^m s \ge 0$ and $(-1)^m t \ge 0$, where m is the number of steps of the algorithm.

Definition 5 (The Extended Euclidean Algorithm). Similar to the common algorithm, one starts with dividing a by b, followed by successive divisions with remainders. One has to keep track of both the quotients q_i and remainders r_i. But the extended algorithm needs two extra columns, which start in row 0 and 1 with the 2×2 unit matrix, whereas the divisions start at row 1 with the calculation of $a : b$. The operation to successively produce these two extra columns is *row two above current row plus quotient × row one above current row ↦ gives current row.*

The algorithm stops when a division has zero remainder for the first time. The greatest common divisor $\gcd(a, b)$ is the last nonzero remainder. In the m-th row, adjacent to the zero remainder r_{m+1}, appear the numbers s_m and t_m such that

$$(\text{*m}) \qquad\qquad (-1)^m \cdot \gcd(a, b) = s_m a - t_m b$$

and hence (I.2.10) follows with $s = (-1)^m s_m$ and $t = (-1)^m t_m$. The optional extra row $m + 1$ does not contain a division, only s_{m+1} and t_{m+1} are calculated. Because of the relations

$$(\text{I.1.6}) \qquad\qquad s_{m+1} = \frac{b}{\gcd(a, b)}, \qquad t_{m+1} = \frac{a}{\gcd(a, b)}$$

shown in Corollary 3 below, we get a convenient check.

Example I.1.2. *Take $a = 42, b = 16$.*

row 0:						1	0
row 1:	42 :	16 =	2	rem	10	0	1
row 2:	16 :	10 =	1	rem	6	1	2
row 3:	10 :	6 =	1	rem	4	1	3
row 4:	6 :	4 =	1	rem	2	2	5
row m=5:	4 :	2 =	2	rem	0	3	8
row m+1:						8	21

Hence $\gcd(42, 16) = 2$ and $3 \cdot 42 - 8 \cdot 16 = -2 = -\gcd(42, 16)$. We get $s = -3, t = -8$. These numbers are negative since the number $m = 5$ of steps of the algorithm is odd in this example.

The optional extra row $m+1$ does no longer involve a division step of the ordinary Euclidean algorithm. The extra calculation of s_{m+1} and t_{m+1} is a convenient check. s_{m+1} and t_{m+1} are calculated. Because of the relations

$$(\text{I.2.1}) \qquad\qquad s_{m+1} = \frac{b}{\gcd(a, b)}, \qquad t_{m+1} = \frac{a}{\gcd(a, b)}$$

shown in Corollary 3 below, we get a convenient check.

Reason for the extended algorithm. Let subscript i count the rows. Starting with the given numbers as remainders $r_0 = a$ and $r_1 = b$, the i-th row of the extended scheme is

$$(\text{i-extend}) \qquad\qquad r_{i-1} : r_i = q_i \quad \text{rem } r_{i+1} \qquad s_i \ t_i$$

The two extra columns start with $s_0 = 1, s_1 = 0$ and $t_0 = 0, t_1 = 1$. The extended Euclidean algorithm calculates the successive remainders r_i and quotients q_i via

$$(\text{I.1.7}) \qquad r_{i+1} = r_{i-1} - q_i r_i \ \text{ and } \ 0 \le r_{i+1} < r_i$$

and uses the recursion to get the sequences s_i and t_i:

$$(\text{I.1.8}) \qquad \begin{aligned} s_{i+1} &= s_{i-1} + q_i s_i \\ t_{i+1} &= t_{i-1} + q_i t_i \end{aligned}$$

Since we get the same two step recursion

$$(-1)^{i+1} r_{i+1} = (-1)^{i-1} r_{i-1} + q_i \, (-1)^i r_i$$

for the alternating remainders $(-1)^i \, r_i$, we show inductively

$$(\text{combine}) \qquad (-1)^i \, r_i = s_i a - t_i b$$

to hold for all rows $i = 0, 1, \ldots m{+}1$. Indeed equation (combine) holds for $i = 0, 1$, and inductively follows for all i, because of formulas (I.1.7) and (I.1.8). The algorithm stops when a zero remainder $r_{m+1} = 0$ appears for the first time in row m. As explained above, the last nonzero remainder is the greatest common divisor.

$$(\text{I.1.9}) \qquad r_m = \gcd(a, b), \ r_{m+1} = 0$$

The identity (combine) with $i = m$ implies

$$(-1)^m \, r_m = s_m a - t_m b$$

hence formula (*m) holds. The numbers s_m, t_m show up next to the zero remainder $r_{m+1} = 0$. For odd number m steps $s = -s_m, t = -t_m$ are negative, for even m, we get $s = s_m, t = t_m$ which are positive. $\qquad \square$

Remark. The above use of the identity (combine) with $i = m$ has been taken from a remark in the article [24].

Problem 2. *Express the greatest common divisor* $\gcd(4321, 1234)$ *as integer combination*

$$(\text{I.2.10}) \qquad \gcd(a, b) = sa - tb$$

of these two given integers.

Answer.

row 0:						1	0
row 1:	$4231:$	$1234 =$	3 rem	619		0	1
row 2:	$1234:$	$619 =$	1 rem	615		1	3
row 3:	$619:$	$615 =$	1 rem	4		1	4
row 4:	$615:$	$4 =$	153 rem	3		2	7
row 5:	$4:$	$3 =$	1 rem	1		307	1 075
row m=6:	$3:$	$1 =$	3 rem	0		309	1 082
row m+1:						1234	4321

Hence $\gcd(4321, 1234) = 1$ and $309 \cdot 4321 - 1\,082 \cdot 1234 = 1 = \gcd(4321, 1234)$. We get $s = 309, t = 1\,082$. These numbers are positive since the number $m = 6$ of steps of the algorithm is even.

Remark (Backtracking "the old fashioned way"). For completeness I mention the backtracking algorithm, which can be used to produce formula (I.2.10), too. At first one does the common Euclidean algorithm, keeping the quotients. But as the price for not planning ahead, one has to remember the two entire sequences of remainders and quotients. Going backwards, one uses the divisions of the algorithm in reversed order and calculates expressions for the greatest common divisor in terms of two successively larger remainders.

Example I.1.3. *Get integers s, t such that $s42 - t16 = \gcd(42, 16)$.*

$$\begin{aligned}
\gcd(42, 16) = 2 = \underline{6} - \underline{4} \qquad &= \underline{6} - (\underline{10} - \underline{6}) \\
= -\underline{10} + 2 \cdot \underline{6} \qquad &= -\underline{10} + 2 \cdot [\underline{16} - \underline{10}] \\
= 2 \cdot \underline{16} - 3 \cdot \underline{10} \quad &= 2 \cdot \underline{16} - 3 \cdot [\underline{42} - 2 \cdot \underline{16}] \\
= -3 \cdot \underline{42} + 8 \cdot \underline{16}
\end{aligned}$$

from which we see one (non unique) solution $s = -3, t = -8$ of formula (I.2.10).

Problem 3. *Use the old fashioned backtracking to express the greatest common divisor $\gcd(4321, 1234) = 1$ as integer combination.*

Answer.

$$\begin{aligned}
\gcd(4321, 1234) &= 1 \\
= \underline{4} - \underline{3} \qquad\qquad &= \underline{4} - [\underline{615} - 153 \cdot \underline{4}] \\
= -\underline{615} + 154 \cdot \underline{4} \qquad &= -\underline{615} + 154 \cdot [\underline{619} - \underline{615}] \\
= 154 \cdot \underline{619} - 155 \cdot \underline{615} \quad &= 154 \cdot \underline{619} - 155 \cdot [\underline{1234} - \underline{619}] \\
= -155 \cdot \underline{1234} + 309 \cdot \underline{619} &= -155 \cdot \underline{1234} + 309 \cdot [\underline{4321} - 3 \cdot \underline{1234}] \\
= 309 \cdot \underline{4321} - 1082 \cdot \underline{1234}
\end{aligned}$$

8

from which we see one (non unique) solution $s = 309, t = 1082$. such that $s4321 - t1234 = \gcd(4321, 1234)$.

Problem 4. *Calculate the greatest common divisor of 765 and 567. Find integers s and t such that* $\gcd(765, 567) = s \cdot 765 - t \cdot 567$.

Answer. The extended Euclidean algorithms is used to calculate the greatest common divisor. In an additional parallel calculation, one gets the greatest common divisor as integer combination of the two given numbers.

row 0:							1	0
row 1:	765	567 $=$	1	rem	198		0	1
row 2:	567 :	198 $=$	2	rem	171		1	1
row 3:	198 :	171 $=$	1	rem	27		2	3
row 4:	171 :	27 $=$	6	rem	9		3	4
row 5:	27 :	9 $=$	3	rem	0		20	27
row 5+1:							63	85

Indeed, $\gcd(765, 567) = 9 = (-20) \cdot 765 + 27 \cdot 567$. Hence $s = -20$ and $t = -27$.

Remark. The optional extra row $5 + 1$ does not contain a division, only s_{M+1} and t_{M+1} are calculated. This is a convenient check, since $63 \cdot 765 - 85 \cdot 567 = 0$.

Problem 5. *Use the last problem to calculate the least common multiple* $lcm(765, 567)$.

Answer.
$$lcm(765, 567) = 765 \cdot \frac{567}{\gcd(765, 567)} = 765 \cdot 63 = 48\,195$$

Problem 6. *Calculate the greatest common divisor of 367 and 47. Find integers s and t such that* $\gcd(367, 47) = s \cdot 367 + t \cdot 47$.

Answer. The extended Euclidean algorithms is used to calculate the greatest common divisor. In an additional parallel calculation, one gets the greatest common divisor as integer combination of the two given numbers. Here is the example:

row 0:							1	0
row 1:	367 :	47 $=$	7	rem	32		0	1
row 2:	47 :	38 $=$	1	rem	9		1	7
row 3:	38 :	9 $=$	4	rem	2		1	8
row 4:	9 :	2 $=$	4	rem	1		5	39
row 5:	2 :	1 $=$	2	rem	0		21	164
row 5+1:							47	367

Indeed, $\gcd(367, 74) = 1 = (-21) \cdot 367 + 164 \cdot 47$. The optional extra row $5 + 1$ does not contain a division, only s_{M+1} and t_{M+1} are calculated. This is a convenient check, since $47 \cdot 367 - 367 \cdot 47 = 0$.

I.1.3 Further Properties of the Extended Algorithm

The extended Euclidean algorithm is of basic importance, and deserves further observations and Corollaries.

Corollary 1. *The linear diophantine equation*

$$(I.2.10) \qquad\qquad \gcd(a, b) = sa - tb$$

has the set of integer solutions s', t'

$$(I.1.10) \qquad\qquad s' = s + \lambda\frac{b}{\gcd(a,b)} \ , \quad t' = t + \lambda\frac{a}{\gcd(a,b)}$$

with arbitrary integer λ.

Reason. It is easy to check that formula (I.1.10) gives integer solutions of equation (I.2.10). Conversely, suppose that both s, t and s', t' are solutions of (I.2.10):

$$sa - tb = \gcd(a, b)$$
$$s'a - t'b = \gcd(a, b)$$

Multiplying the first equation above by t', the second one by t and subtracting yields the first equation below, multiplying the first equation by s', the second one by s and subtracting yields the first equation below. Similarly, we get the second one.

$$(t's - ts')a = (t' - t)\gcd(a, b)$$
$$(-s't + st')b = (s' - s)\gcd(a, b)$$

Hence equation (I.1.10) holds with $\lambda = st' - s't$. $\qquad\qquad\square$

Corollary 2. *The least common multiple of any positive integers a and b is*

$$(I.1.11) \qquad\qquad lcm(a, b) = \frac{a \cdot b}{\gcd(a, b)}$$

Proof. We check that the number

$$(I.1.12) \qquad\qquad l := \frac{a \cdot b}{\gcd(a, b)}$$

on the right hand side satisfies both requirements (a) and (b) from the definition 3 of the least common multiple.

(a) The number l is a multiple of both a and b, since $l = a \cdot \frac{b}{\gcd(a,b)} = \frac{a}{\gcd(a,b)} \cdot b$.

(b) Assume the positive integer k is a multiple of both a and b. We need to check that k is a multiple of l.

$\square$

Check of item (b). Since the positive integer k is assumed to be a multiple of both a and b, there exist integers p and q such that $k = aq = bp$. The greatest common divisor satisfies

$$(I.2.10) \qquad\qquad \gcd(a, b) = sa - tb$$

from the extended Euclidean algorithm. Hence

$$\gcd(a, b) \cdot k = sa \cdot k - tb \cdot k = sa \cdot bp - tb \cdot aq = (a \cdot b) \cdot (sp - tq)$$

$$k = \frac{a \cdot b}{\gcd(a, b)} \cdot (sp - tq) = l \cdot (sp - tq)$$

Hence k is a multiple of l, as to be shown. $\square$

Proposition 1 (The combination and determinant identities).

(combine) $\qquad\qquad (-1)^i r_i = s_i a - t_i b \qquad$ for $i = 0, 1 \ldots m + 1$

(det) $\qquad s_i t_{i+1} - s_{i+1} t_i = (-1)^i$

(det-r) $\qquad r_i t_{i+1} + r_{i+1} t_i = a$ for $i = 0, 1 \ldots m$

Problem 7. *Show the determinant identity* (det) *by induction for* $i = 0, 1, \ldots m$. *Use induction step* "$i - 1 \mapsto i$".

Answer. $s_0 t_1 - s_1 t_0 = 1 \cdot 1 - 0 \cdot 0 = 1$ gives the start. Here is the induction step "$i - 1 \mapsto i$":

$$s_i t_{i+1} - s_{i+1} t_i = s_i[t_{i-1} + q_i t_i] - [s_{i-1} + q_i s_i]t_i = s_i t_{i-1} - s_{i-1} t_i$$
$$= -[s_{i-1} t_i - s_i t_{i-1}] = -(-1)^{i-1} = (-1)^i$$

Too, we get formula (det-r), since

$$r_i t_{i+1} + r_{i+1} t_i = (-1)^i [s_i a - t_i b] t_{i+1} + (-1)^{i+1} [s_{i+1} a - t_{i+1} b] t_i$$
$$= (-1)^i a [s_i t_{i+1} - s_{i+1} t_i] = a$$

Corollary 3. *One may take the calculation of the sequences s_i and t_i one step further up to $i = m+1$ and get*

$$((\text{I.2.1})) \qquad s_{m+1} = \frac{b}{\gcd(a,b)}, \qquad t_{m+1} = \frac{a}{\gcd(a,b)}$$

Reason for Corollary 3. We set up a linear system to calculate the extra values s_{m+1} and t_{m+1}. To this end we use formula (combine) for $i = m+1$, and the determinant identity (det) with $i = m$.

$$(\text{sys}) \qquad \begin{aligned} t_m \, s_{m+1} - s_m \, t_{m+1} &= (-1)^{m+1} \\ a \, s_{m+1} - b \ \ t_{m+1} &= 0 \end{aligned}$$

The determinant of system (sys) is

$$\Delta = -t_m b + s_m a = (-1)^m \cdot \gcd(a,b) \neq 0$$

Hence system (sys) has a unique solution, to be obtained by Cramer's rule. It turns out to be (I.2.1). $\qquad \square$

Remark. Here is a direct calculation to check (I.2.1):

$$\begin{aligned} b &= (-1)^{m+1}[(b s_{m+1}) \cdot t_m - s_m \cdot (b t_{m+1})] \\ &= (-1)^{m+1}[(b s_{m+1}) \cdot t_m - s_m \cdot (a s_{m+1})] \\ &= (-1)^{m+1}[b t_m - a s_m] \cdot s_{m+1} \\ &= \gcd(a,b) \cdot s_{m+1} \end{aligned}$$

$$\begin{aligned} a &= (-1)^{m+1}[(a s_{m+1}) \cdot t_m - s_m \cdot (a t_{m+1})] \\ &= (-1)^{m+1}[(b t_{m+1}) \cdot t_m - s_m \cdot (a t_{m+1})] \\ &= (-1)^{m+1}[b t_m - a s_m] \cdot t_{m+1} = \gcd(a,b) \cdot t_{m+1} \end{aligned}$$

Corollary 4. *Assume $a \neq b$. The solution s, t of equation (I.2.10) constructed by the extended Euclidean algorithm is the <u>unique</u> solution of (I.2.10) satisfying*

$$(\text{**}) \qquad |s| \leq \frac{b}{2 \gcd(a,b)} \quad \text{and} \quad |t| \leq \frac{a}{2 \gcd(a,b)}$$

Reason. The assumption $a \neq b$ implies that the last quotient $q_m \geq 2$. Hence corollary 3 and formula (I.1.8) with $i = m$ imply

$$\frac{b}{\gcd(a,b)} = s_{m+1} = s_{m-1} + q_m s_m \geq 2|s|$$

$$\frac{a}{\gcd(a,b)} = t_{m+1} = t_{m-1} + q_m t_m \geq 2|t|$$

for the solution s, t of (I.2.10) constructed above. Hence s, t satisfy (**). $\square$

Problem 8. *Prove uniqueness for the solutions of* (I.2.10) *and* (**).

Answer. Suppose that s, t and s', t' are different solutions of formulas (I.2.10) and (**). Both

$$|s' - s| = |\lambda| \frac{b}{\gcd(a,b)} \geq \frac{b}{\gcd(a,b)} \quad \text{and} \quad |s' - s| \leq |s| + |s'| \leq 2\frac{b}{2\gcd(a,b)}$$

by (I.1.10) from Corollary 1, and formula (**). Since $s \neq s'$, both inequalities together imply $|s' - s| = |s| + |s'|$ and hence $s = -s' = \pm\frac{b}{2\gcd(a,b)}$. Similarly one gets $t = -t' = \pm\frac{a}{2\gcd(a,b)}$. Now formula (I.2.10) implies

$$\gcd(a,b) = sa - tb = -(s'a - t'b) = -\gcd(a,b) = 0$$

which is impossible.

I.1.4   The Uniqueness of Prime Decomposition

Definition 6 (prime number). A *prime number* is an integer $p \geq 2$, which is divisible only by 1 and itself.

Theorem 1 (Euclid). *There exist infinitely many primes.*

Proof. Put the first k primes into the increasing sequence p_i and let

$$P = \prod_{1 \leq i \leq k} p_i$$

The number $P + 1$ may be a prime or composite. In the first case, we have found a prime larger than p_k. In he second case, the number $P + 1$ may be prime factored. All its prime factors are larger that p_k. In both cases, we have shown that there exist at least $k + 1$ primes. Since this argument holds for all natural numbers k, the number of primes is not finite. $\square$

Euclid—and many other mathematicians—have shown that there exist infinitely many prime numbers. We put them into the increasing sequence

$$p_1 = 2,\ p_2 = 3,\ p_3 = 5,\ p_4 = 7,\dots$$

It is rather easy to see that for every positive integer, there <u>exists</u> a decomposition into prime factors. Let a and b be any positive integers. There exist sequence $\alpha_i \geq 0$ and $\beta_i \geq 0$, with index $i = 1, 2, \dots$ and only finitely many terms nonzero such that

$$\text{(I.1.13)} \qquad a = \prod_{i \geq 1} p_i^{\alpha_i}\,, \quad b = \prod_{i \geq 1} p_i^{\beta_i}$$

The <u>uniqueness</u> of the prime decomposition turns out harder to prove. Astonishingly, the proof depends on Euclid's lemma,

Proposition 2 (From the extended Euclidean algorithm). *Let $a \geq 0$ and $b \geq 0$ not be both equal to zero. There exist unique integers $0 \leq s \leq \frac{b}{2}$ and $0 \leq t \leq \frac{a}{2}$ such that*

$$(-1)^m \gcd(a, b) = a \cdot s - b \cdot t$$

Here m is the number of division with remainder occurring for the extended Euclidean algorithm starting with the division with remainder of a by b.

Proposition 3 (Euclid's Lemma). *If a prime number divides the product of two integers, the prime number divides at least one of the two integers.*

Reason. Let p be the prime number, and the integers be a and b. We assume that p divides the product ab, but p does not divide a. We need to show that p divides b.

Because p does not divide a, the definition of a prime number implies $\gcd(a, p) = 1$. By the extended Euclidean algorithm, there exist integers s, t such that

$$1 = sa + tp$$

Hence

$$\frac{b}{p} = s\frac{ab}{p} + tb$$

Because p divides ab, the right hand side is an integer. Hence p divides b, as to be shown. $\qquad\square$

Corollary 5 (Euclidean property or general Euclid lemma). *If the integer c divides the product ab, and a and c are relatively prime, then c divides b.*

It is rather easy to see that for every positive integer, there <u>exists</u> a decomposition into prime factors. Let a and b be any positive integers. There exist sequence $\alpha_i \geq 0$ and $\beta_i \geq 0$, with index $i = 1, 2, \ldots$ and only finitely many terms nonzero such that

$$(\text{I.1.14}) \qquad a = \prod_{i \geq 1} p_i^{\alpha_i}$$

The <u>uniqueness</u> of the prime decomposition turns out harder to prove. Astonishingly, the proof depends on Euclid's lemma, the proof of which in turn relies on the extended Euclidean algorithm.

Proposition 4 (Uniqueness of prime decomposition). *The prime decomposition of any positive integer is unique.*

Reason. Assume

$$n = \prod_{i \geq 1} p_i^{\alpha_i} \quad \text{and} \quad n = \prod_{i \geq 1} p_i^{\beta_i}$$

we natural exponents $\alpha_i \geq 0, \beta_i \geq 0$. We need to allow that some of the α_i or β_i may be zero, but require that the primes $p_i \neq p_j$ for $i \neq j$ are *different*.

At first we observer that any prime q dividing n is among the primes p_i for which holds $\alpha_i \geq 1$. This can be checked inductively with Euclid's lemma.

Because n divides n, we see that $p_1^{\beta_1} =: c$ divides n. Now we use the general Euclid lemma with

$$a := \prod_{i \geq 2} p_i^{\alpha_i} \quad \text{and} \quad b := p_1^{\alpha_1}$$

The prime p_1 does not divide a by the remark above since it is not among the primes p_i for $i \geq 2$. Too, p_1 is the only prime dividing c. Hence c and a are relatively prime.

Since c divides $a \cdot b = n$ the general Euclid lemma yields that c divides b. We see that $p_1^{\beta_1}$ divides $p_1^{\alpha_1}$. This implies that $\beta_1 \leq \alpha_1$. Similarly we get the reversed inequality $\alpha_1 \leq \beta_1$, and hence holds the equality $\beta_1 = \alpha_1$. Similarly we get $\beta_i = \alpha_i$ for all $i \geq 1$, which confirms the uniqueness of the prime decomposition. $\qquad\square$

Proposition 5 (The little proposition). *I use the convention from mathematica that the greatest common divisor is always positive, except for* $\gcd(0,0) = 0$. *Given are integers* $a \neq 0$ *and* $n, m, k, l \geq 0$. *If k divides l, then $a^k - 1$ divides $a^l - 1$. Moreover*

$(\text{I.1.15}) \quad \gcd(a^m - 1, a^n - 1) = |a^{\gcd(m,n)} - 1| \;$ *in all cases;*

$(\text{I.1.16}) \quad \gcd(a^m - 1, a^n - 1) = a^{\gcd(m,n)} - 1 \;$ *if $a > 0$ or n, m both even;*

$(\text{I.1.17}) \quad \gcd(a^m - 1, a^n - 1) = -a^{\gcd(m,n)} + 1 \;$ *if $a < 0$, and either m or n is odd.*

Definition 7 (Greatest integer or floor function). For any real x, the floor of x is the unique integer such that $n \leq x < n + 1$. The floor of x is denoted by $\lfloor x \rfloor$.

I.2 The Chinese Remainder Theorem

I.2.1 Simultaneous Congruences

Problem 9. *Given that*

$$x \equiv 19 \mod 765 \quad and \quad x \equiv 1 \mod 567$$

Find the smallest positive solution for x.

Answer. One can determine the unknown integer x modulo the least common multiple of $\mathrm{lcm}(765, 567)$. To get a solution, one needs p and q such that $x = 19 + p \cdot 765 = 1 + q \cdot 567$. Since $\frac{19-1}{9} = 2$ is an integer, the congruence is solvable. We need to multiply the result of the problem above by this integer 2 and get

$$9 = (-20) \cdot 765 + 27 \cdot 567$$
$$18 = (-40) \cdot 765 + 54 \cdot 567$$
$$18 + 40 \cdot 765 = 54 \cdot 567$$
$$19 + 40 \cdot 765 = 1 + 54 \cdot 567$$
$$30\,619 = 30\,619$$

We get a solution $x = 30\,619$, which turns out to be the smallest one, in this example.
[1]

Problem 10. *Solve the simultaneous congruences*

$$\text{(I.2.1)} \qquad \begin{aligned} a &\equiv 2 \pmod{121} \\ a &\equiv 5 \pmod{23} \end{aligned}$$

Answer. We want to calculate with a sequence of congruences modulo the least common multiple $m = \mathrm{lcm}[121, 23] = 121 \cdot 23 = 2783$, and use the same operations as done by the Euclidean algorithm to get the greatest common divisor of 121 and 23.

---

[1]One does not always get immediately the smallest solution. The values p and q are not needed any more.

q_i	s_i	r_i	congruence	
	1	121	$121(a-2)$	$\equiv 121(5-2)$
5	0	23	$23(a-2)$	$\equiv 0$
3	1	6	$6(a-2)$	$\equiv 363$
1	3	5	$5(a-2)$	$\equiv (-3)\cdot 363$
5	4	1	$1(a-2)$	$\equiv 4\cdot 363$
0	23	0	$0(a-2)$	$\equiv (-23)\cdot 363$

One gets $a \equiv 2 + 4\cdot 363 = 1454 \pmod{2783}$, which can easily be checked to be a solution of the system (I.2.1).

Problem 11. *Find integers s and t such that $22s + 8t = 2$. Find the least common multiple of 8 and 22.*

Answer. One can get immediately $\gcd(22,8) = 2$. Hence the least common multiple of 22 and 8 is 88. The extended Euclidean algorithm gives the greatest common divisor as linear combination:

$$
\begin{array}{llllllll}
\text{row 0:} & & & & & & 1 & 0 \\
\text{row 1:} & 22: & 8 = & 2 \text{ rem} & 6 & & 0 & 1 \\
\text{row 2:} & 8: & 6 = & 1 \text{ rem} & 2 & & 1 & 2 \\
\text{row 3:} & 6: & 2 = & 3 \text{ rem} & 0 & & 1 & 3 \\
\end{array}
$$

Indeed, $\gcd(22,8) = 2 = (-1)\cdot 22 + 3\cdot 8$.

Problem 12. *Given that*

$$x \equiv 3 \mod 8 \quad and \quad x \equiv 5 \mod 22$$

determine the unknown integer x modulo the least common multiple of 8 and 22.

Answer. One needs s and t such that $3 + 8s = 5 + 22t$. This works for $3 + 8\cdot 3 = 5 + 22\cdot 1 = 27$. Hence both

$$x \equiv 27 \mod 8 \quad \text{and} \quad x \equiv 27 \mod 22$$
$$\text{what implies} \quad x \equiv 27 \mod 88$$

We now turn to the general case of simultaneous congruences

(I.2.2)
$$x \equiv u \pmod a$$
$$x \equiv v \pmod b$$

We want to calculate with a sequence of congruences modulo $m = ab$, and use the same sequence of operations as those occurring in the Euclidean algorithm determining the greatest common divisor $\gcd(a, b)$. For an appropriate new start, the system (ch) is changed to the equivalent equations

$$\text{(I.2.3)} \qquad \begin{aligned} a(x - u) &\equiv a(v - u) \quad (\text{mod } ab) \\ b(x - u) &\equiv 0 \qquad\qquad (\text{mod } ab) \end{aligned}$$

Note that the second equation of system (ch) has been multiplied with a, and the first equation has been multiplied with by b, moreover the equations have been switched. The extended Euclidean algorithm calculates the successive remainders r_i and quotients q_i via

$$\text{(I.1.7)} \qquad r_0 = a\,, \ r_1 = b\,, \ r_{i+1} = r_{i-1} - q_i r_i \ \text{ and } \ 0 \le r_{i+1} < r_i$$

and gets the sequences $s_i\ t_i$ by the recursions

$$\text{(I.2.4)} \qquad \begin{aligned} s_0 &= 1\,, \ s_1 = 0\,, \ s_{i+1} = s_{i-1} + q_i s_i \\ t_0 &= 0\,, \ t_1 = 1\,, \ t_{i+1} = t_{i-1} + q_i t_i \end{aligned}$$

for $i = 1, 2, \ldots, m$. The algorithm stops when a zero remainder $r_{m+1} = 0$ appears for the first time in row m. The last nonzero remainder is the greatest common divisor.

$$\text{(I.2.5)} \qquad r_m = \gcd(a, b)\,, \ r_{m+1} = 0$$

We want to calculate with a sequence of congruences modulo the least common multiple $m = ab$. As already shown in the example, we use the system (I.2.3) in rows with $i = 0$ and $i = 1$ and perform the same sequence of operations as in Euclidean algorithm. In this manner, we get a sequence of congruences

$$\text{(I.2.6)} \qquad r_i(x - u) \equiv a(v - u) \cdot (-1)^i s_i \quad (\text{mod } ab)$$

for $i = 0, 1, 2, \ldots, m+1$. From row $i = m$ we conclude, after division by $r_m = \gcd(a, b)$,

$$\text{(I.2.7)} \qquad x \equiv u + a(v - u) \cdot (-1)^m s_m \quad (\text{mod } \operatorname{lcm}[a, b])$$

We have shown earlier that the recursions (I.2.4) imply

$$s_{m+1} = \frac{b}{\gcd(a, b)}\,, \qquad t_{m+1} = \frac{a}{\gcd(a, b)}$$

Hence the equation from row $i = m + 1$ tells that $0 \equiv a(v - u) \cdot s_{m+1} \pmod{ab}$ and hence

$$(\text{I.2.8}) \qquad\qquad 0 \equiv v - u \pmod{\gcd(a, b)}$$

It is easy to convince oneself that any two successive rows obtained during the Euclidean algorithm are equivalent to the first and second row. Hence the original system (ch) is equivalent to the two congruences (I.2.7) <u>together with</u> (I.2.8). In other words, we have proved

Proposition 6. *The solution of the simultaneous congruences (ch) can be calculated by the extended Euclidean algorithm. One obtains the result (I.2.7). But this is a valid solution if and only if the solvability condition (I.2.8) holds.*

Problem 13. *Convince yourself that in case the solvability condition (I.2.8) not hold, the result (I.2.7) is at least a solution of the relaxed system*

$$(\text{I.2.9}) \qquad \begin{aligned} x &\equiv u \pmod{\dfrac{a}{\gcd(a, b)}} \\[2mm] x &\equiv v \pmod{\dfrac{b}{\gcd(a, b)}} \end{aligned}$$

(Simultaneous congruences program for the TI84). *The program asks for the numbers u, a, v, b from the simultaneous congruences*

$$(\text{ch}) \qquad \begin{aligned} x &\equiv u \quad \bmod a \\ x &\equiv v \quad \bmod b \end{aligned}$$

Three successive remainders r_i are

X,Y,Z

Three successive terms $(-1)^i s_i$ are

R,S,G

In the end, one should have
$r_M = Y$, $(-1)^M s_M = S$, $(-1)^{M+1} s_{M+1} = G$.

Unfortunately, in the version written down, I needed to replace list L_1 by L1, and $\neq$ by not= . Please find out from the context.

PROGRAM:TRY

```
Prompt U,A
Prompt V,B
A -> Y: B -> Z
1 -> S: 0 -> G: 0 -> M
Lbl 1
If Z = 0:Goto 2
Y -> X: Z -> Y
S -> R: G -> S
int(X/Y) -> Q
X-QY -> Z: R-QS -> G: M+1 -> M
Goto 1
Lbl 2
Disp{Y,M}
B/Y -> H: AH -> L
(U-V)/Y -> D
DS -> P
P-int(P/H)H -> P
U-PA -> W
W-int(W/L)L -> W
Lbl 3
{D,W-U-int((W-U)/A)A,W-V-int((W-V)/B)B} ->  L_1
Disp L_1
Disp {W,L}
```

The output of the program consists of these three lists:

{Y,M}

{D,W-U-int((W-U)/A)A,W-V-int((W-V)/B)B}

> *if and only if the congruences are solvable, and two check numbers that are* $0, 0$
> *if and only if the congruences solved correctly;*

{W,L}

> *which yields the actual solution* $x \equiv w \mod l.$

I.2.2   Chinese Remainder (Sun-Ze's) Theorem

At this point, we additionally assume the integers a and b to be <u>relatively prime</u>. The solution of the simultaneous congruences (ch) defines a (total) function $f : \mathbf{Z}_a \times \mathbf{Z}_b \mapsto \mathbf{Z}_{ab}$ since we know the simultaneous congruences to be solvable for any right-hand side $(u, v) \in \mathbf{Z}_a \times \mathbf{Z}_b$, and the solution to be uniquely determined modulo ab.

The inverse function $x \mapsto f^{-1}(x) = (u, v)$ is obtained by reducing any $x \in \mathbf{Z}_{ab}$ to the remainders $u \in \mathbf{Z}_a$ and $v \in \mathbf{Z}_b$; as one naturally has to do when checking the solution of the Chinese remainder problem. Since the reduced remainders are unique, we see that f^{-1} is a function, too, and hence the function f is injective. Moreover, the value $x \in \mathbf{Z}_{ab}$ can be chosen arbitrarily. Hence f^{-1} is a well-defined (total) function, and hence f is surjective.

Sun's-Ze's result was obtained before 850, the statement below was obtained by Chin-Chin Shao about 1250.

Theorem 2 (Chinese Remainder Theorem). *Assume a and b are relatively prime positive integers. Let s, t be integers satisfying*

$$(\text{I.2.10}) \qquad\qquad \gcd(a, b) = sa - tb$$

obtained via the extended Euclidean algorithm. The function

$$f : \mathbf{Z}_a \times \mathbf{Z}_b \mapsto \mathbf{Z}_{ab}$$
$$f(u, v) = v \cdot sa - u \cdot tb$$

is a bijection which yields the solution of the Chinese remainder problem

$$(\text{ch}) \qquad\qquad \begin{aligned} x &\equiv u \quad \mathrm{mod}\ a \\ x &\equiv v \quad \mathrm{mod}\ b \end{aligned}$$

Its inverse $x \mapsto f^{-1}(x) = (u, v)$ is obtained by reducing $x \in \mathbf{Z}_{ab}$ to the remainders $u \in \mathbf{Z}_a$ and $v \in \mathbf{Z}_b$.

I.3   Little Fermat from Iterated Mappings

I.3.1   Moebius Inversion

Definition 8 (Moebius function). The *Moebius function $\mu(n)$* is defined to be
$$(\text{I.3.1})$$

$$\mu(n) = \begin{cases} +1 & \text{if } n \text{ is square-free and has an even number of prime divisors;} \\ -1 & \text{if } n \text{ is square-free and has an odd number of prime divisors;} \\ 0 & \text{if } n \text{ is not square-free.} \end{cases}$$

One puts $\mu(1) = 1$.

Definition 9 (group). A *ring* is a collection of elements $a, b, \ldots$ which contains the unit element $e = 1$ and allows the binary operation $\cdot$ such that for all a, b in the group, the product $a \cdot b = ab$ is an element of the group, too. Moreover for all a, b, c in the ring hold the following arithmetic rules:

unit element $1 \cdot a = a \cdot e = a$

negative element for all a exists the inverse element a^{-1} such that $aa^{-1} = a^{-1}a = e$.

associativity $(ab)c = a(bc)$

The commutative law is not in general required.

Proposition 7 (Moebius inversion). *Let $\mathcal{G}$ be any commutative group, written with addition as group operation. Let $n \in \mathbf{N} \mapsto A_n \in \mathcal{G}$ and $n \in \mathbf{N} \mapsto B_n \in \mathcal{G}$ be any functions. Let $n \geq 1$ be any integer.*

$$\text{If for all divisors } d \mid n \text{ holds } A_d = \sum_{e \mid d} B_e$$

$$\text{then for all divisors } d \mid n \text{ holds } B_d = \sum_{e \mid d} \mu\left(\frac{d}{e}\right) A_e = \sum_{e \mid d} \mu(e) A\left(\frac{d}{e}\right)$$

Remark. Proposition 7 is as useful in an <u>multiplicatively</u> written version, too.

Lemma 1. *Let $\mathcal{G}$ be any commutative group or field. One needs a commutative and associative multiplication, the unit e and the inverse. One puts $g^0 = e$ for all group elements g. Let $n \in \mathbf{N} \mapsto A_n \in \mathcal{G}$ and $n \in \mathbf{N} \mapsto B_n \in \mathcal{G}$ be any functions. Let $n \geq 1$ be any integer.*

$$\text{If for all divisors } d \mid n \text{ holds } A_d = \prod_{e \mid d} B_e$$

$$\text{then for all divisors } d \mid n \text{ holds } B_d = \prod_{e \mid d} (A_e)^{\mu(d/e)} = \prod_{e \mid d} (A_{d/e})^{\mu(e)}$$

Definition 10 (λ-function). We define the function $\lambda : \mathbf{N} \mapsto \mathbf{N}$ by setting

$$\lambda(n) = \begin{cases} p & \text{if } n = p^r \text{ is a prime power;} \\ 1 & \text{if } n = 1 \text{ or } n \text{ has at least two prime divisors.} \end{cases}$$

22

Lemma 2. *For all $n \geq 1$ holds*

$$(I.3.2) \qquad n = \prod_{d|n} \lambda(d)$$

Proof. Indeed that is easy to check for prime-powers. Define the sets

$$D_n = \{d \in \mathbf{N} : 1 \leq d \mid n \text{ and } d \text{ not prime-power}\}$$

Suppose there is a smallest counterexample. Let

$$n = \prod_i p_i^{r_i}$$

be its prime factorization. We already know that n is not a prime power. But simply calculate

$$\prod_{d|n} \lambda(d) = \prod_i \prod_{d=p_i^s|n} \lambda(d) \cdot \prod_{d \in D_n} \lambda(d) = \prod_i p_i^{r_i} = n$$

No smallest counterexample does exists, the assertion holds for all n $\qquad \square$

Let me use the Kronecker delta

$$\delta_{ab} = \begin{cases} 1 & \text{if } a = b; \\ 0 & \text{if } a \neq b. \end{cases}$$

Lemma 3 (Remember Moebius). *The Moebius function satisfies for all $n \geq 1$*

$$(I.3.3) \qquad \sum_{d|n} \mu(d) = \delta_{n1}$$

$$(I.3.4) \qquad \sum_{d|n} \frac{n}{d} \mu(d) = \phi(n)$$

$$(I.3.5) \qquad \prod_{d|n} \left(\frac{n}{d}\right)^{\mu(d)} = \lambda(n)$$

Problem 14. *Obviously holds*

$$\sum_{d|n} \delta_{d1} = 1$$

Use Moebius inversion to get formula (I.3.3).

Problem 15. *By Gauss' proposition 15*

$$\text{(I.5.9)} \qquad\qquad \sum \{\, \phi(d) : d \text{ divides } m \,\} = m$$

Use Moebius inversion to get formula (VI.1.32).

Problem 16. *We have proved above the formula* (I.3.2). *Use multiplicative Moebius inversion to get formula* (I.3.5).

I.3.2 Little Fermat from an Iterated Mapping

Let $a \geq 1$ be an integer. Let $T : [0, 1) \mapsto [0, 1)$ be the transformation

$$T(x) = \lfloor ax \rfloor$$

Remark. One may close the half-open interval $[0, 1)$ to a circle, use the topology for this circle and thus even get a continuous mapping T. That approach is more natural but does not touch the matter dealt with below.

The iterated mappings $T^{[n]}$ are defined recursively for all $n \geq 1$ by setting

$$T^{[1]} := T \ \text{ and } \ T^{[n+1]} := T^{[n]} \circ T \ \text{ for all } n \geq 1.$$

The reader should convince himself that the following facts are true.

Lemma 4. *Let $a \geq 1$. The iterated mapping are*

$$T^{[n]}(x) = \lfloor a^n x \rfloor \ \text{ for all } n \geq 1.$$

Let $a \geq 2$ and $n \geq 1$ be integers. The mapping $T^{[n]}$ has $a^n - 1$ fixed points. They occur at the points

$$x = \frac{l}{a^n - 1} \ \text{ for integers } 0 \leq l < a^n - 1.$$

For any $d \geq 1$ let b_d count the number of periodic <u>orbits</u> of $T^{[d]}$ with the <u>least</u> period d. For each divisor $d \mid n$ such a periodic orbit

$$x, T(x), T^{[2]}(x), \ldots,, T^{[d-1]}(x)$$

gives rise to d fixed points of $T^{[n]}$; Hence

Lemma 5. *Let $a \geq 2$ and $n \geq 1$ be integers. The number of fixed points of the mapping $T^{[n]}$ is*

$$\sum_{d|n} d \cdot b_d$$

Lemma 6 (Thanks to Moebius). *Let $a \geq 1$ and $n \geq 1$ be any integers.*

(I.3.6)
$$\sum_{d|n} d\, b_d = a^n - 1$$

$$\sum_{d|n} \delta_{d1} + d\, b_d = a^n$$

(I.3.7)
$$\sum_{d|n} \mu(d) a^{n/d} = n b_n + \delta_{n1}$$

Corollary 6 (Thanks to Moebius). *For any $a \geq 1$ and $n \geq 1$ the sum*

(I.3.8)
$$\sum_{d|n} \mu(d)\, a^{\frac{n}{d}} \quad \text{is divisible by } n.$$

Too, this sum is positive or zero. It is zero if and only if $n \geq 2$ and no orbit of period n exists.

Corollary 7 (Fermat's Little Theorem). *For any integer a and any prime p holds $p \mid a^p - a$. If $p \nmid a$ then even $p \mid a^{p-1} - 1$ holds.*

Proof. Put $n = p$ and obtain for the sum (I.3.8)

$$\sum_{d|n} \mu(d)\, a^{\frac{n}{d}} = \mu(1)a^n - \mu(p)\, a^{\frac{n}{p}} = a \cdot [a^{p-1} - 1]$$

to be divisible by p. Under the assumption that $p \nmid a$, we conclude from Euclid's lemma that p divides the second factor $a^{p-1} - 1$. $\qquad\square$

One may also get a few less obvious consequences.

Lemma 7. *For any integer a not divisible by p and any prime power p^s holds*

$$p^s \mid a^{(p-1)p^{s-1}} - 1$$

Proof. Put $n = p^s$ and obtain for the sum (I.3.8)

$$\sum_{d|n} \mu(d)\, a^{\frac{n}{d}} = \mu(1)a^n - \mu(p)\, a^{\frac{n}{p}} = a^{\frac{n}{p}}\big[a^{\frac{(p-1)n}{p}} - 1\big]$$

to be divisible by p^s. Under the assumption that $p \nmid a$, we conclude from Euclid's lemma that p^s divides the second factor $a^{(p-1)p^{s-1}} - 1$. $\qquad\square$

Problem 17. *Let $p \neq q$ be two different prime and let a be an integer not divisible by neither p nor q. Prove that*

$$pq \mid a^{(p-1)(q-1)} - 1$$

Proof. Put $n = pq$ and obtain for the sum (I.3.8)

$$\sum_{d|n} \mu(d)\, a^{\frac{n}{d}} = \mu(1)a^n - \mu(p)\, a^{\frac{n}{p}} - \mu(q)\, a^{\frac{n}{q}} - \mu(pq)\, a^{\frac{n}{pq}} = a^{pq} - a^q - a^p + a$$

which is divisible by pq. Since $\gcd(pq, a) = 1$ we get from the sum above, and from Little Fermat

$$\begin{aligned}
pq &\mid a^{pq-1} - a^{p-1} - a^{q-1} + 1 \\
pq &\mid (a^{p-1} - 1)(a^{q-1} - 1) = a^{p+q-2} - a^{p-1} - a^{q-1} + 1 \\
pq &\mid a^{pq-1} - a^{p+q-2} = a^{p+q-2} \cdot \big[a^{pq-p-q+1} - 1\big] \\
pq &\mid a^{pq-p-q+1} - 1 = a^{(p-1)(q-1)} - 1
\end{aligned}$$

as to be shown. $\qquad\square$

I.4 Euler's Totient Function

I.4.1 The Euler Group

Definition 11 (Unit group). For any commutative ring $\mathcal{R}$, its *unit group* consists of the elements having an inverse.

Problem 18. *Convince yourself that the unit group of any ring with unit is indeed a group.*

Definition 12 (Euler group). The *Euler group* consists of the remainder classes of a modulo m which are relatively prime to m. The modular multiplication is the group operation.

Lemma 8. *Fix some integer $m \geq 2$. The Euler group modulo m is the unit group of* $\mathbf{Z}_m$.

Proof. Let a be a unit in the ring $\mathbf{Z}_m$. There exists an inverse $b = a^{-1}$ for which holds $ab \equiv 1 \pmod{m}$. Hence

$$ab + km = 1$$

which implies that $\gcd(a, m) = 1$ and a is an element of the Euler group.

Conversely, let a be an element of the Euler group. By definition holds $\gcd(a, m) = 1$. By the extended Euclidean algorithm there exist integers b and k such that

$$ab + km = 1$$

Hence $ab \equiv 1 \pmod{m}$ and $b = a^{-1}$ in the ring $\mathbf{Z}_m$. Thus a is a unit in the ring $\mathbf{Z}_m$. $\qquad\square$

We denote the Euler group by G_m^* or $\mathcal{U}(\mathbf{Z}_m)$. The order of the Euler group

$$|\mathcal{U}(\mathbf{Z}_m)| = \phi(m)$$

is the Euler totient function.

I.4.2   The Euler Totient Function

Definition 13. For any $m > 1$, the *Euler totient function* $\phi(m)$ counts the number of integers among $1, 2, \ldots, m - 1$ which are relatively prime to m. Equivalently, one can count residue classes. One puts $\phi(1) = 1$.

Moreover, consider for the Chinese remainder problem

$$\text{(ch)} \qquad x \equiv u \pmod{a} \quad \text{and} \quad x \equiv v \pmod{b}$$

the case that $\gcd(u, a) = \gcd(v, b) = 1$, and $\gcd(a, b) = 1$ as was assumed before. Under these assumptions, $\gcd(f(u, v), ab) = 1$ holds for the solution $f(u, v)$ of problem (ch). The converse is true, too: $\gcd(f(u, v), ab) = 1$ implies $\gcd(u, a) = \gcd(v, b) = 1$. Hence the restriction of the solution function $f : (u, v) \mapsto x$ to the residue classes $(u, v) \in \mathcal{U}(\mathbf{Z}_a) \times \mathcal{U}(\mathbf{Z}_b)$ leads to an interesting result about the structure of the group $\mathcal{U}(\mathbf{Z}_{ab})$:

Corollary 8. *For any relatively prime a and b we have a group isomorphism,*

$$\mathcal{U}(\mathbf{Z}_a) \times \mathcal{U}(\mathbf{Z}_b) \simeq \mathcal{U}(\mathbf{Z}_{ab})$$

obtained via a restriction the bijection from the Chinese Remainder Theorem 2. Hence the Euler totient function is multiplicative. In other words,

$$\phi(ab) = \phi(a)\phi(b) \tag{I.4.1}$$

holds for any relatively prime integers a and b.

Definition 14 (Multiplicative function). A number theoretic function $n \in \mathbf{N} \mapsto f(n) \in \mathbf{Z}$ is called *multiplicative* if and only if

$$f(ab) = f(a) \cdot f(b) \tag{I.4.2}$$

holds for any relatively prime a and b.

Problem 19. *Convince yourself that the Moebius function μ given by definition 8 is multiplicative.*

Problem 20. *Convince yourself that the Lambda function λ given by definition 10 is <u>not</u> multiplicative.*

Problem 21. *Show that for any prime power p^r the Euler totient function takes the value $\phi(p^r) = p^{r-1}(p-1)$.*

Proposition 8 (Formula for the totient function). *The Euler totient function takes the values:*

$$\phi(1) = \phi(2) = 1\,, \ \ \phi(2^r) = 2^{r-1} \quad \text{for } r \geq 1,$$
$$\phi(p^r) = p^{r-1}(p-1) \quad \text{for any odd prime } p \text{ and } r \geq 1$$
$$\phi(p_1^{r_1} \cdot p_2^{r_2} \cdots p_s^{r_s}) = \phi(p_1^{r_1}) \cdot \phi(p_2^{r_2}) \cdots \phi(p_s^{r_s})$$
$$= p_1^{r_1-1}(p_1-1) \cdot p_2^{r_2-1}(p_2-1) \cdots p_s^{r_s-1}(p_s-1)$$

where $p_1, p_2, \ldots p_s$ are any different primes.

Lemma 9. *The Euler totient function satisfies $\phi(n) \geq \sqrt{n}$ for all $n \neq 2, 6$. Equality holds only for $n = 1$ and $n = 4$.*

Proof. In the case that $n = p^r$ where $r \geq 1$ and $p \geq 5$ is an odd prime

$$\frac{\phi(n)^2}{n} = (p-1)^2 p^{r-2} \geq \frac{(p-1)^2}{2p} \geq 1.6$$

since $(p-1)^2 - 3 \cdot 2p = p^2 - 5 \cdot 2p + 1 \geq 0$ for $p \geq 5$. In the case that $n = 2^s p^r$ where $s \geq 1, r \geq 1$ and $p \geq 5$ is an odd prime

$$\frac{\phi(n)^2}{n} = 2^{s-2}(p-1)^2 p^{r-2} \geq \frac{(p-1)^2}{2p} = \frac{\phi(2p)^2}{2p} = \frac{(p-1)^2}{2p} \geq 1.6$$

since $(p-1)^2 - 3 \cdot 2p = p^2 - 5 \cdot 2p + 1 \geq 0$ for $p \geq 5$.
In the case that $n = 2^r$ with $r \geq 2$

$$\frac{\phi(n)^2}{n} = 2^{2(r-1)-r} \geq 1$$

In the case that $n = 3^r$ with $r \geq 1$

$$\frac{\phi(n)^2}{n} = 2^2 \cdot 3^{2(r-1)-r} \geq \frac{4}{3} > 1$$

In the case that $n = 2^s \cdot 3$ with $s \geq 2$

$$\frac{\phi(n)^2}{n} = 2^{2(s-1)+2-s} \cdot 3^{-1} \geq \frac{4}{3} > 1$$

In the case that $n = 2^s 3^r$ with $s \geq 1, r \geq 2$

$$\frac{\phi(n)^2}{n} = 2^{2(s-1)+2-s} \cdot 3^{2(r-1)-r} \geq 2 > 1$$

The remaining cases may be covered by multiplicativity (I.4.1). $\qquad\square$

I.4.3 Euler's Theorem

Definition 15. Let $m > 1$ and $\gcd(a, m) = 1$. The *order of an integer a modulo m* is the smallest positive integer ω such that

$$a^\omega \equiv 1 \pmod{m}$$

We denote the order of an integer a modulo m by $\operatorname{ord}_m(x)$.

Note that $a, a^2, \ldots a^\omega \equiv 1$ are the elements of a cyclic subgroup of $\mathcal{U}(\mathbf{Z}_m)$. By Lagrange's Theorem, the order of a subgroup is a divisor of the order of the group. Hence we conclude

Proposition 9. *Let $m > 1$ and $\gcd(a, m) = 1$. The order of an integer a modulo m is a divisor of the Euler totient function $\phi(m)$.*

Theorem 3 (Euler's Theorem).

$$a^{\phi(m)} \equiv 1 \quad (\bmod\ m)$$

for any $m > 1$ provided that $\gcd(a, m) = 1$.

Independent proof of Euler's Theorem. Let $b_1, b_2, \ldots, b_{\phi(m)}$ be representatives for all the residue classes relatively prime to m and put

$$A := b_1 \cdot b_2 \cdots b_{\phi(m)}$$

For any integer a with $\gcd(a, m) = 1$, the numbers $ab_1, ab_2, \ldots, ab_{\phi(m)}$ are again representatives for all the residue classes relatively prime to m. In other words, as elements of $\mathbf{Z}_m$, they are a permutation of $b_1, b_2, \ldots, b_{\phi(m)}$. We conclude

$$a^{\phi(m)} A = ab_1 \cdot ab_2 \cdots ab_{\phi(m)} \equiv b_1 \cdot b_2 \cdots b_{\phi(m)} = A$$
$$(a^{\phi(m)} - 1)A \equiv 0$$

Since A is relatively prime to m, Euclid's Lemma implies m divides $a^{\phi(m)} - 1$ and hence $a^{\phi(m)} \equiv 1$ $(\bmod\ m)$ as to be shown. $\square$

A second proof of Euler's Theorem. For the special case that $m = p^s$ is a prime power we may use lemma 7. Since the integer a is assumed to be relatively prime to p^s, it is not divisible by p. Hence

$$p^s \mid a^{(p-1)p^{s-1}} - 1 = a^{\phi(p^s)} - 1$$

Let any relatively prime integers a and m with $\gcd(a, m) = 1$ be given. By the little proposition 5 we know that $d \mid \phi(m)$ implies $a^d - 1 \mid a^{\phi(m)} - 1$. Let

$$m = \prod_i p_i^{r_i}$$

be the prime factorization. Since for any index i we know that $\phi(p_i^{s_i}) \mid \phi(m)$, one gets for all i

$$p_i^{s_i} \mid a^{\phi(p_i^{s_i})} - 1 \mid a^{\phi(m)} - 1$$

Since these prime powers are relatively prime, we may conclude

$$m = \prod_i p_i^{s_i} \mid a^{\phi(m)} - 1$$

as to be shown. $\square$

Proposition 10. *Given are any integer $a \neq 1, -1$, integer $m > 1$ and a prime p. If $m \mid a^p - 1$ then either $p \mid \phi(m)$ or $m \mid a - 1$.*

Proof. By Euler's theorem $m \mid a^{\phi(m)} - 1$. Hence the little proposition 5 yields

$$m \mid \gcd(a^{\phi(m)} - 1, a^p - 1) = a^{\gcd(\phi(m), p)} - 1$$

In the case that $p \nmid \phi(m)$ we get $\gcd(\phi(m), p) = 1$ since p is prime. Hence $m \mid a - 1$. $\square$

Proposition 11. *Given are any integer $a \neq 1, -1$, and two primes p and q. If $q \mid a^p - 1$ then either $q = 1 + kp$ or $m \mid a - 1$. If p and q are odd, we conclude even $q = 1 + 2kp$ or $m \mid a - 1$.*

I.5 Reviewing some Number Theory

I.5.1 Legendre symbols

Definition 16 (Quadratic residue and nonresidue). Let p be a prime. The remainder class of a modulo p is called a *quadratic residue* if the congruence $x^2 \equiv a$ (mod p) is solvable.
The remainder class of a modulo p is called a *quadratic nonresidue* if the congruence $x^2 \equiv a$ (mod p) is not solvable.

Proposition 12 (Euler's criterium). *Let p be an odd prime and suppose that $p \nmid a$ is not a divisor of a. The quadratic congruence*

$$x^2 \equiv a \quad (\text{mod } p) \qquad \text{is solvable iff} \quad a^{\frac{p-1}{2}} \equiv 1 \quad (\text{mod } p)$$

$$x^2 \equiv a \quad (\text{mod } p) \quad \text{is not solvable iff} \quad a^{\frac{p-1}{2}} \equiv -1 \quad (\text{mod } p)$$

Proof. Let

$$\mathbf{Z}_p^* = \{a \in \mathbf{Z}_p \; : \; p \nmid a\}$$

The mapping $x \in \mathbf{Z}_p^* \mapsto x^2 \in \mathbf{Z}_p^*$ is two to one. Indeed $x^2 \equiv y^2$ (mod p) implies that either $x = y \equiv 0$ (mod p) which is excluded from the domain, or one gets two different solutions $x \equiv y$ (mod p) and $x \equiv -y$ (mod p). Hence the range

$$K = \{a \in \mathbf{Z}_p^* \; : \; x^2 \equiv a \quad (\text{mod } p) \text{ is solvable}\}$$

has $\frac{p-1}{2}$ elements. Next we use Lagrange's Theorem 7. By Fermat's Little Theorem, the polynomial $x^{p-1} - 1$ has exactly $p - 1$ zeros, indeed these are the remainder classes

$1, 2, \ldots p - 1$ modulo p. This turns out to be the maximal possible number of zeros. In the polynomial factoring

$$x^{p-1} - 1 = (x^{\frac{p-1}{2}} - 1)(x^{\frac{p-1}{2}} + 1)$$

each factor on the right-hand side has at most $\frac{p-1}{2}$ zeros but their product has $p - 1$ zeros. Hence the factors $x^{\frac{p-1}{2}} - 1$ and $x^{\frac{p-1}{2}} + 1$ have both exactly $\frac{p-1}{2}$ zeros. Thus the set

$$H = \{a \in \mathbf{Z}_p^* : a^{\frac{p-1}{2}} \equiv 1 \pmod{p}\}$$

turns out to have $\frac{p-1}{2}$ elements, too.

Assume the quadratic congruence $x^2 \equiv a \pmod{p}$ to be solvable. The assumption $p \nmid a$ implies $p \nmid x$. Hence Fermat's Little Theorem implies

$$a^{\frac{p-1}{2}} \equiv x^{p-1} \equiv 1 \pmod{p}$$

as claimed. In other words, we get the inclusion $K \subseteq H$. Since we know already that both sets K and H have the same number of elements, the equality $K = H$ occurs. Thus Euler's criterium is shown to be valid. $\square$

Problem 22. *Give a simple proof of Euler's criterium assuming existence of a primitive root.*

Answer. Assume the quadratic congruence $x^2 \equiv a \pmod{p}$ to be solvable. The assumption $p \nmid a$ implies $p \nmid x$. Hence Fermat's Little Theorem implies

$$a^{\frac{p-1}{2}} \equiv x^{p-1} \equiv 1 \pmod{p}$$

as claimed.

Assume the quadratic congruence $x^2 \equiv a \pmod{p}$ not to be solvable. There exists s primitive root r modulo p. For this root holds

$$r^{\frac{p-1}{2}} \equiv -1 \pmod{p}$$

since the powers $\{r, r^2, \ldots r^{p-1}\}$ exhaust all $p - 1$ congruence classes modulo p which are not divisible by p. The assumption $p \nmid a$ implies $a \equiv r^s$ to hold to some integer s. For even s, it is easy to see that $x = r^{s/2}$ solves the quadratic congruence. Hence s is odd and hence

$$a^{\frac{p-1}{2}} \equiv r^{\frac{s(p-1)}{2}} \equiv (-1)^s \equiv -1 \pmod{p}$$

$\square$

For a thorough treatment of Legendre and Jacob symbols and gapless proofs, the reader is referred to the literature. Too, the relevant material is included in the volume about number theory *Essays about Number Theory* by Rothe.

Definition 17 (Legendre symbol). Let p be an odd prime and a any integer. The *Legendre symbol* is defined to be

$$(\text{I.5.1}) \qquad \left(\frac{a}{p}\right) = \begin{cases} 1, & \text{if } x^2 \equiv a \pmod{p} \text{ is solvable and } p \nmid a; \\ -1, & \text{if } x^2 \equiv a \pmod{p} \text{ is not solvable and } p \nmid a; \\ 0, & \text{if } p \mid a. \end{cases}$$

Theorem 4 (The basic formula for the Legendre symbol). *Let p be an odd prime and a any integer. The Legendre symbol is given by Euler's formula*

$$(\text{I.5.2}) \qquad \left(\frac{a}{p}\right) \equiv a^{\frac{p-1}{2}} \pmod{p}$$

Assume additionally that p does not divide a. Then the Legendre symbol, equals $+1$ if a is a quadratic residue, and -1 if a is a quadratic non-residue modulo p.

Proposition 13. *Let $m, n \geq 3$ be odd integers. Let a be an integer such that $\gcd(a, m) = 1$. The parity count has the following properties*

$$(\text{I.5.3}) \qquad (1/m) = 1$$
$$(\text{I.5.4}) \qquad (a/m) = (b/m) \quad \text{if } a \equiv b \pmod{m}$$
$$(\text{I.5.5}) \qquad (2a/m) = (2/m)(a/m)$$
$$(\text{I.5.6}) \qquad (2/m) = (-1)^{\frac{m^2-1}{8}}$$
$$(\text{I.5.7}) \qquad (n/m)(m/n) = (-1)^{\frac{n-1}{2}\frac{m-1}{2}} \quad \text{if } n, m \geq 3 \text{ are odd and relatively prime.}$$

Let

$$K = \{a \in G_m^* : x^2 \equiv a \pmod{m} \text{ is solvable }\}$$

denote the remainder classes of a modulo m which are relatively prime to m, and for which the quadratic congruence

$$x^2 \equiv a \pmod{m}$$

is solvable. The mapping $h : x \in G_m^* \mapsto x^2 \in K$ is a group homomorphism. The image of h is , $K \subset G_m^*$. This is hence a subgroup. The kernel of h is

$$\text{kernel } h = \{x \in G_m^* : x^2 \equiv 1 \pmod{m}\} =: Q$$

By the main theorem of group theory the image of h is isomorphic to the quotient of the domain by the kernel:

$$K \simeq G_m^* / \text{kernel } h$$

From here we get the subgroup index

$$(\text{I.5.8}) \qquad \text{index } [K, G_m^*] = \dim \text{ kernel } h$$

Proposition 14. *Let $m = 2^s$ be a power of 2. For $s \geq 3$, the quadratic residues are a cyclic subgroup with index $[K, G_m^*] = 4$.*

The cases $m = 2, 4$ are exceptional:
$G_2^* = K = \{1\}$ with index $[K, G_2^*] = 1$; whereas
$G_4^* = \{1, 3\} \sim \mathbf{Z}_2$ with $K = \{1\}$ and index $[K, G_4^*] = 2$

A simple proof. The kernel $\dim \text{ kernel } h$ consists of the four elements $\{1, -1, 1 + \frac{m}{2}, -1 + \frac{m}{2}\}$. I may leave to the reader that these four elements are in the kernel. Conversely, Let $x = 1 + a$ and solve $x^2 \equiv 1 \pmod{m}$. One gets $a(2 + a) \equiv 0 \pmod{m}$ which has four solutions:

$$a = 0, x = 1,$$
$$a = -2, x = -1,$$
$$a = \frac{m}{2}, x = 1 + \frac{m}{2},$$
$$a = -2 + \frac{m}{2}, x = -1 + \frac{m}{2}$$

Indeed, either one factor is divisible by m, or $a \equiv 2 \pmod 4$, $2 + a = m/2$ or $2 + a \equiv 2 \pmod 4$, $a = m/2$. All four solution are different since $m \geq 8$. By equation (I.5.8) index $[K, G_m^*] = \dim \text{ kernel } h = 4$. $\qquad \square$

I.5.2   Primitive Roots

Theorem 5. *Let $m > 1$. There exists a primitive root modulo m in the following cases:*

- *m is $2, 4$ or p^r or $2p^r$ where p is an odd prime and $r \geq 1$.*

There exists no primitive root modulo m in the following cases:

- *m is divisible by 8;*

- *m is divisible by 4 and an odd prime;*

- *m is divisible by two different odd primes.*

34

Gauss has proved that for any prime number there exists a primitive root. See theorem 6 below.

Problem 23. *Assume that no primitive root exists modulo $m > 1$. Show that the Euler totient function $\phi(m)$ is divisible by 4. Is the converse true?*

Answer. A primitive root exists modulo the integer m in the cases where m is $2, 4$ or p^r or $2p^r$ where p is an odd prime and $r \geq 1$. The opposite case that no primitive exists occurs if the integer m is either divisible by 8, or by two different odd primes $p \neq q$. In these cases, it is easy to check that $\phi(m)$ is divisible by 4.

The converse is not true. Let p be any prime with $p \equiv 1 \pmod 4$ and put $m = p^r$ or $2p^r$. In this case, a primitive root exists and $\phi(m) = p^{r-1}(p-1)$ is divisible by 4, nevertheless.

Proposition 15 (Gauss). *For any $m \geq 1$, the sum of the values $\phi(d)$ over all divisors $d \mid m$ equals m:*

$$(\text{I.5.9}) \qquad \sum \{ \phi(d) \ : \ d \text{ divides } m \ \} = m$$

Problem 24. *Check the formula* (I.5.9) *for a prime power $m = p^r$.*

Answer.

$$\sum \{ \phi(d) \ : \ d \mid p^r \} = \sum_{t=0}^{r} \phi(p^t) = 1 + \sum_{t=1}^{r} p^{t-1}(p-1) = 1 + \frac{p^r - 1}{p - 1} \cdot (p - 1) = p^r$$

Lemma 10. *Assume the integers a and b are relatively prime, and assume that the formula* (I.5.9) *holds for a and b. Then the formula* (I.5.9) *holds for the product ab, too.*

Proof. Since $\gcd(a, b) = 1$, any divisor $d \mid ab$ of the product can be factored

$$d = \gcd(ab, d) = \gcd(a, d) \cdot \gcd(b, d) =: e \cdot f$$

Conversely, the product ef of any divisors $e \mid a$ and $f \mid b$ is a divisor of ab. From the multiplicativity of the ϕ-function one gets

$$\sum \{ \phi(d) \ : \ d \mid ab \} = \sum \{ \phi(ef) \ : \ e \mid a \text{ and } f \mid b \}$$
$$= \sum \{ \phi(e)\phi(f) \ : \ e \mid a \text{ and } f \mid b \}$$
$$= \sum \{ \phi(e) \ : \ e \mid a \} \cdot \sum \{ \phi(f) \ : \ f \mid b \} = a \cdot b$$

as to be shown. $\qquad \square$

End of the proof of Proposition 15. The reasoning uses induction on the number s of different prime factors of $m = p_1^{r_1} \cdot p_2^{r_2} \cdots p_s^{r_s}$. Problem 24 gives the induction start for $s = 1$, and Lemma 10 provides the induction step. $\qquad\square$

Gauss' independent proof of Proposition 15. We partition the integers $1, 2, , \ldots, m$ into disjoint classes

$$C_d = \{ j \; : \; 1 \le j \le m \text{ and } \gcd(j, m) = d \}$$

for all divisors $d \mid m$. Since $\gcd(j, m) = d$ if and only if $\gcd(j/d, m/d) = 1$, the possible values for j/d are just the integers among $1, 2, \ldots m/d$ which are relatively prime to m/d. Hence $|C_d| = \phi(m/d)$ and counting yields

$$m = \sum \{ |C_d| : d \text{ divides } m \} = \sum \{ \phi(\tfrac{m}{d}) : d \text{ divides } m \}$$
$$= \sum \{ \phi(d) : d \text{ divides } m \}$$

$\qquad\square$

Problem 25. *As an example, put* $m = 12$ *and determine the classes* C_d.

Answer.

divisor d	C_d	$\phi\left(\frac{12}{d}\right)$
1	$\{1, 5, 7, 11\}$	$\phi(12) = 4$
2	$\{2, 10\}$	$\phi(6) = 2$
3	$\{3, 9\}$	$\phi(4) = 2$
4	$\{4, 8\}$	$\phi(3) = 2$
6	$\{6\}$	$\phi(2) = 1$
12	$\{12\}$	$\phi(1) = 1$

Theorem 6 (Gauss). *For any prime number there exists a primitive root.*

Proof. Let p be an odd prime number, and let the integer a represent any of the residue classes modulo p among the numbers $1, 2, \ldots, p - 1$.

The Little Fermat Theorem tells that $a^{p-1} \equiv 1 \pmod{p}$. But $p - 1$ need not be the smallest exponent with that property. According to Definition 15, the order of a is the smallest positive integer ω such that

$$(\text{I.5.10}) \qquad\qquad a^\omega \equiv 1 \pmod{p}$$

36

All divisors $\omega \mid p-1$ can occur as the order of the integer a.

Let $c(\omega)$ count, among the numbers $1, 2, \ldots, p-1$, the residue classes modulo p having the order ω . Counting all different cases yields

$$(\text{I.5.11}) \qquad p - 1 = \sum \{\, c(\omega) \ : \ \omega \mid p - 1 \,\}$$

Next we show the inequality

$$(\text{I.5.12}) \qquad c(\omega) \leq \phi(\omega)$$

In the case that $c(\omega) = 0$, we are obviously ready. Assume that $c(\omega) > 0$ and let h be a residue class of order ω. It is not hard to check that all powers

$$(\text{I.5.13}) \qquad h^j \ \text{ with } 1 \leq j < \omega \text{ and } \gcd(j, \omega) = 1$$

are $\phi(\omega)$ different residue class of order ω. Can there be any more residue classes of the same order ω? From Lagrange's Theorem 7, we conclude that the equation (I.5.10) has at most ω roots. It is easy to see that these roots are $h, h^2, \ldots h^\omega$. Hence we have obtained indeed all roots. But among them only the roots given by formula (I.5.13) actually have the order ω, the order of the remaining roots is a proper divisor of ω. Hence we conclude that there are exactly $\phi(\omega)$ different residue class of order ω, confirming the estimate (I.5.12).

The "coup de grâce" is now relating the counting formula (I.5.11) and estimate (I.5.12) with equation (I.5.9) from Gauss' Proposition,—to finally obtain

$$p - 1 = \sum \{\, c(\omega) \ : \ \omega \mid p - 1 \,\} \leq \sum \{\, \phi(\omega) \ : \ \omega \mid p - 1 \,\} = p - 1$$

These chain of inequalities can only be true with equality everywhere. Hence we conclude that $c(\omega) = \phi(\omega)$ for all divisors $\omega \mid p - 1$. We get the Corollary 9 below and especially the existence of a primitive root. $\qquad \square$

Corollary 9. *Let p be any odd prime number.*

- *There exists residue classes for all orders $\omega \mid p - 1$ dividing $p - 1$;*

- *for each divisor there exist exactly $\phi(\omega)$ residue classes of order ω;*

- *especially there exist exactly $\phi(p - 1)$ primitive roots modulo the prime number p.*

I.6   The Product over the Euler Group

Recall that by definition 12 the *Euler group* G_m^* consists of the remainder classes of a modulo m which are relatively prime to m. The modular multiplication is the group operation. In the familiar proof I.4.3 of Euler's theorem 3 the product of all elements of the Euler group comes up, but its value is not obtained. We calculate now its value modulo m. It turns out to be either -1 or 1, depending on existence or nonexistence of a primitive root.

Problem 26. *Show that in the cases where a primitive root exists, the product*

$$A := b_1 \cdot b_2 \cdots b_{\phi(m)}$$

occurring in the proof of Euler's Theorem satisfies $A \equiv -1 \pmod{m}$.

Answer. The cases $m = 2, 4$ can be checked directly Let g be a primitive root modulo $m = p^r$ (odd prime power) or $m = 2p^r$. The Euler totient function $\phi(m)$ is even. The product A is

$$A = g^1 \cdot g^2 \cdots g^{\phi(m)} = g^{1+2+\cdots+\phi(m)} = g^{\frac{\phi(m)(\phi(m)+1)}{2}} \equiv g^{\frac{\phi(m)}{2}} =: h \pmod{m}$$

But $h^2 \equiv 1 \pmod{m}$ by Euler's theorem and $h \not\equiv 1 \pmod{m}$ since g is a primitive root. Since as primitive root is assumed to exist, the equation $h^2 \equiv 1 \pmod{m}$ has only two solutions. Hence $h \equiv -1 \pmod{m}$ as claimed.

Problem 27. *Show that in the cases where no primitive root exists, the product*

$$A := b_1 \cdot b_2 \cdots b_{\phi(m)}$$

occurring in the proof of Euler's Theorem satisfies $A \equiv 1 \pmod{m}$.

Here is a solution of this problem. The set

$$Q := \{a \in G_m^* : a^2 \equiv 1 \quad \mathrm{mod}\ m\}$$

is the kernel of the squaring homomorphism $h : x \in G_m^* \mapsto x^2$ and hence a subgroup of G_m^*. Under the involution $a_i \mapsto a_i^{-1}$, defined by taking inverses, the elements of Q are the fixed points and the elements of $G \setminus H$ are paired. Hence

$$(\text{I.6.1}) \qquad\qquad \prod_{a \in G_m^*} a \equiv \prod_{a \in Q} a \pmod{m}$$

Lemma 11. *The mapping $a \mapsto i(a) = m - a$ is a second (different) involution $G_m^* \mapsto G_m^*$ with restriction $Q \mapsto Q$. For $m \geq 3$, the involution i has no fixed points.*

Proof. Indeed we calculate modulo m and obtain from the assumption $i(a) = a \in G_m^*$: $m - a \equiv a$ implies $2a \equiv m$ and hence $a \mid m$. Hence $\gcd(a, m) = 1$ implies $a = 1$ and finally $2 = 2a \equiv m \equiv 0 \pmod{m}$ implies $m = 2$. But this special case has been excluded by assuming $m \geq 3$. $\qquad\square$

Now the group Q can be partitioned into pairs $\{a, i(a)\}$. For each pair holds $a(m - a) \equiv -1 \pmod{m}$ and hence we obtain

$$(I.6.2) \qquad \prod_{a \in Q} a \equiv (-1)^{|Q|/2} \quad \bmod\ m$$

From equations (I.6.1) and (I.6.2) we see

Lemma 12. *The product of the elements of the Euler group is*

$$(I.6.3) \qquad \prod_{a \in G_m^*} a \equiv (-1)^{|Q|/2} \quad \bmod\ m$$

and hence either $+1$ or -1 modulo m.

Proposition 16. *Excluding the case $m = 2$ we see: the alternative*

$$\prod_{a \in G_m^*} a = \prod_{a \in Q} a \equiv -1 \quad \bmod\ m$$

is equivalent to $Q = \{\pm 1\}$ which is in turn equivalent to existence of a primitive root. The alternative

$$\prod_{a \in G_m^*} a = \prod_{a \in Q} a \equiv +1 \quad \bmod\ m$$

occurs if and only if the order $|Q|$ is divisible by 4 which is in turn equivalent to nonexistence of a primitive root.

Lemma 13 (The kernel of the squaring homomorphism). *Fix some integer $m \geq 2$. Let t denote the number of its different prime factors. The kernel of the squaring homomorphism $h : x \in G_m^* \mapsto x^2$ has the dimension*

(i) dim kernel $h = 2^t$ *if m is either odd or $m \equiv 4 \pmod 8$;*

(ii) dim kernel $h = 2^{t-1}$ *if m is even and $m \equiv 2 \pmod 4$;*

(iii) dim kernel $h = 2^{t+1}$ *if m is even and $m \equiv 0$* (mod 8).

Proof. Let

$$m = \prod_{1 \leq i \leq t} p_i^{r_i}$$

be the prime factorization of m The kernel of the squaring homomorphism h can be determined via the Chinese remainder theorem. Indeed any solution of $x^2 \equiv 1$ (mod m) is obtained by solving at first

(I.6.4) $$x_i^2 \equiv 1 \quad (\text{mod } p_i^{r_i})$$

for all $1 \leq i \leq t$ separately, then solving

(I.6.5) $$x \equiv x_i \quad (\text{mod } p_i^{r_i}) \quad \text{for } 1 \leq i \leq t$$

which is a standard Chinese remainder problem. The equation (I.6.4) has two solutions $x_i = \pm 1$ if either p_i is odd or $p_i^{r_i} = 4$. By lemma 14 the equation (I.6.4) has four solutions if $8 \mid p_i^{r_i}$. But there exists only one solution if $p_i^{r_i} = 2$ Because of equation (I.6.5) these solutions may be combined arbitrarily. Hence the dimension of the kernel is obtained by taking the product of all $1 \leq i \leq t$ and comes out as claimed. $\qquad\square$

Proof. It is easy to see that $|Q|$ is divisible by 4 unless $Q = \{\pm 1\}$. Indeed, if there exists $a \in Q \setminus \{\pm 1\}$, the set $\{1, -1, a, -a\}$ is a subgroup of Q of order four, and hence $|Q|$ is divisible by 4.

Too, the order of kernel Q of squaring homomorphism $h : x \in G_m^* \mapsto x^2$ has already be determined by lemma 13. Let t denote the number of its different prime factors of $m \geq 3$. From there we see that dim $Q = 2$ occurs for the cases

(i) dim $Q = 2^t = 2$ if m is either an odd prime power or $m = 4$;

(ii) dim $Q = 2^{t-1} = 2$ if $m \equiv 2$ (mod 4) and $t = 2$ and hence $m = 2p^r$.

Because of theorem 5 these turn just out to be the cases where a primitive root exists. $\qquad\square$

Remark. Proposition 16 is indeed a theorem of Gauss, who sketched the proof in his Disquisitiones Arithmeticae. Too, a detailed proof can be found in Ore's book [] p.263-267.

It seems rather peculiar that case (1-) does occur if and only if there exists a primitive root. It would be nice to establish the equivalence of (1-) with existence of a primitive root even more directly. How does (1-) imply the existence of a primitive root?

I.7 Fermat Primes

I.7.1   A Short Paragraph about Fermat Numbers

Definition 18 (Fermat number, Fermat prime). The numbers

$$F_n = 2^{2^n} + 1$$

for any integer $n \geq 0$ are called *Fermat numbers*. A Fermat number which is prime is called a *Fermat prime*.

Lemma 14 (Fermat). *If p is any odd prime, and $p-1$ is a power of two, then $p = F_n$ is a Fermat prime.*

Proof. Assume $p = 1 + 2^a$ and $a \geq 3$ odd. Because of the factorization $a = 2^n \cdot (2b+1)$ one can factor

$$p = 1 + 2^{2^n \cdot (2b+1)} = (1 + 2^{2^n})(1 - 2^{2^n} + 2^{2^n \cdot 2} - \cdots + 2^{2^n \cdot 2b})$$

If p is an odd prime, the only possibility is $b = 0$. Hence the exponent a cannot have any odd prime factor. Hence we are left with $p = F_n$ as the only possibility. $\qquad\square$

The only <u>known</u> Fermat primes are F_n for $n = 0, 1, 2, 3, 4$.
They are $3, 5, 17, 257$, and 65537, see[2].

Fermat numbers and Fermat primes were first studied by Pierre de Fermat, who conjectured 1654 in a letter to Blaise Pascal that all Fermat numbers are prime, but told he had not been able to find a proof. Indeed, the first five Fermat numbers are easily shown to be prime. However, Fermat's conjecture was refuted by Leonhard Euler in 1732 when he showed, as one of his first number theoretic discoveries:

$$F_5 = 2^{2^5} + 1 = 2^{32} + 1 = 4294967297 = 641 \cdot 6700417.$$

Later, Euler proved that every prime factor p of F_n must have the form $p = i \cdot 2^{n+1} + 1$. Almost hundred years later, Lucas showed that even $p = j \cdot 2^{n+2} + 1$. These results give a vague hope that one could factor Fermat numbers and perhaps settle Fermat's conjecture to the negative.

There are no other known Fermat primes F_n with $n > 4$. However, little is known about Fermat numbers with large n. In fact, each of the following is an open problem:

1. Is F_n composite for all $n > 4$? By this *anti-Fermat hypothesis* $3, 5, 17, 257$, and 65537 would be the only Fermat primes.

---

[2]sequence A019434 in OEIS

2. Are there infinitely many Fermat primes? (Eisenstein 1844)

3. Are there infinitely many composite Fermat numbers?

4. Are all Fermat numbers square free?

As of 2012, the next twenty-eight Fermat numbers, F_5 through F_{32}, are known to be composite. As of February 2012, only F_0 to F_{11} have been completely factored. For complete information, see Fermat factoring status by Wilfrid Keller, on the internet at

`http://www.prothsearch.net/fermat.html`

Main Theorem 1 (Gauss-Wantzel Theorem). *A regular polygon with n sides is constructible if and only if*

$$n = 2^h p_1 \cdot p_2 \cdots p_s$$

where $p_1 \cdots p_s$ is a product of different Fermat primes.

If the anti-Fermat hypothesis is true, there are exactly five Fermat primes, and hence exactly 31 regular constructible polygons with an odd number of sides.

Problem 28. *Prove that the totient function $\phi(n)$ is a power of two if and only if*

$$n = 2^r \cdot \prod F_i$$

where the F_i are different Fermat primes. Hence a regular polygon with n sides is constructible if and only if the totient function $\phi(n)$ is a power of two.

Problem 29. *Determine all solutions of $\phi(n) = 1\,024$. There are 12 solutions. Prime factor these solution. Why do we see from the Gauss-Wantzel Theorem that the corresponding regular n-gons are constructible with compass and straightedge.*

Answer. The natural numbers for which $\phi(n) = 1\,024$. are

$$(1285,\ 2048,\ 2056,\ 2176,\ 2560,\ 2570,\ 2720,\ 3072,\ 3084,\ 3264,\ 3840,\ 4080)$$

Here they are prime factored

$$1285 = 5 \cdot 257$$
$$2048 = 2^{11}$$
$$2056 = 2^3 \cdot 257$$
$$2176 = 2^7 \cdot 17$$
$$2560 = 2^9 \cdot 5$$
$$2570 = 2 \cdot 5 \cdot 257$$
$$2720 = 2^5 \cdot 5 \cdot 17$$
$$3072 = 2^{10} \cdot 3$$
$$3084 = 2^2 \cdot 3 \cdot 257$$
$$3264 = 2^6 \cdot 3 \cdot 17$$
$$3840 = 2^8 \cdot 3 \cdot 5$$
$$4080 = 2^4 \cdot 3 \cdot 5 \cdot 17$$

These number are all products of some power of two with different Fermat primes. By the Gauss-Wantzel Theorem, the corresponding regular n-gons are constructible with compass and straightedge. $\square$

Corollary 10 (Gauss-Wantzel). *For all natural $n \geq 3$ are equivalent*

(a) *The regular polygon with n sides is constructible with compass and straightedge.*

(b) *The totient function $\phi(n)$ is a power of two.*

(c) $n = 2^h p_1 \cdot p_2 \cdots p_s$ *where* $p_2 \cdots p_s$ *is a product of* different *Fermat primes.*

(a) *implies* (b). We assume that the regular n-gon is constructible with compass and straightedge. Hence the roots of the polynomial equation $\Phi_n(z) = 1$ are in a field extension of the rationals the dimension of which is a power of two. A full proof of necessity was given by Pierre Wantzel in 1837. (More details are given in Hartshorne's book *Euclid and Beyond.*) Hence the totient function $\phi(n) = \deg \Phi_n$ is a power of two. $\square$

(b) *implies* (c). We assume that the totient function $\phi(n)$ is a power of two. Hence the prime decomposition of n is a power of two multiplied with simple odd prime factors F_k for which $\phi(F_k)$ is a power of two. Hence $F_k = 1 + 2^s$ and by lemma 14 the number F_k is a Fermat prime. $\square$

(c) *implies* (a). We assume that n has the prime decomposition

$$n = 2^h p_1 \cdot p_2 \cdots p_s$$

where the primes p_k are Fermat primes. The construtibility of the regular n-gon follows from the constructibility of the F_k-gons for the Fermat primes, which is Gauss' main result (see corollary 12 below), and some further simple transfers and bisection of angles. These steps are possible for any given angle with compass and straightedge.

$\square$

	n	$factored$
1	3	F_0
2	5	F_1
3	15	$3 \cdot 5$
4	17	F_2
5	51	$3 \cdot 17$
6	85	$5 \cdot 17$
7	255	$3 \cdot 5 \cdot 17$
8	257	F_3
9	771	$3 \cdot 257$
10	1 285	$5 \cdot 257$
11	3 855	$3 \cdot 5 \cdot 257$
12	4 369	$17 \cdot 257$
13	13 107	$3 \cdot 17 \cdot 257$
14	21 845	$5 \cdot 17 \cdot 257$
15	65 535	$3 \cdot 5 \cdot 17 \cdot 257$
16	65 537	F_4

	n	$factored$
17	196 611	$3 \cdot 65\,537$
18	327 685	$5 \cdot 65\,537$
19	983 055	$3 \cdot 5 \cdot 65\,537$
20	1 114 129	$17 \cdot 65\,537$
21	3 342 387	$3 \cdot 17 \cdot 65\,537$
22	5 570 645	$5 \cdot 17 \cdot 65\,537$
23	16 711 935	$3 \cdot 5 \cdot 17 \cdot 65\,537$
24	16 843 009	$257 \cdot 65\,537$
25	50 529 027	$3 \cdot 257 \cdot 65\,537$
26	84 215 045	$5 \cdot 257 \cdot 65\,537$
27	252 645 135	$3 \cdot 5 \cdot 257 \cdot 65\,537$
28	286 331 153	$17 \cdot 257 \cdot 65\,537$
29	858 993 459	$3 \cdot 17 \cdot 257 \cdot 65\,537$
30	1 431 655 765	$5 \cdot 17 \cdot 257 \cdot 65\,537$
31	4 294 967 295	$3 \cdot 5 \cdot 17 \cdot 257 \cdot 65\,537$

Part II

Construction of Regular Polygons

II.0.1 The 17-gon Construction

The regular heptadecagon is a constructible polygon (that is, one that can be constructed using a compass and unmarked straightedge), as was shown by Carl Friedrich Gauss in 1796 at the age of nineteen. This proof represented the first progress in regular polygon construction in over 2000 years. It can be found in his *Disquisitiones Arithmeticae* and is reprinted in his Werke (1870-77), vol. I. Gauss's proof of the constructibility of the relies on the fact that constructibility is equivalent to expressibility of the trigonometric functions of the angle $360°/17$ in terms of arithmetic operations and square root extractions.

We use radian measure for angles in this section. According to theorems from geometry the number $2\cos(2\pi/17)$ is constructible if and only if there exists a tower of dimension extensions all of dimension two

$$\mathbf{Q} = \mathbf{F}_0 \subset \mathbf{F}_1 \subset \mathbf{F}_2 \subset \cdots \subset \mathbf{F}_n \ni 2\cos\frac{2\pi}{17}$$

Since 17 is a prime, Proposition 38 shows that the polynomial

$$\Phi_{17}(z) = \frac{z^{17} - 1}{z - 1}$$

is irreducible over the integers. We use Proposition 25 about an extension generated by one element. Since $\deg \Phi_{17} = 16$, we conclude that

$$\left[\mathbf{Q}\left(\exp\frac{2\pi i}{17}\right) : \mathbf{Q}\right] = 16$$

We give an explicit construction of the tower with three two dimensional extensions

$$(\text{II.0.1}) \qquad \mathbf{Q} = \mathbf{F}_0 \subset \mathbf{F}_1 \subset \mathbf{F}_2 \subset \mathbf{Q}\left(\cos\frac{2\pi}{17}\right)$$

One further quadratic extension is needed to adjoin the imaginary parts $i\sin\frac{2\pi}{17}$, the calculation of which uses simply $\sin\alpha = \pm\sqrt{1 - \cos^2\alpha}$.

Here I explain my own semi empirical approach which I call the "pairing method". In this way, one avoids the primitive roots from number theory. Define

$$c_j = 2\cos\frac{2\pi j}{17}$$

Clearly c_j as a function of the integer j is symmetric and has period 17. We get the roots in question from $j = 1\ldots 8$. Moreover the addition theorem for cosine implies

$$(\text{II.0.2}) \qquad 2\cos\alpha\cos\beta = \cos(\alpha + \beta) + \cos(\alpha - \beta)$$

$$c_i \cdot c_j = c_{i+j} + c_{i-j}$$

The tower (II.0.1) in now reconstructed backwards. The third step in the tower constructs the c_i from four sums of pairs of the c_i. How can one find these four pairs? To get two roots c_i and c_j by means of a quadratic equation, we need to know both the sum $c_i + c_j$ and the product $c_i c_j = c_{i+j} + c_{i-j}$. To have the product available, the four pairs have to be chosen according to the following rules:

If c_i and c_j are paired, then c_{i+j} and c_{i-j} are paired, too.

If c_i and c_j are paired, then c_{2i} and c_{2j} are paired, too.

Problem 30. *Convince yourself that the only way to get four pairs from the numbers* $\{1, 2, 3, 4, 5, 6, 7, 8\}$ *is*

$$\{c_1, c_4\} \quad \{c_2, c_8\} \quad \{c_3, c_5\} \quad \{c_6, c_7\}$$

Solution. We see that 1 and 2 cannot be paired. Otherwise we would get the $2, 4$ which is contradictory to the rule. Can $1, 3$ be a pair? No, one would get the further pairs $2, 4$ and $2, 6$,—again contradictory to the rule. The next case $1, 4$ leads to the above success. $\qquad\square$

Now we guess that

$$\mathbf{F}_2 = \mathbf{Q}(c_1 + c_4) = \mathbf{Q}(c_2 + c_8) = \mathbf{Q}(c_3 + c_5) = \mathbf{Q}(c_6 + c_7)$$

To get the previous extension $\mathbf{F}_2/\mathbf{F}_1$, we need to find out which two pairs have to be added.

A bid of calculation shows

$$(c_1 + c_4)(c_2 + c_8) = \sum_{i=1}^{8} c_i \quad \text{and} \quad (c_3 + c_5)(c_6 + c_7) = \sum_{i=1}^{8} c_i$$

whereas in other such products some c_i occur more than once. We guess that

$$\mathbf{F}_1 = \mathbf{Q}(c_1 + c_4 + c_2 + c_8) = \mathbf{Q}(c_3 + c_5 + c_6 + c_7)$$

For the first extension we need still the product

$$(c_1 + c_4 + c_2 + c_8)(c_3 + c_5 + c_6 + c_7) = 4 \sum_{i=1}^{8} c_i$$

and the total sum $\sum_{i=1}^{8} c_i$. Before going on, recapitulate what we have obtained so far:

Lemma 15. *Let c_j be any real valued function of the integer j with the properties:*

$$c_j = c_{-j} = c_{17-j} \quad and \quad c_i \cdot c_j = c_{i+j} + c_{i-j}$$

for all integer i and j. Then

$$c_1 c_4 = c_3 + c_5 \quad and \quad c_3 c_5 = c_2 + c_8 \quad and \quad c_2 c_8 = c_6 + c_7 \quad and \quad c_6 c_7 = c_1 + c_4$$

$$(c_1 + c_4)(c_2 + c_8) = \sum_{i=1}^{8} c_i \quad and \quad (c_3 + c_5)(c_6 + c_7) = \sum_{i=1}^{8} c_i$$

$$(c_1 + c_4 + c_2 + c_8)(c_3 + c_5 + c_6 + c_7) = 4 \sum_{i=1}^{8} c_i$$

From a known $\sum_{i=1}^{8} c_i$, one can calculate all c_i with a tree of quadratic equations.

The sum of all eight c_i can be obtained from a complex geometric series

$$1 + \sum_{i=1}^{8} c_i = \sum_{k=0}^{16} \exp \frac{2\pi i\, k}{17} = \frac{\exp \frac{2\pi i\, 17}{17} - 1}{\exp \frac{2\pi i}{17} - 1} = 0 \quad \text{hence} \quad \sum_{i=1}^{8} c_i = -1$$

We can now write down the quadratic equations for the extensions. We check numerically which sign of the square root in their solutions corresponds to which partial sum of c_i.

1. The polynomial with the roots

$$x_1 = c_1 + c_4 + c_2 + c_8 \quad \text{and} \quad x_2 = c_3 + c_5 + c_6 + c_7 \quad \text{is}$$
$$(x - x_1)(x - x_2) = x^2 + x - 4 \quad \text{and the roots are}$$
$$x_1 = \frac{-1 + \sqrt{17}}{2} \quad \text{and} \quad x_2 = \frac{-1 - \sqrt{17}}{2}$$

We have to check numerically that the signs of $\sqrt{17}$ occur in that arrangement.

2. The polynomial with the roots

$$y_1 = c_1 + c_4 \quad \text{and} \quad y_2 = c_2 + c_8 \quad \text{is}$$
$$(y - y_1)(y - y_2) = y^2 - x_1 y - 1 \quad \text{and the roots are}$$
$$y_{1,2} = \frac{x_1 \pm \sqrt{x_1^2 + 4}}{2} \quad \text{hence}$$
$$y_1 = \frac{-1 + \sqrt{17} + \sqrt{34 - 2\sqrt{17}}}{4} \quad \text{and} \quad y_2 = \frac{-1 + \sqrt{17} - \sqrt{34 - 2\sqrt{17}}}{4}$$

We have to check numerically that the signs of the longer root occur in that arrangement.

3. The polynomial with the roots

$$y_3 = c_3 + c_5 \quad \text{and} \quad y_4 = c_6 + c_7 \quad \text{is}$$
$$(y - y_3)(y - y_4) = y^2 - x_2 y - 1 \quad \text{and the roots are}$$
$$y_{3,4} = \frac{x_2 \pm \sqrt{x_2^2 + 4}}{2} \quad \text{hence}$$
$$y_3 = \frac{-1 - \sqrt{17} + \sqrt{34 + 2\sqrt{17}}}{4} \quad \text{and} \quad y_4 = \frac{-1 - \sqrt{17} - \sqrt{34 + 2\sqrt{17}}}{4}$$

We have to check numerically that the signs of the longer root occur in that arrangement.

4. The polynomial with the roots

$$z_1 = c_1 \quad \text{and} \quad z_2 = c_4 \quad \text{is}$$
$$(z - z_1)(z - z_2) = z^2 - y_1 z + y_3 \quad \text{and the roots are}$$
$$z_{1,2} = \frac{y_1 \pm \sqrt{y_1^2 - 4y_3}}{2}$$

We have to check numerically that the signs occur in that arrangement.

5. The polynomial with the roots

$$z_3 = c_2 \quad \text{and} \quad z_4 = c_8 \quad \text{is}$$
$$(z - z_3)(z - z_4) = z^2 - y_2 z + y_4 \quad \text{and the roots are}$$
$$z_{3,4} = \frac{y_2 \pm \sqrt{y_2^2 - 4y_4}}{2}$$

We have to check numerically that the signs occur in that arrangement.

6. The polynomial with the roots

$$z_5 = c_3 \quad \text{and} \quad z_6 = c_5 \quad \text{is}$$
$$(z - z_5)(z - z_6) = z^2 - y_3 z + y_2 \quad \text{and the roots are}$$
$$z_{1,2} = \frac{y_3 \pm \sqrt{y_3^2 - 4y_2}}{2}$$

We have to check numerically that the signs occur in that arrangement.

7. The polynomial with the roots

$$z_7 = c_6 \quad \text{and} \quad z_8 = c_7 \quad \text{is}$$
$$(z - z_7)(z - z_8) = z^2 - y_4 z + y_1 \quad \text{and the roots are}$$
$$z_{3,4} = \frac{y_4 \pm \sqrt{y_4^2 - 4y_1}}{2}$$

We have to check numerically that the signs occur in that arrangement.

For the root z_1, I have finished the calculation and have obtained

$$16 \cos \frac{2\pi}{17} = -1 + \sqrt{17} + \sqrt{34 - 2\sqrt{17}}$$
$$+ \sqrt{\left(-1 + \sqrt{17} + \sqrt{34 - 2\sqrt{17}}\right)^2 + 16 + 16\sqrt{17} - 16\sqrt{34 + 2\sqrt{17}}}$$

Additionally to the above obtained result, computations with mathematica confirm the following result

Proposition 17 (The 17-gon with mathematica). *With $\pm l = 1 \ldots 8$ and $s_1, s_2, s_3 = \pm 1$, we obtain all eight values*

$$16 \cos \frac{2\pi l}{17} = -1 + s_1\sqrt{17} + s_2\sqrt{34 - 2s_1\sqrt{17}}$$
$$+ s_2 s_3 \sqrt{68 + 12 s_1\sqrt{17} - 2 s_2\sqrt{34 - 2s_1\sqrt{17}} + A + B}$$

with $A := +2 s_1 s_2 \sqrt{17(34 - 2s_1\sqrt{17})}$ and $B := -16 s_1 s_2 \sqrt{34 + 2s_1\sqrt{17}}$

in the following order

l	s_1	s_2	s_3	$(k\,j\,i)_2$	$3^{4k+2j+i} \pmod{17}$
1	$+1$	$+1$	$+1$	0	1
-15	$+1$	-1	-1	$110_2 = 6$	15
3	-1	$+1$	$+1$	1	3
-13	$+1$	$+1$	-1	$100_2 = 4$	13
5	-1	$+1$	-1	$101_2 = 5$	5
-11	-1	-1	-1	$111_2 = 7$	11
-10	-1	-1	$+1$	$10_2 = 3$	10
-9	$+1$	-1	$+1$	$10_2 = 2$	9

This order follows the rule

$$(\text{II.0.3}) \qquad |l| = 3^{4k+2j+i} \pmod{17} \quad with \ k = \frac{1-s_3}{2}, j = \frac{1-s_2}{2}, i = \frac{1-s_1}{2}$$

Again with l from the rule (II.0.3), there hold the following formulas given by Gauss' book Disquisitiones Arithmeticae—*in modern notation:*

$$(\text{II.0.4}) \quad 16\cos\frac{2\pi l}{17} = -1 + s_1\sqrt{17} + s_2\sqrt{34 - 2s_1\sqrt{17}}$$

$$+ 2s_2 s_3\sqrt{17 + 3s_1\sqrt{17} - s_2\sqrt{34 - 2s_1\sqrt{17}} - 2s_1 s_2\sqrt{34 + 2s_1\sqrt{17}}}$$

as well as the following additional less known formula

$$(\text{II.0.5}) \qquad 16\cos\frac{2\pi l}{17} = -1 + s_1\sqrt{17} + s_2\sqrt{34 - 2s_1\sqrt{17}}$$

$$+ 2s_2 s_3\sqrt{17 + 3s_1\sqrt{17} - s_1 s_2\sqrt{2(85 + 19s_1\sqrt{17})}}$$

Checks for proposition 17. This is a numerical check with mathematica, as well as a way to obtain formulas (III.2.2) and (II.0.5).

```
In[1]:= n = 2; Fermat = Function[n, 2^(2^n) + 1]; Fermat[n]
Out[1]= 17

In[2]:= Table[4 k + 2 j + i, {k, 0, 1}, {j, 0, 1}, {i, 0, 1}]
Out[2]= {{{0, 1}, {2, 3}}, {{4, 5}, {6, 7}}}

In[3]:= cos16 = -1 + Sqrt[17] s1 + Sqrt[34 - 2 Sqrt[17] s1] s2 +
    s2 *s3 \[Sqrt](68 + 12 Sqrt[17] s1
       - 2 Sqrt[34 - 2 Sqrt[17] s1] s2 +
         2 Sqrt[17] s1 Sqrt[34 - 2 Sqrt[17] s1] s2 -
         16 s1 Sqrt[34 + 2 Sqrt[17] s1] s2) ;

In[4]:= Table[mysigns
   = {s1 -> 1 - 2 i, s2 -> 1 - 2 j, s3 -> 1 - 2 k};
  {s1, s2, s3,
   N[(cos16 - 16 Cos[2 Pi*Mod[3^(4 k + 2 j + i), 17]/17])]} /.
  mysigns,  {k, 0, 1}, {j, 0, 1}, {i, 0, 1}]
```

```
Out[4]= {{{{1, 1, 1, -3.55271*10^-15}, {-1, 1, 1, 0.}},
          {{1, -1, 1, 0.}, {-1, -1, 1, -1.77636*10^-15}}},
         {{{1, 1, -1,  8.88178*10^-16},
           {-1, 1, -1, -8.88178*10^-16}},
          {{1, -1, -1, 3.55271*10^-15},
           {-1, -1, -1, 8.88178*10^-16}}}}

In[5]:= cos16g = -1 + s1 Sqrt[17] + s2 Sqrt[34 - 2 s1 Sqrt[17]] +
    2 *s2*s3 \[Sqrt](17 + 3 s1 Sqrt[17] -
        s2 Sqrt[34 - 2 s1 Sqrt[17]]
        - 2 s1*s2 Sqrt[34 + 2 s1 Sqrt[17]]);

In[6]:= Table[
 mysigns = {s1 -> 1 - 2 i, s2 -> 1 - 2 j, s3 -> 1 - 2 k};
    {s1, s2, s3,
 N[(cos16g - 16 Cos[2 Pi*Mod[3^(4 k + 2 j + i), 17]/17])]}
   /. mysigns, {k, 0, 1}, {j, 0, 1}, {i, 0, 1}]

Out[6]= {{{{1, 1, 1, -1.77636*10^-15}, {-1, 1, 1, 0.}},
          {{1, -1, 1, 0.}, {-1, -1, 1, -1.77636*10^-15}}},
         {{{1, 1, -1, 0.}, {-1, 1, -1, -8.88178*10^-16}},
          {{1, -1, -1, 0.}, {-1, -1, -1,  8.88178*10^-16}}}}

In[7]:= cos16new = -1 + s1 Sqrt[17] + s2 Sqrt[34
   - 2 s1 Sqrt[17]] + 2 s2*s3 \[Sqrt](17 + 3 s1 Sqrt[17]
        - s2*s1 Sqrt[2 (85 + 19 s1 Sqrt[17])]);

In[8]:= Table[mysigns
        = {s1 -> 1 - 2 i, s2 -> 1 - 2 j, s3 -> 1 - 2 k};
 {s1, s2, s3,
 N[(cos16new - 16 Cos[2 Pi*Mod[3^(4 k + 2 j + i), 17]/17])]}
 /.  mysigns, {k, 0, 1}, {j, 0, 1}, {i, 0, 1}]

Out[8]= {{{{1, 1, 1, -1.77636*10^-15}, {-1, 1, 1, 0.}},
    {{1, -1, 1, 0.}, {-1, -1, 1, 0.}}}, {{{1, 1, -1, 0.},
    {-1, 1, -1, -1.77636*10^-15}}, {{1, -1, -1, 0.},
    {-1, -1, -1, -8.88178*10^-16}}}}
```

```
In[9]:= Solve[Sqrt[17] s1 Sqrt[34 - 2 Sqrt[17] s1] -
    4 s1 Sqrt[34 + 2 Sqrt[17] s1]
          == -Sqrt[34 - 2 Sqrt[17] s1], {s1}]
Out[9]= {{s1 -> -1}, {s1 -> 1}}


In[10]:= RootReduce[(34 + 6 Sqrt[17] - Sqrt[34 - 2 Sqrt[17]]
+ Sqrt[17 (34 - 2 Sqrt[17])] - 8 Sqrt[34 + 2 Sqrt[17]])/8]


Out[10]= Root[17 - 85 #1 + 68 #1^2 - 17 #1^3 + #1^4 &, 3]


In[11]:= MinimalPolynomial[
 Root[17 - 85 #1 + 68 #1^2 - 17 #1^3 + #1^4 &, 3], x]
Out[11]= 17 - 85 x + 68 x^2 - 17 x^3 + x^4


In[12]:= Solve[17 - 85 x + 68 x^2 - 17 x^3 + x^4 == 0, x]
Out[12]= {{x -> 1/4 (17 - 3 Sqrt[17]
                    - Sqrt[2 (85 - 19 Sqrt[17])])},
{x -> 1/4 (17 - 3 Sqrt[17] + Sqrt[2 (85 - 19 Sqrt[17])])},
{x -> 1/4 (17 + 3 Sqrt[17] - 2 Sqrt[85/2 + (19 Sqrt[17])/2])},
{x -> 1/4 (17 + 3 Sqrt[17] + 2 Sqrt[85/2 + (19 Sqrt[17])/2])}}


In[14]:= Solve[
 Sqrt[2] s1 Sqrt[85 + 19 Sqrt[17] s1] ==
   Sqrt[34 - 2 Sqrt[17] s1]
        + 2 s1 Sqrt[34 + 2 Sqrt[17] s1], {s1}]
Out[14]= {{s1 -> -1}, {s1 -> 1}}
```

$\square$

Remark. The primitive root 3 has not been used,— and even is not need to be known—. I have no good explanation for formula (II.0.3). It seems to be an accidental coincidence that the above procedure agrees with Gauss' procedure using 3 as a primitive root.

Problem 31. *Check that the rule (II.0.3) holds similarly for all primitive roots of 17, but of course only after modification of the meaning of s_i.*

Problem 32. *Show by symbolic manipulation that indeed*

$$s_1\sqrt{17}\sqrt{34 - 2s_1\sqrt{17}} - 4s_1\sqrt{34 + 2s_1\sqrt{17}} = -\sqrt{34 - 2s_1\sqrt{17}}$$

holds for $s_1 = \pm 1$.

Problem 33. *Show by symbolic manipulation that indeed*

$$s_1 \sqrt{2(85 + 19 s_1 \sqrt{17})} = \sqrt{34 - 2 s_1 \sqrt{17}} + 2 s_1 \sqrt{34 + 2 s_1 \sqrt{17}}$$

holds for $s_1 = \pm 1$.

Of course, all these empirical checks of signs are still unsatisfactory. More important, one wants to set up a more systematic procedure at the very beginning. It is the merit of Gauss to have found satisfactory answers to the latter problem. Thus he opens up the possibility to construct the polygons for other Fermat primes.

The signs $s_1, s_2, s_3 = \pm$ correspond to the digits $i, j, k = 0$ or 1. We refer to the binary tree from the proof of Gauss' theorem in the next section. While building the tree from its root, the signs s_1, s_2, s_3 determine the branches and the signs $s_.$ are combined to the binary number $(k\,j\,i)_2 = 4k + 2j + i \in \mathcal{B}$.

The list of the seventeen remainders of 3^k modulo 17 for $k = 0, 1, \ldots, 15$ is

$$[\,1, 3, -8, -7\,,\ -4, 5, -2, -6\,,\ -1, -3, 8, 7\,,\ 4, -5, 2, 6\,]$$

We see they are all different. Hence one calls 3 a *primitive root* modulo 17. Consequently the sixteen numbers

$$d_r = \exp\left(\frac{2\pi i}{17} 3^r\right)$$

are a permutation of the roots of $\frac{z^{17}-1}{z-1} = 0$. After pairing conjugate complex roots d_r and d_{r+8}, we get the ordered list

$$[\,c_1, c_3, c_8, c_7\,,\ c_4, c_5, c_2, c_6\,] = d_r + d_{r+8} \quad \text{for } r = 0, 1, \ldots, 7$$

The pairs formed at the beginning of the empirical procedure, and sums of four in the next step are

$$c_1 + c_4 = d_0 + d_4 + d_8 + d_{12}$$
$$c_2 + c_8 = d_2 + d_6 + d_{10} + d_{14}$$
$$c_3 + c_5 = d_1 + d_5 + d_9 + d_{13}$$
$$c_6 + c_7 = d_3 + d_7 + d_{11} + d_{15}$$
$$c_1 + c_4 + c_2 + c_8 = d_0 + d_2 + d_4 + d_6 + d_8 + d_{10} + d_{12} + d_{14}$$
$$c_3 + c_5 + c_6 + c_7 = d_1 + d_3 + d_5 + d_7 + d_9 + d_{11} + d_{13} + d_{15}$$

One can see from sums of d_r much more easily how to set up the tree of quadratic equations. Indeed in Proposition 17

$$l \equiv (3^r \bmod 17) \quad \text{and} \quad r = \frac{1}{2}(1 - s_1) + (1 - s_2) + 2(1 - s_3)$$

These hints must be enough towards a guess how to generalize the procedure for other Fermat primes. Results which I have obtained by putting together Gauss' ideas with numerical computation are given in the next two sections.

II.0.2   Gauss' Polygons

Let $F_n \geq 5$ be a Fermat prime. The known cases are only $5, 17, 257$ and $65\,537$. The construction of the regular F_n-gon by ruler and compass involves the solution of the equation

$$(\text{II.0.6}) \qquad\qquad \Phi_{F_n}(z) = 0$$

by means of a binary tree of quadratic equations. Of course, Euler's formula $e^{it} = \cos t + i \sin t$ gives the analytic solutions

$$(\text{II.0.7}) \qquad\qquad z_k = \exp \frac{2k\pi i}{F_n} \quad \text{with } k = 1, 2, \ldots, F_n - 1.$$

Gauss' method proceeds by finding at first suitable sums of z_k. It turns out that these can be obtained from <u>quadratic equations</u> only. At first we remark that

$$s := \sum_{1 \leq k < F_n} z_k = -1$$

since the z_k and 1 are the roots of $z^{F_n} - 1 = 0$, and by Viëta's formula these roots have the sum zero.

Corollary 11 (A primitive root). *Especially $a = 3$ is a primitive root for the known Fermat primes $5, 17, 257$ and $65\,537$.*

Proof. Let h denote the order of a modulo F_n. This number h is defined to be the minimal number for which holds

$$a^h \equiv 1 \quad \bmod F_n$$

By Little Fermat one gets

$$h \mid F_n - 1 = 2^{2^n}$$

This implies $h = 2^t$ for some integer t. Now the assumption

$$a^{(F_n-1)/2} \not\equiv 1 \mod F_n$$

implies $t = 2^n$ and $h = F_n - 1$. Hence a is a primitive root. $\square$

By corollary 11, we get the primitive root $a = 3$ for any Fermat prime $F_n \geq 5$. Hence all $k \in \mathcal{U}(F_n)$ are the powers

$$k \equiv 3^r \mod F_n \quad \text{with } 0 \leq r < F_n - 1.$$

Since F_n is a prime holds $\phi(F_n) = F_n - 1 = 2^{2^n}$. Indeed we have obtained a bijection $r \in \mathbf{Z}(\phi(F_n)) \mapsto k \in \mathcal{U}(F_n)$. To determine $k \mod F_n$, it is enough to know $r \mod \phi(F_n)$. One puts this formula into equation (II.0.7) to obtain

$$(\text{II.0.8}) \qquad d_r := \exp \frac{2\pi i \cdot 3^r}{F_n} \quad \text{with } 0 \leq r < F_n - 1.$$

Hence it is convenient to put the exponent r into binary representation

$$(\text{II.0.9}) \qquad r = \sum_{0 \leq q < 2^n} b_q 2^q \text{ with the binary digits } b_q \in \{0, 1\}$$

Only 2^n binary digits are needed to determine $3^r \mod F_n$. We see that

$$\mathcal{B} = \{(b_{2^n-1} \ldots b_0) : b_q \in \{0, 1\} \text{ for } 0 \leq q < 2^n \}$$

is the entire set of binary digit-tuples involved. The corresponding binary integers are the set

$$(\text{II.0.10}) \qquad \mathcal{R}(n, 0, 0) = \{ \sum_{0 \leq q < 2^n} b_q 2^q : b_q \in \{0, 1\} \text{ for } 0 \leq q < 2^n \}$$

Formula (II.0.8) is now used to determine the 2^{2^n} vertices of the regular F_n-gon different from the exceptional vertex 1. Let $1 \leq q \leq 2^n$. For any fixed lowest digits $l_{q-1} \ldots l_0$ in formula (II.0.9) the low digits yield

$$(\text{II.0.11}) \qquad l = \sum_{0 \leq i < q} l_i 2^i \text{ with the binary digits } l_i \in \{0, 1\}$$

which is a number $0 \leq l < 2^q$. We define the subsets $\mathcal{S}(n, q, l) \subseteq \mathcal{U}(F_n)$ as well as sums of corresponding polygon vertices $S(n, q, l) \in \mathbb{C}$ by setting

$$\mathcal{S}(n, q, l) = \{3^r \in \mathcal{U}(F_n) : r = \sum_{0 \leq i < q} l_i 2^i + \sum_{q \leq j < 2^n} h_j 2^j \text{ and } h_j \in \{0, 1\}\} \quad \text{or}$$

$$(\text{II.0.12}) \qquad \mathcal{R}(n,q,l) = \{r \; : \; r = l + x2^q \text{ with } 0 \le x < 2^{2^n - q}\}$$

$$(\text{II.0.13}) \qquad \mathcal{S}(n,q,l) = \{3^r \in \mathcal{U}(F_n) \; : \; r \in \mathcal{R}(n,q,l)\}$$

$$(\text{II.0.14}) \qquad S(n,q,l) = \sum \{\exp \frac{2\pi i \cdot k}{F_n} \; : \; k \in \mathcal{S}(n,q,l)\}$$

At each vertex of the rooted binary tree with 2^n levels of descendents is put a suitable sum $S(n,q,l)$. Here $1 \le q \le 2^n$ is the level in the tree and $0 \le l < 2^q$ enumerates the vertices in a given level, say from left to right. At each tree vertex, the low digits (II.0.11) are fixed. The root of the binary tree has level 0. At the root is put the sum $s = -1$, which is the sum of all d_r.

At the vertices of the q-th level of the tree are put those sums for which the q lowest digits $l_{q-1} \dots l_0$ are fixed, and the sum is taken over all choices for the high digits $h_{2^n-1} \dots h_q$. The node $[l_{q-1} \dots l_0]$ gets the two children $[1, l_{q-1} \dots l_0]$ and $[0, l_{q-1} \dots l_0]$. Increasing the level by one, one more low digit becomes fixed, whereas the sums over the high digits get shorter by one digit. The numbers (II.0.8) are obtained at the 2^{2^n} leaves of the tree.

The sums $S(n,q,l)$ are needed during this process. They may be computated by means of quadratic equations. How does one get the sums on the $q+1$-st level from those on the q-th level? To get the two children' sums from its (one!) [3] parent by means of a quadratic equation, one needs to know their sum and product. The sum is easy:

$$P := S(n, q+1, l) + S(n, q+1, l + 2^q) = S(n, q, l)$$

To determine the product

$$Q := S(n, q+1, l) \cdot S(n, q+1, l + 2^q)$$

I need god, Gauss and good guesses. Clearly, the multiplication leads to addition of exponents.

$$S(n, q+1, l) \cdot S(n, q+1, l + 2^q)$$

$$(\text{II.0.15}) \quad = \sum_{k \in \mathcal{S}(n,q+1,l)} \exp \frac{2\pi i \cdot k}{F_n} \cdot \sum_{k' \in \mathcal{S}(n,q+1,l+2^q)} \exp \frac{2\pi i \cdot k'}{F_n}$$

$$(\text{II.0.16}) \quad = \sum \{\exp \frac{2\pi i \cdot (k + k')}{F_n} \; : \; (k, k') \in \mathcal{S}(n, q+1, l) \times \mathcal{S}(n, q+1, l + 2^q)\}$$

The sums $k + k'$ occurs over all pairs (k, k') from the Cartesian product set $\mathcal{S}(n, q+1, l) \times \mathcal{S}(n, q+1, l + 2^q)$. Not too convenient!

---

[3] In any rooted tree a node has always <u>only one</u> parent. But a parent node of a binary tree can have up to two children.

Finally the sums $S(n, q+1, l)$ and $S(n, q+1, l+2^q)$ can in the the next step be determined from the quadratic equation $x^2 - Px + Q = 0$ with the coefficients

$$P = S(n, q+1, l) + S(n, q+1, l+2^q)$$
$$Q = S(n, q+1, l) \cdot S(n, q+1, l+2^q)$$

Lemma 16. *From the root to the $2^n - 1$-st level all sums $S(n, q, l)$ turn out to be real.*

Reason. Let

$$(\text{II.0.17}) \qquad\qquad h = \frac{F_n - 1}{2} = 2^{2^n - 1}$$

Since $3^h \equiv -1 \mod F_n$, we get from equation (II.0.8) conjugate complex pairs $d_{r+h} = \overline{d_r}$ and

$$(\text{II.0.18}) \qquad\qquad d_r + d_{r+h} = 2\cos\frac{2\pi \cdot 3^r}{F_n}$$

is real for all r. Indeed, for $q < 2^n$ each set $\mathcal{S}(n, q, l)$ contains with any member 3^r a second member $3^{r+h} \equiv -3^r \mod F_n$. Since $d_{r+h} = \overline{d_r}$ are conjugate complex to each other, the sums $S(n, q, l) \in \mathbb{R}$ are sums of cosinus terms (II.0.18), and hence real. $\quad\square$

Lemma 17. $1 \leq q \leq 2^n - 1$. *To produce the $q+1$-st level sums from the q-th level sums one needs the product from equation (II.0.16). The special case $k + k' \equiv 0 \mod F_n$ occurs only for the maximal last level $q = 2^n - 1$. In this special case always hold $k + k' \equiv 0 \mod F_n$ and $S(n, q+1, l) \cdot S(n, q+1, l+2^q) = 1$.*

Reason. Let $k \in \mathcal{S}(n, q+1, l)$ and $k' \in \mathcal{S}(n, q+1, l+2^q)$. Let $k = 3^r \mod F_n$ and assume $k + k' \equiv 0 \mod F_n$. Hence $k' = 3^{r+h}$ with $h = 2^{2^n - 1}$. By definition (II.0.12) $r = l + x2^{q+1}$ and $0 \leq x < 2^{2^n - q - 1}$. Similarly $r + h = l + 2^q + y2^{q+1}$ and $0 \leq y < 2^{2^n - q - 1}$.

Calculation in the double exponent are done modulo $F_n - 1 = 2^{2^n}$, in other words with n binary digits. One gets

$$2^{2^n - 1} = h \equiv 2^q + (y - x)2^{q+1} \pmod{2^{2^n}} \text{ and } q \leq 2^n - 1$$

For $q < 2^n - 1$ we get a contradiction. Hence $k + k' \equiv 0$ is impossible. For the second last level $q = 2^n - 1$, because of equation (II.0.17) the congruence holds indeed. The node sets are conjugate complex pairs. Their sum gives the double cosins (II.0.18). $\quad\square$

I use the notation

$$\oplus A \times B := \text{Multiset}\{a + b : a \in A \text{ and } b \in B\}$$

mainly for the multiset

$$\oplus \mathcal{S}(n, q+1, l) \times \mathcal{S}(n, q+1, l+2^q)$$
$$= \text{Multiset } \{3^r + 3^{r'} : r \in \mathcal{R}(n, q+1, l) \text{ and } r' \in \mathcal{R}(n, q+1, l+2^q)\}$$

Lemma 18. *We now explain how to go to the last level. With $q = 2^n - 1$ and $h = 2^q$ we have to solve the quadratic equations $x^2 - Px + Q = 0$ with the coefficients*

$$P = S(n, 2^n - 1, l) = S(n, 2^n, l) + S(n, 2^n, l+h) = 2\cos\frac{2\pi \cdot 3^l}{F_n}$$

$$Q = S(n, 2^n, l) \cdot S(n, 2^n, l + 2^{2^n - 1}) = 1$$

As expected, one gets the solutions

$$S(n, 2^n, l), \ S(n, 2^n, l+h) = \exp \pm\frac{2\pi i \cdot 3^l}{F_n}$$

Reason. With $q = 2^n - 1$ and $2^q = h$ holds

$$(\text{II.0.19}) \qquad \mathcal{S}(n, 2^n - 1, l) = \{3^r \mod F_n : r = l + xh \text{ and } 0 \le x < 2\} = \{\pm 3^l\}$$

$$(\text{II.0.20}) \qquad S(n, 2^n - 1, l) = \sum \{\exp\frac{2\pi i \cdot k}{F_n} : k \in \{\pm 3^l\}\} = 2\cos\frac{2\pi \cdot 3^l}{F_n}$$

$$\oplus \mathcal{S}(n, 2^n, l) \times \mathcal{S}(n, 2^n, l+h)$$
$$= \text{Multiset } \{3^r + 3^{r'} : 3^r = 3^l \text{ and } 3^{r'} = -3^l\} = \{0\}$$
$$S(n, 2^n, l) \cdot S(n, 2^n, l+h) = 1$$

$$\square$$

Lemma 19. *To go to the first level one gets the equation $x^2 + x - 2^{n-2} = 0$ hence the solutions*

$$S(n, 1, 0) = \frac{-1 + \sqrt{F_n}}{2} \quad and \quad S(n, 1, 1) = \frac{-1 - \sqrt{F_n}}{2}$$

or exchanged. The signs of the roots are harder to know.

Remark. Formula (II.0.38) for $q = 0$ is

$$-\frac{F_n - 1}{4} = S(n, 1, 0) \cdot S(n, 1, 1) = e[n, 0, 0]S(n, 0, 0) = -e[n, 0, 0]$$
$$e[n, 0, 0] = 2^{2^n - 2}$$

The same result follows from lemma 26 below.

Reason. The set $\mathcal{S}(n,1,0)$ consists of the $h = 2^{2^n-1}$ quadratic residues, and the set $\mathcal{S}(n,1,1)$ of the 2^{2^n-1} quadratic non-residues. Assume that

$$a = 3^{2s} + 3^{2s'+1} \in \oplus \mathcal{S}(n,1,0) \times \mathcal{S}(n,1,1)$$

is a quadratic residue,— as for example $3^0 + 3^1 = 4$. Then $3^{2t}a = 3^{2(s+t)} + 3^{2(s'+t)+1}$ are quadratic residues for all t. By mapping $(s,s') \mapsto (s+t, s'+t)$ for $0 \le t < h$ one produces in this way h different quadratic residues. But $3^{2t-1}a = 3^{2(s'+t)} + 3^{2(s+t-1)+1}$ are quadratic non-residues for all t. By $(s,s') \mapsto (s'+t, s+t-1)$ for $0 \le t < h$ one obtains in this way h different quadratic non-residues. The entire set $\mathcal{S}(n,0,0) = \mathcal{U}(F_n)$ is covered, indeed $\frac{h}{2}$ times as one sees choosing $s' - s = 1 \ldots \frac{h}{2}$.

Remark. We claim that $3^{2s} + 3^{2s'+1}$ is a quadratic residue if $s + s'$ is even, as for example $3^0 + 3^2$. Moreover $3^{2s} + 3^{2s'+1}$ is a quadratic non-residue if $s + s'$ is odd, as for example $3^0 + 3^1$.

In other word, the product multiset $\oplus \mathcal{S}(n,1,0) \times \mathcal{S}(n,1,1)$ contains $\frac{h^2}{2}$ quadratic residues and $\frac{h^2}{2}$ quadratic non-residues and

$$\oplus \mathcal{S}(n,1,0) \times \mathcal{S}(n,1,1) = \tfrac{h}{2}\mathcal{S}(n,0,0) = 2^{2^n-2}\mathcal{U}(F_n)$$

The quadratic equation $x^2 - Px + Q = 1$ with the coefficients

$$P = S(n,1,0) + S(n,1,1) = S(n,0,0) = -1$$
$$Q = S(n,1,0) \cdot S(n,1,1) = -2^{2^n-2}$$

has the solutions $S(n,1,0)$, $S(n,1,1)$ given above. $\qquad\square$

The hard part is to set up the quadratic equations for the levels $0 < q < 2^n - 1$. We do not know much more than

$$\mathcal{S}(n,0,0) = \mathcal{U}(F_n)$$

$$\#\mathcal{S}(n,q,l) = 2^{2^n-q}$$

The sum $k + k'$ in formula (II.0.16) occurs over all pairs (k,k') from the Cartesian product set $\mathcal{S}(n,q+1,l) \times \mathcal{S}(n,q+1,l+2^q)$! This Cartesian product has $2^{2^n-q-1} \cdot 2^{2^n-q-1} = 2^{2(2^n-q-1)}$ members, many of which result in equal sums $k + k'$ since there are only 2^{2^n} choices.

Conjecture 1. *The first guess is that always $2^{2^n - 2q - 2}$ terms are equal and all 2^{2^n} possible values for $k + k'$ occur. But we see that is only possible for $q \le 2^{n-1} - 1$. One could get*

$$\oplus \mathcal{S}(n, q+1, l) \times \mathcal{S}(n, q+1, l+2^q) = 2^{2^n - 2q - 2} \mathcal{U}(F_n) \quad \text{for } q \le 2^{n-1} - 1.$$

The second guess is

$$\oplus \mathcal{S}(n, q+1, l) \times \mathcal{S}(n, q+1, l+2^q) = 2^{2^n - q - 2} \mathcal{S}(n, q, l^*) \quad \text{for } 2^{n-1} \le q < 2^n - 1.$$

and $l^* = (l+1) \mod 2^q$. The latter multiset has $2^{2^n - q}$ members, and each one occurs $2^{2^n - q - 2}$ times.

Conjecture 2. *The third guess is that for $2^{n-1} \le q < 2^n - 1$, always $2^{2^n - 2q - 2 + q'}$ terms are equal and all $2^{2^n - q'}$ terms occur*

$$\oplus \mathcal{S}(n, q+1, l) \times \mathcal{S}(n, q+1, l+2^q) = 2^{2^n - 2q - 2 + q'} S(n, q', l^*) \quad \text{for } 2^{n-1} \le q < 2^n - 1.$$

and $l^* = (l+1) \mod 2^{q'}$. The latter multiset has $2^{2^n - q'}$ members, and each one occurs $2^{2^n - q - 2 + q'} = 1$ times.

$$q' = 2q + 2 - 2^n, \; 2 \le q' < 2^n$$

Proposition 18 (The 17-gon is solved). *The checked results for the 17-gon from the previous section are:*

(II.0.21) $$S(2, 1, 0) \cdot S(2, 1, 1) = -4$$
(II.0.22) $$S(2, 2, 0) \cdot S(2, 2, 2) = -1$$
(II.0.23) $$S(2, 2, 1) \cdot S(2, 2, 3) = -1$$
(II.0.24) $$S(2, 3, 0) \cdot S(2, 3, 4) = S(2, 2, 1)$$
(II.0.25) $$S(2, 3, 1) \cdot S(2, 3, 5) = S(2, 2, 2)$$
(II.0.26) $$S(2, 3, 2) \cdot S(2, 3, 6) = S(2, 2, 3)$$
(II.0.27) $$S(2, 3, 3) \cdot S(2, 3, 7) = S(2, 2, 0)$$

This confirms both conjectures 1 and 2 to hold for $n = 2$ and $F_2 = 17$.

Proof. The results from the last section are in the present notation

$$(c_1 + c_4 + c_2 + c_8)(c_3 + c_5 + c_6 + c_7) = -4$$
$$S(2, 1, 0) \cdot S(2, 1, 1)$$
$$= (d_0 + d_2 + d_4 + d_6 + d_8 + d_{10} + d_{12} + d_{14})$$
$$\cdot (d_1 + d_3 + d_5 + d_7 + d_9 + d_{11} + d_{13} + d_{15})$$
$$= -4$$

$$(c_1 + c_4)(c_2 + c_8) = -1$$
$$S(2,2,0) \cdot S(2,2,2)$$
$$= (d_0 + d_4 + d_8 + d_{12})(d_2 + d_6 + d_{10} + d_{14}) = -1$$
$$(c_3 + c_5)(c_6 + c_7) = -1$$
$$S(2,2,1) \cdot S(2,2,3)$$

$$= (d_1 + d_5 + d_9 + d_{13})(d_3 + d_7 + d_{11} + d_{15}) = -1$$
$$c_i \cdot c_j = c_{i+j} + c_{i-j}$$
$$c_1 \cdot c_4 = c_3 + c_5$$
$$c_2 \cdot c_8 = c_6 + c_7$$
$$c_3 \cdot c_5 = c_2 + c_8$$
$$c_6 \cdot c_7 = c_1 + c_4$$

$$S(2,3,0) \cdot S(2,3,4) = (d_0 + d_8)(d_4 + d_{12}) = c_1 \cdot c_4 = c_3 + c_5$$
$$= (d_1 + d_5 + d_9 + d_{13}) = S(2,2,1)$$
$$S(2,3,1) \cdot S(2,3,5) = (d_1 + d_9)(d_5 + d_{13}) = c_3 \cdot c_5 = c_2 + c_8$$
$$= (d_2 + d_6 + d_{10} + d_{14}) = S(2,2,2)$$
$$S(2,3,2) \cdot S(2,3,6) = (d_2 + d_{10})(d_6 + d_{14}) = c_2 \cdot c_8 = c_6 + c_7$$
$$= (d_3 + d_7 + d_{11} + d_{15}) = S(2,2,3)$$
$$S(2,3,3) \cdot S(2,3,7) = (d_3 + d_{11})(d_7 + d_{15}) = c_7 \cdot c_6 = c_1 + c_4$$
$$= (d_0 + d_4 + d_8 + d_{12}) = S(2,2,0)$$

In short, we have checked

$$S(n, q+1, l) \cdot S(n, q+1, l+2^q) = -2^{2^n - 2q - 2} \quad \text{for } q \leq 2^{n-1} - 1.$$

hence $q = 0$ and $q = 1$. This turns out to be (II.0.21) (II.0.22) and (II.0.23).
The second guess is that for $2^{n-1} \leq q < 2^n$ hence $q = 2$,

$$S(2, 3, l) \cdot S(2, 3, l+4) = S(2, 2, l^*) \quad \text{for } q = 2.$$

This holds with $l^* = (l+1) \mod 4$ and turns out to be (II.0.24) (II.0.25) (II.0.26)
and (II.0.27). $\qquad \square$

The above conjectures turn out to work for the 17-gon, but are not correct for
the 257-gon. To save the day and prove at least some theoretical principles, I now
borrow an idea from Fourier analysis.

64

Lemma 20. *Let $0 \leq q < 2^n - 1$. Let $\mathcal{E}(n,q)$ be any finite multiset built from multiple copies of elements from $\mathcal{U}(F_n)$ such that elements 3^r and 3^s occur with equal multiplicities if $2^q \mid r - s$. Any such multiset is a linear combination*

$$\mathcal{E}(n,q) = \bigcup_{0 \leq k < 2^q} f_{[q,k]} \mathcal{S}(n,q,k)$$

with non-negative coefficients $c_{[q,k]}$. Moreover these coefficients are obtained with a Fourier-typ analysis. One gets

$$2^{2^n - q} \cdot f_{[q,k]} = \#\mathcal{E}(n,q) \cap \mathcal{S}(n,q,k)$$

for all $0 \leq k < 2^q$.

Proof. The linear combinations (20) exhaust the above specified multisets $\mathcal{E}(n,q)$. The orthogonality relation

$$\#\mathcal{S}(n,q,l) \cap \mathcal{S}(n,q,k) = 2^{2^n - q} \, \delta_{lk}$$

for $0 \leq l < 2^q$ and $0 \leq k < 2^q$ makes the Fourier analysis feasible. $\qquad\square$

Lemma 21. *Let $0 \leq q \leq 2^n - 1$. we may extend the definition*

(II.0.12) $$\mathcal{R}(n,q,l) = \{r \,:\, r = l + x2^q \text{ with } 0 \leq x < 2^{2^n - q}\}$$
(II.0.13) $$\mathcal{S}(n,q,l) = \{3^r \in \mathcal{U}(F_n) \,:\, r \in \mathcal{R}(n,q,l)\}$$
(II.0.14) $$S(n,q,l) = \sum \{\exp \frac{2\pi i \cdot k}{F_n} \,:\, k \in \mathcal{S}(n,q,l)\}$$

to all integer l. In that case holds

$$\mathcal{R}(n,q,l + 2^q) = \mathcal{R}(n,q,l) \quad (\text{mod } F_n - 1)$$
$$\mathcal{S}(n,q,l + 2^q) = \mathcal{S}(n,q,l)$$

Proof. Since $3^{F_n - 1} \equiv 1 \mod F_n$ one may simplify

$$\mathcal{R}(n,q,l + 2^q) = \{r \,:\, r = l + (x+1)2^q \text{ with } 0 \leq x < 2^{2^n - q}\}$$
$$= \{r \,:\, r = l + x2^q \text{ with } 1 \leq x \leq 2^{2^n - q}\}$$
$$= \{r \,:\, r = l + x2^q \text{ with } 1 \leq x < 2^{2^n - q}\} \cup \{l + 2^{2^n}\}$$
$$= \{r \,:\, r = l + x2^q \text{ with } 0 \leq x < 2^{2^n - q}\} \quad (\text{mod } (F_n - 1))$$
$$\mathcal{S}(n,q,l + 2^q) = \mathcal{S}(n,q,l)$$

$\qquad\square$

Lemma 22. *Let $0 \leq q < 2^n - 1$. we may extend the definition*

(II.0.28)
$$\mathcal{M}(n, q, l) := \oplus \mathcal{S}(n, q+1, l) \times \mathcal{S}(n, q+1, l+2^q)$$
$$= \text{Multiset } \{3^r + 3^{r'} : r \in \mathcal{R}(n, q+1, l) \text{ and } r' \in \mathcal{R}(n, q+1, l+2^q)\}$$

to all integer l. In that case holds

$$\mathcal{M}(n, q, l + 2^q) = \mathcal{M}(n, q, l)$$

Proof. One may simplify in the same way as in the previous lemma 21 and go on

$$\mathcal{S}(n, q, l + 2^q) = \mathcal{S}(n, q, l)$$
$$\mathcal{M}(n, q, l + 2^q) = \oplus \mathcal{S}(n, q+1, l+2^q) \times \mathcal{S}(n, q+1, l+2^{q+1})$$
$$= \oplus \mathcal{S}(n, q+1, l+2^q) \times \mathcal{S}(n, q+1, l)$$
$$= \mathcal{M}(n, q, l)$$

$\square$

Lemma 23. *Let $0 \leq q < 2^n - 1$ and $0 \leq l < 2^q$. The above Fourier-typ analysis has to be used for the delicate multiset (II.0.28). Any two elements $3^s, 3^t \in \mathcal{M}(n, q, l)$ with $2^q \mid s - t$ occur with the same multiplicity. (They need not be equal.) Moreover there exist nonnegative coefficients $c_{[q,l,k]}$ for $0 \leq l < 2^q$, $0 \leq k < 2^q$ such that*

(II.0.29)
$$\oplus \mathcal{S}(n, q+1, l) \times \mathcal{S}(n, q+1, l+2^q) = \bigcup_{0 \leq k < 2^q} c_{[q,l,k]} \mathcal{S}(n, q, k)$$

(II.0.30)
$$S(n, q+1, l) \cdot S(n, q+1, l+2^q) = \sum_{0 \leq k < 2^q} c_{[q,l,k]} S(n, q, k)$$

Remark. We can only prove that any two elements $3^s, 3^t \in \mathcal{M}(n, q, l)$ with $2^q \mid s - t$ occur with the same multiplicity, but they need not be equal.

Proof. Take any elements 3^s and 3^t with $2^q \mid s - t$ and assume 3^s occurs in the multiset $\mathcal{M}(n, q, l)$. By assumption $s = l + 2^q x$ and $t = l + 2^q y$ holds for some $0 \leq x, y < 2^{2^n - q}$ for which we may assume $y > x$. We may write

$$3^t = 3^{(l + (y-x)2^q) + 2^q x} \in \mathcal{M}(n, q, l + (y-x)2^q) = \mathcal{M}(n, q, l)$$

to let a wonder happen. $\square$

Lemma 24. *The coefficients from the last lemma 23 are*

$$2^{2^n - q} \cdot c_{[q,l,k]} = \#\mathcal{M}(n, q, l) \cap \mathcal{S}(n, q, k)$$
$$= \#\mathcal{S}(n, q, k) \cap [\oplus \mathcal{S}(n, q + 1, l) \times \mathcal{S}(n, q + 1, l + 2^q)]$$

Hence $2^{2^n - q} \cdot c_{[q,l,k]}$ is the number of solutions (r, r', s) of the congruence

$$(\text{II.0.31}) \qquad\qquad 3^r + 3^{r'} \equiv 3^s \quad (\text{mod } F_n)$$

with $r \in \mathcal{R}(n, q + 1, l)$, $r' \in \mathcal{R}(n, q + 1, l + 2^q)$ and $s \in \mathcal{R}(n, q, k)$.

Definition 19. Let $0 \le q < 2^n$ and $a \equiv 3^{2^q} \pmod{F_n}$ and fix the number d. We define the coefficient $e[n, q, d]$ as the number of solutions of the congruence

$$(\text{II.0.32}) \qquad a^{2x} + a^{1+2y} \equiv 3^d \pmod{F_n} \text{ with } 0 \le x, y < 2^{2^n - q - 1}$$

We define the coefficient $ex0[n, q, d]$ as the number of solutions of the congruence

$$(\text{II.0.33}) \qquad 1 + a^{1+2y} \equiv 3^d \pmod{F_n} \text{ with } 0 \le y < 2^{2^n - q - 1}$$

Lemma 25. *I go on to prepare the calculation of the coefficients $e[n, q, d]$.*

(i)

$$(\text{II.0.34}) \qquad e[n, q, d] = \sum_{0 \le x < 2^{2^n - q - 1}} ex0[n, q, d + x * 2^{q+1}]$$

(ii) *The coefficient $e[n, q, d]$ only depends on $d \mod 2^q$ and on $x \mod 2^{2^n - q - 1}$ and $y \mod 2^{2^n - q - 1}$*

(iii)

$$(\text{II.0.35}) \qquad \sum_{0 \le d < 2^q} e[n, q, d] = 2^{2^n - q - 2}$$

Proof. Fix n and define

$$t[x, y, q, d] := \left(3^{2x \cdot 2^q} + 3^{(1+2y) \cdot 2^q} - 3^d\right) \pmod{F_n}$$

Because of the Little Fermat Theorem, the quantity $t[x, y, q, d]$ depends only on x mod 2^{2^n-q-1} and y mod 2^{2^n-q-1} and d mod 2^{2^n}. The reader should check the identities

$$t[x, y, q, d] \cdot 3^{2^q} = t[y, x+1, d+2^q]$$
$$t[x, y, q, d] \cdot 3^{2^{q+1}} = t[x+1, y+1, d+2^{q+1}]$$
$$t[x, y, q, d] \cdot 3^{z2^{q+1}} = t[x+z, y+z, d+z2^{q+1}]$$

In terms of the characteristic function

$$\chi(P) = \begin{cases} 1 & \text{if P is true} \\ 0 & \text{if P is false} \end{cases}$$

equations (II.0.33) and (II.0.32) become

$$(\text{II.0.36}) \qquad ex0[n, q, d] = \sum_{0 \le y < 2^{2^n-q-1}} \chi(t[0, y, q, d] \equiv 0 \pmod{F_n})$$

$$e[n, q, d] = \sum_{0 \le x < 2^{2^n-q-1} \text{ and } 0 \le y < 2^{2^n-q-1}} \chi(t[x, y, q, d] \equiv 0 \pmod{F_n})$$

In the following proof all congruences are understood modulo F_n.

(i) The sum from item (i) extends over the entire period. Hence substitutions $x \mapsto x + z$ and $x \mapsto -x$ leave them invariant. Finally, we use the identities for t and get

$$\sum_{0 \le x < 2^{2^n-q-1}} ex0[n, q, d + x * 2^{q+1}]$$

$$= \sum_{0 \le x < 2^{2^n-q-1} \text{ and } 0 \le y < 2^{2^n-q-1}} \chi(t[0, y, d + x * 2^{q+1}] \equiv 0)$$

$$= \sum_{0 \le x < 2^{2^n-q-1} \text{ and } 0 \le y < 2^{2^n-q-1}} \chi(t[0, y - x, d - x * 2^{q+1}] \equiv 0)$$

$$= \sum_{0 \le x < 2^{2^n-q-1} \text{ and } 0 \le y < 2^{2^n-q-1}} \chi(t[x, y, q, d] \equiv 0)$$

$$= e[n, q, d]$$

as claimed.

(ii) To confirm the shorter period 2^q for the variable d, we use the fact that the substitution $x, y \mapsto y, x + 1$ is allowed for the sums over the entire periods. Finally, we use the identities for t and get

$$
\begin{aligned}
e[n, q, d + 2^q] &= \sum_{0 \leq x < 2^{2^n-q-1} \text{ and } 0 \leq y < 2^{2^n-q-1}} \chi(t[x, y, d + 2^q] \equiv 0) \\
&= \sum_{0 \leq x < 2^{2^n-q-1} \text{ and } 0 \leq y < 2^{2^n-q-1}} \chi(t[y, x + 1, d + 2^q] \equiv 0) \\
&= \sum_{0 \leq x < 2^{2^n-q-1} \text{ and } 0 \leq y < 2^{2^n-q-1}} \chi(t[x, y, q, d] \equiv 0) \\
&= e[n, q, d]
\end{aligned}
$$

confirming period 2^q.

(iii) Since all these 3^d are inequivalent, for fixed x and y the first sum equals 1. Periodicity 2^q implies the last two lines

$$
\sum_{0 \leq d < 2^{2^n}} \chi(t[x, y, q, d] \equiv 0) = 1
$$

$$
\sum_{0 \leq d < 2^{2^n} \text{ and } 0 \leq x < 2^{2^n-q-1} \text{ and } 0 \leq y < 2^{2^n-q-1}} \chi(t[x, y, q, d] \equiv 0) = 2^{2(2^n-q-1)}
$$

$$
\sum_{0 \leq d < 2^{2^n}} e[n, q, d] = 2^2(2^n - q - 1)
$$

$$
2^{2^n-q} \sum_{0 \leq d < 2^q} e[n, q, d] = 2^{2(2^n-q-1)}
$$

$$
\sum_{0 \leq d < 2^q} e[n, q, d] = 2^{2^n-q-2}
$$

$$\square$$

I go on with some counting.

Lemma 26. *For $0 \leq q < 2^n - 1$ and $0 \leq d < 2^q$, the counting (II.0.35) has the following consequences:*

(a) *If for some fixed value of q all $e[n, q, d]$ are equal, then $e[n, q, d] = 2^{2^n-2q-2}$. Especially we get $e[n, 0, 0] = 2^{2^n-2}$. But this simple outcome is only possible for $q \leq 2^{n-1} - 1$.*

(b) *For the range $2^{n-1} \leq q \leq 2^n - 2$ at most 2^{2^n-q-2} among the coefficients $e[n,q,d]$ are nonzero. Note that*

$$2^q > 2^{2^{n-1}-2} \geq 2^{2^n-q-2} \geq 1$$

If the number of coefficients $e[n,q,d] \neq 0$ is maximal, then all these coefficients are zero or one.

Especially for $q = 2^n - 2$, there exists a unique index $d*$ such that $e[n, 2^n - 2, d] = \delta_{d,d*}$

Lemma 27. *Let $0 \leq q < 2^n$*

(i) *Let $0 \leq l, k < 2^q$ and $d \equiv (k - l) \pmod{2^q}$. The number of solutions (r, r', s) of the congruence*

$$(\text{II.0.31}) \qquad\qquad 3^r + 3^{r'} \equiv 3^s \pmod{F_n}$$

with $r \in \mathcal{R}(n, q+1, l)$, $r' \in \mathcal{R}(n, q+1, l+2^q)$ and $s \in \mathcal{R}(n, q, k)$ equals $2^{2^n-q} \cdot e[n, q, d]$.

(ii) *$c_{[q,l,k]} = e[n, q, d]$ holds for $d \equiv (k - l) \mod 2^q$. Especially the coefficient $e[n, q, d]$ depends only on $d \mod 2^q$. Formula (II.0.29) simplifies to*

$$(\text{II.0.37}) \quad \oplus \mathcal{S}(n, q+1, l) \times \mathcal{S}(n, q+1, l+2^q) = \bigcup_{0 \leq d < 2^q} e[n, q, d]\mathcal{S}(n, q, l+d)$$

$$(\text{II.0.38}) \qquad S(n, q+1, l) \cdot S(n, q+1, l+2^q) = \sum_{0 \leq d < 2^q} e[n, q, d]S(n, q, l+d)$$

for $0 \leq l < 2^q$.

Proof. **(i)** Let $0 \leq l, k < 2^q$ and $d \equiv (k - l) \mod 2^q$. The number of solutions (r, r', s) of the congruence

$$(\text{II.0.31}) \qquad\qquad 3^r + 3^{r'} \equiv 3^s \pmod{F_n}$$

has to be counted under the restrictions $r \in \mathcal{R}(n, q+1, l)$, $r' \in \mathcal{R}(n, q+1, l+2^q)$ and $s \in \mathcal{R}(n, q, k)$. These restriction are

$$r = l + (2x)2^q \text{ with } 0 \leq 2x < 2^{2^n-q}$$
$$r' = l + (1 + 2y)2^q \text{ with } 0 \leq 2y < 2^{2^n-q}$$
$$s \equiv l + d + z2^q \pmod{2^q} \text{ with } 0 \leq z < 2^{2^n-q}$$

and the congruence (II.0.31) is equivalent to

$$3^l \cdot a^{2x} + 3^l \cdot a^{1+2y} \equiv 3^l \cdot 3^{d+\epsilon 2^q} \cdot a^{z-\epsilon} \pmod{F_n}$$

Here

$$\epsilon = \begin{cases} 0 & \text{if } d \geq 0; \\ 1 & \text{if } d < 0. \end{cases}$$

We need to simplify separately in four cases.

- Assume $d \geq 0$ and $z = 2u$. One gets

$$a^{2(x-u)} + a^{1+2(y-u)} \equiv 3^d \pmod{F_n}$$

- Assume $d \geq 0$ and $z = -1 + 2u$. One gets

$$a^{1+2(x-u)} + a^{2(y-u+1)} \equiv 3^d \pmod{F_n}$$

- Assume $d < 0$ and $z = 2u$. One getts

$$a^{1+2(x-u)} + a^{2(y-u+1)} \equiv 3^{d+2^q} \pmod{F_n}$$

- Assume $d < 0$ and $z = 1 + 2u$. One gets

$$a^{2(x-u)} + a^{1+2(y-u)} \equiv 3^{d+2^q} \pmod{F_n}$$

In each one of the four cases, the number of solutions for <u>fixed</u> z is equal to $e[n, q, d]$. Since z may assume $2^{2^n - q}$ values, counting both odd and even cases, the total number of solutions of equation (II.0.31) is $2^{2^n - q} \cdot e[n, q, d]$.

(ii) This item follows by using item (i) and the last lemma 24.

$\square$

Corollary 12 (Gauss' main result). *For all Fermat primes, the regular F_n-gon is constructible by ruler and compass.*

Remark. The first explicit constructions of a regular 257-gon were given by Magnus Georg Paucker (1822) [18] and Friedrich Julius Richelot (1832) [19]. A construction for a regular 65537-gon was first given by Johann Gustav Hermes (1894) [11]. The construction is very complex; Hermes spent 10 years completing the 200-page manuscript.

Proof. We just recapitulate the procedure outlined above. The regular F_n-gon is put into the complex plane with all vertices on the unit circle, and one vertex at $z = 1$. The remaining vertices are obtained from the formula

$$(\text{II.0.8}) \qquad d_r := \exp \frac{2\pi i \cdot 3^r}{F_n} \quad \text{with } 0 \le r < F_n - 1.$$

At each vertex of the rooted binary tree with 2^n levels of descendents is put the set

$$(\text{II.0.12}) \qquad \mathcal{R}(n, q, l) = \{r \ : \ r = l + x2^q \text{ with } 0 \le x < 2^{2^n - q}\}$$

At first are calculated the sums of vertices

$$(\text{II.0.14}) \qquad S(n, q, l) = \sum \{\exp \frac{2\pi i \cdot k}{F_n} \ : \ k \in \mathcal{S}(n, q, l)\}$$
$$= \sum \{\exp \frac{2\pi i \cdot 3^r}{F_n} \ : \ r \in \mathcal{R}(n, q, l)\}$$

This binary tree has a root which gets the level $q = 0$. At the root is put the sum $s = -1$, which is the sum of all d_r. The corresponding binary integers are the set

$$(\text{II.0.10}) \qquad \mathcal{R}(n, 0, 0) = \{ \sum_{0 \le q < 2^n} b_q 2^q \ : \ b_q \in \{0, 1\} \text{ for } 0 \le q < 2^n \}$$

Recursively are computated the sums $S(n, q, l)$ at the 2^q vertices of the q-th level. Here $0 \le q \le 2^n$ is the level in the tree and $0 \le l < 2^q$ enumerates the vertices in a given level, say from left to right. Assume that the sums $S(n, q, l)$ for the q-th level are already known. For the next level, the sums $S(n, q + 1, l)$ and $S(n, q + 1, l + 2^q)$ are the roots of the quadratic equation $x^2 - Px + Q = 0$ with the coefficients

$$P = S(n, q + 1, l) + S(n, q + 1, l + 2^q) = S(n, q, l)$$
$$(\text{II.0.30}) \qquad Q = S(n, q + 1, l) \cdot S(n, q + 1, l + 2^q) = \sum_{0 \le d < 2^q} e[n, q, d] S(n, q, l + d)$$

The coefficients P and Q indeed depend only on the sums of the q-th level and are already recursively known. Indeed, by lemma 27 the coefficient $e[n, q, d]$ for $0 \le q < 2^n - 1$ and $0 \le d < 2^q$ is the number of solutions of the congruence

$$(\text{II.0.32}) \qquad a^{2x} + a^{1+2y} \equiv 3^d \pmod{F_n}$$

with

$$a \equiv 3^{2^q} \pmod{F_n} \text{ and } 0 \le x, y < 2^{2^n - q - 1}$$

From the well-known formula for the roots of a quadratic equation, we obtain

$$(\text{II.0.39}) \qquad \Delta(n,q,l) = S(n,q,l)^2 - 4 \sum_{0 \le d < 2^q} e[n,q,d] S(n,q,l+d)$$

$$(\text{II.0.40}) \qquad S(n,q+1,l) = \frac{S(n,q,l) + \sigma(n,q,l)\sqrt{\Delta(n,q,l)}}{2}$$

$$(\text{II.0.41}) \qquad S(n,q+1,l+2^q) = \frac{S(n,q,l) - \sigma(n,q,l)\sqrt{\Delta(n,q,l)}}{2}$$

with the sign $\sigma(n,q,l) = \pm 1$, which has to be fixed numerically. For all levels $q < 2^n - 1$, all quantities are real. One obtains the result

$$(\text{II.0.42}) \qquad 2\cos\frac{2\pi \cdot 3^l}{F_n} = S(n, 2^n - 1, l) \quad \text{for all } 0 \le l < h$$

with

$$(\text{II.0.17}) \qquad h = \frac{F_n - 1}{2} = 2^{2^n - 1}$$

As explained by lemma 18, the last level with $q = 2^n - 1$ yields the complex solutions

$$S(n, 2^n, l), \; S(n, 2^n, l+h) = \exp \pm \frac{2\pi i \cdot 3^l}{F_n}$$

$\square$

Remark. Theoretically, the corollary 12 solves the construction problem <u>for all</u> possible F_n-gons, even in the case that some huge still unknown Fermat prime should exist.

On the other hand, so many calculations are involved that the result is totally impractical. One needs to work towards further simplifications to go beyond the 17-gon.

With the next corollary 13 I skip ahead a bid and use the numerical results explained below. Especially, this means that corollary 13 and remark II.0.2 only refer to the <u>known</u> Fermat primes, but not to any further hugh Fermat prime in case it might exist.

Corollary 13. *After having done numerical calculations for $F_2 = 17$, $F_3 = 257$, and $F_4 = 65\,537$, one gets the following results confirming the above lemma:*

(a) *For $q = 0$ and $q = 1$, all coefficients $e[n, q, d]$ are equal. Hence we get $e[n, 0, 0] = 2^{2^n - 2}$ and $e[n, 1, 0] = e[n, 1, 1] = 2^{2^n - 4}$. This simple outcome does not occur for any $q \geq 2$.*

(b) *For $n = 2, q \geq 2$ and $n = 3, q \geq 4$ and $n = 4, q \geq 11$, the number of coefficients $e[n, q, d] \neq 0$ is maximal, and hence all these coefficients are equal to one.*

Especially for $q = 2^n - 2$, there exists a unique index $d*$ such that $e[n, 2^n - 2, d] = \delta_{d,d*}$. One gets $d = 1$ for $n = 2$, and $d* = 56$ for $n = 3$, and $d* = 3072$ for $n = 4$.*

(c) *For $n = 3, q = 2, 3$ and $n = 4, 2 \leq q \leq 10$, the coefficients $e[n, q, d]$ take much more random values than I originally did expect. I cannot see any general simple pattern.*

Remark. With the values from Corollary 13 one gets for $q = 0$ as already obtained in lemma 19

$$Q(n, 0, 0) = e[n, 0, 0]S(n, 0, 0) = -2^{2^n - 2}$$
$$\Delta(n, 0, 0) = S(n, 0, 0)^2 - 4Q(n, 0, 0) = F_n$$
$$S(n, 1, 0) = \frac{-1 + \sigma(n, 0, 0)\sqrt{F_n}}{2}$$
$$S(n, 1, 1) = \frac{-1 - \sigma(n, 0, 0)\sqrt{F_n}}{2}$$

For $q = 1$ we obtain for $n = 2, 3, 4$

$$4S(2, 2, 0) = -1 + \sqrt{17} + \sqrt{34 - 2\sqrt{17}}$$

$$4S(3, 2, 0) = -1 + \sqrt{257} + \sqrt{514 - 2\sqrt{257}}$$

$$4S(4, 2, 0) = -1 + \sqrt{65\,537} - \sqrt{131\,074 - 2\sqrt{65\,537}}$$

The first value agrees with the result obtained earlier since

$$4S(2, 2, 0) = 4c_1 + 4c_4 = 8\cos\frac{2\pi}{17} + 8\cos\frac{8\pi}{17} = -1 + \sqrt{17} + \sqrt{34 - 2\sqrt{17}}$$

Problem 34. *Assume that $e[n, 1, 0] = e[n, 1, 1] = 2^{2^n - 4}$ and $\sigma[n, 0, 0] = 1$ hold ,—not only for $n = 2, 3, 4$ as we shall check numerically,—but for all n for which F_n turns* *out to be a Fermat prime. Check the resulting formulas*

$$S(n, 2, 0) = \frac{-1 + \sqrt{F_n} + \sigma(n, 1, 0)\sqrt{2F_n - 2\sqrt{F_n}}}{4}$$

$$S(n, 2, 2) = \frac{-1 + \sqrt{F_n} - \sigma(n, 1, 0)\sqrt{2F_n - 2\sqrt{F_n}}}{4}$$

$$S(n, 2, 1) = \frac{-1 - \sqrt{F_n} + \sigma(n, 1, 1)\sqrt{2F_n + 2\sqrt{F_n}}}{4}$$

$$S(n, 2, 2) = \frac{-1 - \sqrt{F_n} - \sigma(n, 1, 1)\sqrt{2F_n + 2\sqrt{F_n}}}{4}$$

Solution.

$$S(n, 1, 0) = \frac{-1 + \sqrt{F_n}}{2} \quad \text{and} \quad S(n, 1, 1) = \frac{-1 - \sqrt{F_n}}{2}$$

$$Q(n, 1, l) = e[n, 1, 0]S(n, 1, l) + e[n, 1, 1]S(n, 1, l + 1)$$
$$= 2^{2^n - 4}(S(n, 1, l) + S(n, 1, l + 1)) = -2^{2^n - 4}$$

$$\Delta(n, 1, 0) = S(n, 1, 0)^2 - 4Q(n, 1, 0) = \frac{1 + F_n - 2\sigma(n, 0, 0)\sqrt{F_n}}{4} + 2^{2^n - 2} = \frac{F_n - \sqrt{F_n}}{2}$$

$$\Delta(n, 1, 1) = \frac{F_n + \sqrt{F_n}}{2}$$

$$S(n, 2, 0) = \frac{S(n, 1, 0) + \sigma(n, 1, 0)\sqrt{\Delta(n, 1, 0)}}{2}$$
$$= \frac{-1 + \sqrt{F_n} + 2\sigma(n, 1, 0)\sqrt{\Delta(n, 1, 0)}}{4}$$
$$= \frac{-1 + \sqrt{F_n} + \sigma(n, 1, 0)\sqrt{2F_n - 2\sqrt{F_n}}}{4}$$

$$S(n, 2, 2) = \frac{-1 + \sqrt{F_n} - \sigma(n, 1, 0)\sqrt{2F_n - 2\sqrt{F_n}}}{4}$$

$$S(n, 2, 1) = \frac{-1 - \sqrt{F_n} + \sigma(n, 1, 1)\sqrt{2F_n + 2\sqrt{F_n}}}{4}$$

$$S(n, 2, 2) = \frac{-1 - \sqrt{F_n} - \sigma(n, 1, 1)\sqrt{2F_n + 2\sqrt{F_n}}}{4}$$

$\square$

These beginnings are still too simple to guess how to go on. Here are some steps I did to gather confidence for the heavy numerics to follow. Next I try to get more information from my pairing method earlier used successfully for the 17-gon. Take the next example with $F_n = 257$, thus $n = 3$ and $h = 128$. Let $\alpha = \frac{2\pi}{257}$. For the level $q = 2^n - 2 = 6$ occur the 64 pairs of double cosines $2\cos k\alpha$. One of them is

$$3^0,\ 3^{64},\ 3^{128},\ 3^{192} \equiv 1,\ -16,\ -1,\ 16 \quad (\mathrm{mod}\ 257)$$

Further ones are obtained from the addition theorem (II.0.2) of the cosin function to be

$$\{1, 16\},\ \{15, 17\},\ \{2, 32\},\ \{30, 34\},\ \{4, 64\},\ \{60, 68\},\ \{8, 128\},\ \{120, 136\}$$

The next pair $\{16, 256\} \equiv \{1, 16\}$ closes the ring. Hence there exist 8 such rings of each 8 cosine pairs.

Translate into Gauss' approach with the primitive root.

Remark. The pairs $(3^r \mod 257.r)$ are

$((1 . 0)$ $(2 . 48)$ $(3 . 1)$ $(4 . 96)$ $(5 . 55)$ $(6 . 49)$ $(7 . 85)$ $(8 . 144)$ $(9 . 2)$ $(10 . 103)$ $(11 . 196)$ $(12 . 97)$ $(13 . 106)$ $(14 . 133)$ $(15 . 56)$ $(16 . 192)$ $(17 . 120)$ $(18 . 50)$ $(19 . 125)$ $(20 . 151)$ $(21 . 86)$ $(22 . 244)$ $(23 . 28)$ $(24 . 145)$ $(25 . 110)$ $(26 . 154)$ $(27 . 3)$ $(28 . 181)$ $(29 . 94)$ $(30 . 104)$ $(31 . 242)$ $(32 . 240)$ $(33 . 197)$ $(34 . 168)$ $(35 . 140)$ $(36 . 98)$ $(37 . 219)$ $(38 . 173)$ $(39 . 107)$ $(40 . 199)$ $(41 . 19)$ $(42 . 134)$ $(43 . 207)$ $(44 . 36)$ $(45 . 57)$ $(46 . 76)$ $(47 . 61)$ $(48 . 193)$ $(49 . 170)$ $(50 . 158)$ $(51 . 121)$ $(52 . 202)$ $(53 . 89)$ $(54 . 51)$ $(55 . 251)$ $(56 . 229)$ $(57 . 126)$ $(58 . 142)$ $(59 . 118)$ $(60 . 152)$ $(61 . 138)$ $(62 . 34)$ $(63 . 87)$ $(64 . 32)$ $(65 . 161)$ $(66 . 245)$ $(67 . 100)$ $(68 . 216)$ $(69 . 29)$ $(70 . 188)$ $(71 . 163)$ $(72 . 146)$ $(73 . 44)$ $(74 . 11)$ $(75 . 111)$ $(76 . 221)$ $(77 . 25)$ $(78 . 155)$ $(79 . 22)$ $(80 . 247)$ $(81 . 4)$ $(82 . 67)$ $(83 . 15)$ $(84 . 182)$ $(85 . 175)$ $(86 . 255)$ $(87 . 95)$ $(88 . 84)$ $(89 . 102)$ $(90 . 105)$ $(91 . 191)$ $(92 . 124)$ $(93 . 243)$ $(94 . 109)$ $(95 . 180)$ $(96 . 241)$ $(97 . 167)$ $(98 . 218)$ $(99 . 198)$ $(100 . 206)$ $(101 . 75)$ $(102 . 169)$ $(103 . 201)$ $(104 . 250)$ $(105 . 141)$ $(106 . 137)$ $(107 . 31)$ $(108 . 99)$ $(109 . 187)$ $(110 . 43)$ $(111 . 220)$ $(112 . 21)$ $(113 . 66)$ $(114 . 174)$ $(115 . 83)$ $(116 . 190)$ $(117 . 108)$ $(118 . 166)$ $(119 . 205)$ $(120 . 200)$ $(121 . 136)$ $(122 . 186)$ $(123 . 20)$ $(124 . 82)$ $(125 . 165)$ $(126 . 135)$ $(127 . 81)$ $(128 . 80)$ $(129 . 208)$ $(130 . 209)$ $(131 . 7)$ $(132 . 37)$ $(133 . 210)$ $(134 . 148)$ $(135 . 58)$ $(136 . 8)$ $(137 . 72)$ $(138 . 77)$ $(139 . 38)$ $(140 . 236)$ $(141 . 62)$ $(142 . 211)$ $(143 . 46)$ $(144 . 194)$ $(145 . 149)$ $(146 . 92)$ $(147 . 171)$ $(148 . 59)$ $(149 . 227)$ $(150 . 159)$ $(151 . 9)$ $(152 . 13)$ $(153 . 122)$ $(154 . 73)$ $(155 . 41)$ $(156 . 203)$ $(157 . 78)$ $(158 . 70)$ $(159 . 90)$ $(160 . 39)$ $(161 . 113)$ $(162 . 52)$ $(163 . 237)$ $(164 . 115)$ $(165 . 252)$ $(166 . 63)$ $(167 . 233)$ $(168 . 230)$ $(169 . 212)$ $(170 . 223)$ $(171 . 127)$ $(172 . 47)$ $(173 . 54)$ $(174$

76

. 143) (175 . 195) (176 . 132) (177 . 119) (178 . 150) (179 . 27) (180 . 153) (181 . 93) (182 . 239) (183 . 139) (184 . 172) (185 . 18) (186 . 35) (187 . 60) (188 . 157) (189 . 88) (190 . 228) (191 . 117) (192 . 33) (193 . 160) (194 . 215) (195 . 162) (196 . 10) (197 . 24) (198 . 246) (199 . 14) (200 . 254) (201 . 101) (202 . 123) (203 . 179) (204 . 217) (205 . 74) (206 . 249) (207 . 30) (208 . 42) (209 . 65) (210 . 189) (211 . 204) (212 . 185) (213 . 164) (214 . 79) (215 . 6) (216 . 147) (217 . 71) (218 . 235) (219 . 45) (220 . 91) (221 . 226) (222 . 12) (223 . 40) (224 . 69) (225 . 112) (226 . 114) (227 . 232) (228 . 222) (229 . 53) (230 . 131) (231 . 26) (232 . 238) (233 . 17) (234 . 156) (235 . 116) (236 . 214) (237 . 23) (238 . 253) (239 . 178) (240 . 248) (241 . 64) (242 . 184) (243 . 5) (244 . 234) (245 . 225) (246 . 68) (247 . 231) (248 . 130) (249 . 16) (250 . 213) (251 . 177) (252 . 183) (253 . 224) (254 . 129) (255 . 176) (256 . 128))

$$3^0,\ 3^{64},\ 3^{128},\ 3^{192} \equiv 1,\ -16,\ -1,\ 16 \quad (\text{mod } 257)$$
$$3^{56},\ 3^{120},\ 3^{184},\ 3^{248} \equiv 15,\ 17,\ -15,\ -17 \quad (\text{mod } 257)$$
$$3^{48},\ 3^{112},\ 3^{176},\ 3^{240} \equiv 2,\ -32,\ -2,\ 32 \quad (\text{mod } 257)$$
$$3^{40},\ 3^{104},\ 3^{168},\ 3^{232} \equiv -34,\ 30,\ 34,\ -30 \quad (\text{mod } 257)$$
$$3^{32},\ 3^{96},\ 3^{160},\ 3^{224} \equiv 64,\ 4,\ -64,\ -4 \quad (\text{mod } 257)$$
$$3^{24},\ 3^{88},\ 3^{152},\ 3^{216} \equiv -60,\ -68,\ 60,\ 68 \quad (\text{mod } 257)$$
$$3^{16},\ 3^{80},\ 3^{144},\ 3^{208} \equiv -8,\ 128,\ 8,\ -128 \quad (\text{mod } 257)$$
$$3^8,\ 3^{72},\ 3^{136},\ 3^{200} \equiv 136,\ -120,\ -136,,\ 120 \quad (\text{mod } 257)$$
$$3^0,\ 3^{64},\ 3^{128},\ 3^{192} \equiv 1,\ -16,\ -1,\ 16 \quad (\text{mod } 257)$$

We may now determine $d*$ in lemma 26. As

$$\mathcal{S}(3,6,0) = \{3^0,\ 3^{64},\ 3^{128},\ 3^{192} \quad \text{mod } 257\} = \{1,\ -16,\ -1,\ 16\}$$

is subdivided into the two pairs $S(3,7,0) = \{\pm 1\}$ and $\mathcal{S}(3,7,2^6) = \{\pm 16\}$ we need (for the setup of the quadratic equations) to get

$$\oplus \mathcal{S}(3,7,0) \times \mathcal{S}(3,7,2^6) = \oplus\{1,-1\} \times \{-16,16\} = \{\pm 15, \pm 17\}$$
$$= \{3^{56},\ 3^{120},\ 3^{184},\ 3^{248} \quad \text{mod } 257\} = \mathcal{S}(3,6,56)$$

Hence $d* = 56$ in lemma 26.

We go back one level to $q = 5$. There occur 32 groups of 8. We take the one with $l = 0$.

$$S(3,5,0) = \{3^0,\ 3^{32},\ 3^{64},\ 3^{96},\ 3^{128},\ 3^{160},\ 3^{192},\ 3^{224} \quad \text{mod } 257\}$$
$$= \{1,\ 64,\ -16,\ 4,\ -1,\ -64,\ 16,\ -4\}$$

It is subdivided into $\mathcal{S}(3,6,0) = \{\pm 1, \pm 16\}$ and $\mathcal{S}(3,6,2^5) = \{\pm 4, \pm 64\}$ We have to realize the formula (II.0.37), it is enough with say $l = 0$,

$$\mathcal{M}(n,q) := \oplus \mathcal{S}(n, q+1, 0) \times \mathcal{S}(n, q+1, 2^q) = \bigcup_{0 \le d < 2^q} e[n, q, d]\mathcal{S}(n, q, d)$$

$$\oplus \mathcal{S}(3,6,0) \times \mathcal{S}(3,6,2^5) = \bigcup_{0 \le d < 32} e[3, 5, d]\mathcal{S}(3, 5, d)$$

$$\log_3 \mathcal{M}(n,q) = \bigcup_{0 \le d < 2^q} e[3, q, d]\mathcal{R}(n, q, d)$$

$$\frac{\log_3 \mathcal{M}(n,q) \cap \mathcal{R}(n,q,d)}{2^{2^n - q}} = e[3, q, d]$$

and extract from the present instance the coefficients $e[3, 5, d]$.

$$\oplus \mathcal{S}(3,6,0) \times \mathcal{S}(3,6,2^5) = \oplus \{\pm 1, \pm 16\} \times \{\pm 4, \pm 64\}$$
$$= \{\pm 3, \pm 5, \pm 63, \pm 65, \pm 12, \pm 20, \pm 48, \pm 80 \quad \mathrm{mod}\ 257\}$$
$$= 3^{\{1,55,87,161,97,151,193,247\}(+128)}$$
$$= 3^{\{1,129,55,183,87,215,161,33,97,225,151,23,193,65,247,119\}}$$
$$= 3^{\{23,33,55,65,87,97,119,129,151,161,183,193,215,225,247,257\}}$$
$$= 3^{\{23,55,87,119,151,183,215,247\} \cup 10 + \{23,55,87,119,151,183,215,247\}}$$
$$= S(3,5,23) \cup S(3,5,33)$$

Hence $e[3, 5, 23] = e[3, 5, 33] = 1$ and all other $e[3, 5, d] = 0$.

Remark. By lemma 27 item(iii) the coefficient $e[n, q, d]$ depends only on $d \mod 2^q$. Hence we have obtained $= e[3, 5, 1] = e[3, 5, 23] = 1$ which agrees with the numerical result of proposition 20 below.

For the level $q = 4$ there occur 16 groups of 16. One of them is

$$3^0, 3^{16}, 3^{32}, 3^{48}, 3^{64}, 3^{80}, 3^{96}, 3^{112}, 3^{128}, 3^{144}, 3^{160}, 3^{176}, 3^{192}, 3^{208}, 3^{224}, 3^{240}$$
$$\equiv 1, -8, 64, 2, -16, 128, 4, -32, -1, 8, -64, -2, 16, -128, -4, 32 \quad (\mathrm{mod}\ 257)$$

For the level $q = 3$ there occur 8 groups of 32. One of them is

$$3^0,\, 3^8,\, 3^{16},\, 3^{24},\, 3^{32},\, 3^{40},\, 3^{48},\, 3^{56},\, 3^{64},\, 3^{72},\, 3^{80},\, 3^{88},\, 3^{96},\, 3^{104},\, 3^{112},\, 3^{120},$$
$$3^{128},\, 3^{136},\, 3^{144},\, 3^{152},\, 3^{160},\, 3^{168},\, 3^{176},\, 3^{184},$$
$$3^{192},\, 3^{200},\, 3^{208},\, 3^{216},\, 3^{224},\, 3^{232},\, 3^{240},\, 3^{248}$$
$$\equiv 1,\, -121,\, -8,\, -60,\, 64,\, 34,\, 2,\, 15,\, -16,\, 137,\, 128,\, -68,\, 4,\, 30,\, -32,\, 17,$$
$$-1,\, 121,\, 8,\, 60,\, -64,\, 34,\, -2,\, -15,$$
$$16,\, 120,\, -128,\, 68,\, -4,\, -30,\, 32,\, -17 \quad (\mathrm{mod}\ 257)$$

I have made further guesses for the 257-gon, but they continued to turn out wrong. Finally, it doomed to me that there is no easy to guess pattern for the neither the 257-gon nor the $65\,537$-gon. Here is the computer program and the result obtained with the help of DrRacket:

```
(require math/base)
(require math/number-theory)

(define (Fermat n)(add1(expt 2(expt 2 n))))
(define proot 3)

;a^{2x} + a^{1+2y} \equiv 3^{d} \pmod{F_n}
;a \equiv  3^{2^q} \pmod {F_n} \,\text{ and } \;
;0\le x,y  < 2^{2^n-q-1}%\,,\; 0\le y < 2^{2^n-q-1}

(define (eshort n q)
  (define 2q (expt  2 q))
  (define large(expt 2(sub1(-(expt 2 n)q))))
  (define a(with-modulus(Fermat n)
                    (modexpt proot (expt 2 q))))
  (define (iter x y d incid result)
  (cond [(equal? d 2q) result]
        [(equal? y large)
          (if(equal? incid 0)
             (iter 0 0 (add1 d) 0 result)
             (iter 0 0 (add1 d) 0 (cons(cons d incid)result)))]
       [(equal? x large) (iter 0 (add1 y) d incid result)]
       [else (let*[(axy(with-modulus(Fermat n)
                (mod+(modexpt a(* 2 x))
```

```
                    (modexpt a(add1(* 2 y))))))
            (3d (with-modulus(Fermat n) (modexpt proot d)))
                (try(with-modulus(Fermat n)(mod- axy 3d)))]
          (if (equal? try 0)
              (iter(add1 x)y d (add1 incid) result)
              (iter(add1 x)y d incid result)))])))
    ; body of eshort n q calls iter:
    (reverse (iter 0 0 0 0 null)))

(define (R n q low)
   (let [(r(lambda(x)
       (with-modulus(sub1(Fermat n))
         (mod+ low(mod* x(modexpt 2 q))))))
        (large(expt 2(- (expt 2 n)q)))]
              (build-list large (lambda(i)(r i)))))

(define (S n q l)
   (map (lambda(r)
       (with-modulus (Fermat n) (modexpt proot r)))
       (R n q l)))

(define (Sum n q l)
   (let*[(alpha(/(* 2 pi)(Fermat n)))
         (lmod (with-modulus(expt 2 q) l))
         (anglelist(map(lambda(k)
             (\cos(* k alpha)))(S n q lmod)))]
       (foldl + 0 anglelist)))

(define (Q n q l)
  (define Qterms (map(lambda(eshortitem)
       (let*[(d(car eshortitem))
             (multi(cdr eshortitem))]
        (* multi(Sum n q (+ 1 d)))))
     (eshort n q)))
   (foldl + 0 Qterms))

(define (sigma n q l)
  (let*[(S (Sum n q l))
```

```
                   (Delta (-(sqr S)(* 4 (Q n q 1))))
                   (top(-(* 2 (Sum n (add1 q)1))S))]
                   (inexact->exact(round(/ top(sqrt Delta)))) ))

(displayln (let[(n 3)]
    (map (lambda(q)
          (let* [(enq(eshort n q))]
          (for-each (lambda(l)
                      (let[(item (list-ref enq l))]
                        (printf " e[~a,~a,~a]=" n q (car item))
                        (display (cdr item )))
                        (printf ",") null)
          (build-list(length enq) values)))
        (printf "\n")"end")
        (build-list(-(expt 2 n)1)values))))

(displayln (let[(n 3)]
    (map (lambda(q)
      (for-each (lambda(l)(printf " sigma(~a,~a,~a)="  n q l)
            (display (sigma n q l))null)
        (build-list(expt 2 q)values)) (displayln "")"end")
      (build-list(-(expt 2 n)1)values))))

(displayln (let[(n 4)]
    (map (lambda(q)
          (let* [(enq(eshort n q))]
          (for-each (lambda(l)
                      (let[(item (list-ref enq l))]
                        (printf " e[~a,~a,~a]=" n q (car item))
                        (display (cdr item )))
                        (printf ",") null)
          (build-list(length enq) values)))
        (printf "\n")"end")
        (build-list(-(expt 2 n)1)values))))

(displayln (let[(n 4)]
    (map (lambda(q)
      (for-each (lambda(l)(printf " sigma(~a,~a,~a)="  n q l)
```

```
      (display (sigma n q l))null)
    (build-list(expt 2 q)values)) (displayln "")"end")
  (build-list(-(expt 2 n)1)values)))))
```

Proposition 19. *Let $n = 2$ and $F_2 = 17$. As always, with primitive root 3, one obtains the nonzero structural constants $e[2, q, d] \neq 0$ with $q = 0, 1, 2$ and $0 \leq d < 2^q$ and the signs $\sigma(n, q, l)$ to be*

$$e[2, 0, 0] = 4,$$
$$e[2, 1, 0] = 1, e[2, 1, 1] = 1,$$
$$e[2, 2, 1] = 1,$$
$$\sigma(2, 0, 0) = 1,$$
$$\sigma(2, 1, 0) = 1, \; \sigma(2, 1, 1) = 1,$$
$$\sigma(2, 2, 0) = 1, \; \sigma(2, 2, 1) = 1, \; \sigma(2, 2, 2) = -1, \; \sigma(2, 2, 3) = -1$$

Proposition 20. *Let $n = 3$ and $F_3 = 257$. The nonzero structural constants $e[3, q, d] \neq 0$ with $q = 0, 1, 2, 3, 4, 5, 6$ and $0 \leq d < 2^q$ are the following ones*

$$e[3, 0, 0] = 64, \; e[3, 1, 0] = e[3, 1, 1] = 16,$$
$$e[3, 2, 0] = 2, \; e[3, 2, 1] = 5, \; e[3, 2, 2] = 4, \; e[3, 2, 3] = 5,$$
$$e[3, 3, 0] = 2, \; e[3, 3, 2] = 2, \; e[3, 3, 4] = 1, \; e[3, 3, 5] = 2, \; e[3, 3, 6] = 1,$$
$$e[3, 4, 0] = e[3, 4, 1] = e[3, 4, 2] = e[3, 4, 5] = 1,$$
$$e[3, 5, 1] = e[3, 5, 23] = 1, \; e[3, 6, 56] = 1$$

Remark. Many of the coefficients $e[3, q, d]$ depend on the choice of the primitive root a, as is shown by the following.

$$proot = 3$$
$$e[3, 0, 0] = 64, e[3, 1, 0] = 16, e[3, 1, 1] = 16,$$
$$e[3, 2, 0] = 2, e[3, 2, 1] = 5, e[3, 2, 2] = 4, e[3, 2, 3] = 5,$$
$$e[3, 3, 0] = 2, e[3, 3, 2] = 2, e[3, 3, 4] = 1, e[3, 3, 5] = 2, e[3, 3, 6] = 1,$$
$$e[3, 4, 0] = 1, e[3, 4, 1] = 1, e[3, 4, 2] = 1, e[3, 4, 5] = 1,$$
$$e[3, 5, 1] = 1, e[3, 5, 23] = 1, e[3, 6, 56] = 1$$

$$proot = 5$$
$$e[3,0,0] = 64, e[3,1,0] = 16, e[3,1,1] = 16,$$
$$e[3,2,0] = 2, e[3,2,1] = 5, e[3,2,2] = 4, e[3,2,3] = 5,$$
$$e[3,3,0] = 2, e[3,3,2] = 1, e[3,3,3] = 2, e[3,3,4] = 1, e[3,3,6] = 2,$$
$$e[3,4,0] = 1, e[3,4,3] = 1, e[3,4,7] = 1, e[3,4,14] = 1,$$
$$e[3,5,1] = 1, e[3,5,7] = 1, e[3,6,8] = 1$$

$$proot = 7$$
$$e[3,0,0] = 64, e[3,1,0] = 16, e[3,1,1] = 16,$$
$$e[3,2,0] = 2, e[3,2,1] = 5, e[3,2,2] = 4, e[3,2,3] = 5,$$
$$e[3,3,0] = 2, e[3,3,1] = 2, e[3,3,2] = 2, e[3,3,4] = 1, e[3,3,6] = 1,$$
$$e[3,4,0] = 1, e[3,4,1] = 1, e[3,4,10] = 1, e[3,4,13] = 1,$$
$$e[3,5,27] = 1, e[3,5,29] = 1, e[3,6,24] = 1$$

$$proot = 10$$
$$e[3,0,0] = 64, e[3,1,0] = 16, e[3,1,1] = 16,$$
$$e[3,2,0] = 2, e[3,2,1] = 5, e[3,2,2] = 4, e[3,2,3] = 5,$$
$$e[3,3,0] = 2, e[3,3,2] = 2, e[3,3,4] = 1, e[3,3,5] = 2, e[3,3,6] = 1,$$
$$e[3,4,0] = 1, e[3,4,1] = 1, e[3,4,2] = 1, e[3,4,5] = 1,$$
$$e[3,5,7] = 1, e[3,5,17] = 1, e[3,6,56] = 1$$

$$proot = 14$$
$$e[3,0,0] = 64, e[3,1,0] = 16, e[3,1,1] = 16,$$
$$e[3,2,0] = 2, e[3,2,1] = 5, e[3,2,2] = 4, e[3,2,3] = 5,$$
$$e[3,3,0] = 2, e[3,3,1] = 2, e[3,3,2] = 2, e[3,3,4] = 1, e[3,3,6] = 1,$$
$$e[3,4,0] = 1, e[3,4,1] = 1, e[3,4,10] = 1, e[3,4,13] = 1,$$
$$e[3,5,11] = 1, e[3,5,13] = 1, e[3,6,24] = 1,$$

Remark. We may now check the entire procedure. It is straightforward to calculate the sums $S(n, q, l)$ numerically. We solve equations (II.0.39) and (II.0.40) for the sign

$\sigma(n, q, l)$.

$$Q(n, q, l) = \sum_{0 \le d < 2^q} e[n, q, d] S(n, q, l + d)$$

$$\Delta(n, q, l) = S(n, q, l)^2 - 4Q(n, q, l)$$

$$\sigma(n, q, l) = \frac{2S(n, q+1, l) - S(n, q, l)}{\sqrt{\Delta(n, q, l)}}$$

These signs should turn out to be approximately ± 1, with small numerical defects. Indeed I have done this procedure successfully for $F_2 = 17$, which is rather easy, and already explained in the previous section.

More remarkably, this procedure is successful for $F_3 = 257$, too. As always, with primitive root 3, one obtains

$$\sigma(3, 0, 0) = 1 \quad \sigma(3, 1, 0) = 1 \quad \sigma(3, 1, 1) = 1$$
$$\sigma(3, 2, 0) = 1 \quad \sigma(3, 2, 1) = -1 \quad \sigma(3, 2, 2) = 1 \quad \sigma(3, 2, 3) = -1$$

$$\sigma(3, 3, 0) = 1 \quad \sigma(3, 3, 1) = 1 \quad \sigma(3, 3, 2) = 1 \quad \sigma(3, 3, 3) = 1$$
$$\sigma(3, 3, 4) = 1 \quad \sigma(3, 3, 5) = 1 \quad \sigma(3, 3, 6) = -1 \quad \sigma(3, 3, 7) = 1$$

$$\sigma(3, 4, 0) = 1 \quad \sigma(3, 4, 1) = 1 \quad \sigma(3, 4, 2) = 1 \quad \sigma(3, 4, 3) = 1$$
$$\sigma(3, 4, 4) = 1 \quad \sigma(3, 4, 5) = 1 \quad \sigma(3, 4, 6) = -1 \quad \sigma(3, 4, 7) = -1$$
$$\sigma(3, 4, 8) = -1 \quad \sigma(3, 4, 9) = -1 \quad \sigma(3, 4, 10) = 1 \quad \sigma(3, 4, 11) = -1$$
$$\sigma(3, 4, 12) = 1 \quad \sigma(3, 4, 13) = -1 \quad \sigma(3, 4, 14) = -1 \quad \sigma(3, 4, 15) = 1$$

$$\sigma(3, 5, 0) = 1 \quad \sigma(3, 5, 1) = 1 \quad \sigma(3, 5, 2) = -1 \quad \sigma(3, 5, 3) = 1$$
$$\sigma(3, 5, 4) = 1 \quad \sigma(3, 5, 5) = 1 \quad \sigma(3, 5, 6) = 1 \quad \sigma(3, 5, 7) = -1$$
$$\sigma(3, 5, 8) = -1 \quad \sigma(3, 5, 9) = -1 \quad \sigma(3, 5, 10) = -1 \quad \sigma(3, 5, 11) = -1$$
$$\sigma(3, 5, 12) = 1 \quad \sigma(3, 5, 13) = -1 \quad \sigma(3, 5, 14) = -1 \quad \sigma(3, 5, 15) = 1$$
$$\sigma(3, 5, 16) = -1 \quad \sigma(3, 5, 17) = -1 \quad \sigma(3, 5, 18) = -1 \quad \sigma(3, 5, 19) = -1$$
$$\sigma(3, 5, 20) = -1 \quad \sigma(3, 5, 21) = -1 \quad \sigma(3, 5, 22) = 1 \quad \sigma(3, 5, 23) = 1$$
$$\sigma(3, 5, 24) = -1 \quad \sigma(3, 5, 25) = -1 \quad \sigma(3, 5, 26) = 1 \quad \sigma(3, 5, 27) = 1$$
$$\sigma(3, 5, 28) = 1 \quad \sigma(3, 5, 29) = -1 \quad \sigma(3, 5, 30) = 1 \quad \sigma(3, 5, 31) = -1$$

84

$$\sigma(3,6,0) = 1 \quad \sigma(3,6,1) = 1 \quad \sigma(3,6,2) = 1 \quad \sigma(3,6,3) = 1$$
$$\sigma(3,6,4) = -1 \quad \sigma(3,6,5) = 1 \quad \sigma(3,6,6) = 1 \quad \sigma(3,6,7) = -1$$
$$\sigma(3,6,8) = -1 \quad \sigma(3,6,9) = -1 \quad \sigma(3,6,10) = -1 \quad \sigma(3,6,11) = 1$$
$$\sigma(3,6,12) = 1 \quad \sigma(3,6,13) = 1 \quad \sigma(3,6,14) = 1 \quad \sigma(3,6,15) = -1$$
$$\sigma(3,6,16) = 1 \quad \sigma(3,6,17) = 1 \quad \sigma(3,6,18) = 1 \quad \sigma(3,6,19) = 1$$
$$\sigma(3,6,20) = -1 \quad \sigma(3,6,21) = -1 \quad \sigma(3,6,22) = -1 \quad \sigma(3,6,23) = 1$$
$$\sigma(3,6,24) = 1 \quad \sigma(3,6,25) = -1 \quad \sigma(3,6,26) = 1 \quad \sigma(3,6,27) = -1$$
$$\sigma(3,6,28) = 1 \quad \sigma(3,6,29) = 1 \quad \sigma(3,6,30) = -1 \quad \sigma(3,6,31) = -1$$
$$\sigma(3,6,32) = -1 \quad \sigma(3,6,33) = -1 \quad \sigma(3,6,34) = -1 \quad \sigma(3,6,35) = 1$$

$$\sigma(3,6,36) = 1 \quad \sigma(3,6,37) = -1 \quad \sigma(3,6,38) = -1 \quad \sigma(3,6,39) = -1$$
$$\sigma(3,6,40) = -1 \quad \sigma(3,6,41) = -1 \quad \sigma(3,6,42) = -1 \quad \sigma(3,6,43) = -1$$
$$\sigma(3,6,44) = 1 \quad \sigma(3,6,45) = 1 \quad \sigma(3,6,46) = -1 \quad \sigma(3,6,47) = -1$$
$$\sigma(3,6,48) = 1 \quad \sigma(3,6,49) = 1 \quad \sigma(3,6,50) = 1 \quad \sigma(3,6,51) = 1$$
$$\sigma(3,6,52) = -1 \quad \sigma(3,6,53) = 1 \quad \sigma(3,6,54) = -1 \quad \sigma(3,6,55) = 1$$
$$\sigma(3,6,56) = 1 \quad \sigma(3,6,57) = 1 \quad \sigma(3,6,58) = -1 \quad \sigma(3,6,59) = -1$$
$$\sigma(3,6,60) = 1 \quad \sigma(3,6,61) = -1 \quad \sigma(3,6,62) = -1 \quad \sigma(3,6,63) = -1$$

From here it is routine to build the formula with the box square roots for the coordinates of the 257-gon vertices,—at least in principle!

For more convincing news, I take up mathematica.

II.0.3 The 257-gon with mathematica

This is a mathematica program that computates the numbers $e[3,q,d]$ and $\sigma[3,q,d]$ exactly; next computates numerically the Gaussian sums $S[3,q,d]$ by means of above obtained formulas

$$((\text{II.0.39})) \qquad \Delta(n,q,l) = S(n,q,l)^2 - 4 \sum_{0 \leq d < 2^q} e[n,q,d] S(n,q,l+d)$$

$$((\text{II.0.40})) \qquad S(n,q+1,l) = \frac{S(n,q,l) + \sigma(n,q,l)\sqrt{\Delta(n,q,l)}}{2}$$

$$((\text{II.0.41})) \qquad S(n,q+1,l+2^q) = \frac{S(n,q,l) - \sigma(n,q,l)\sqrt{\Delta(n,q,l)}}{2}$$

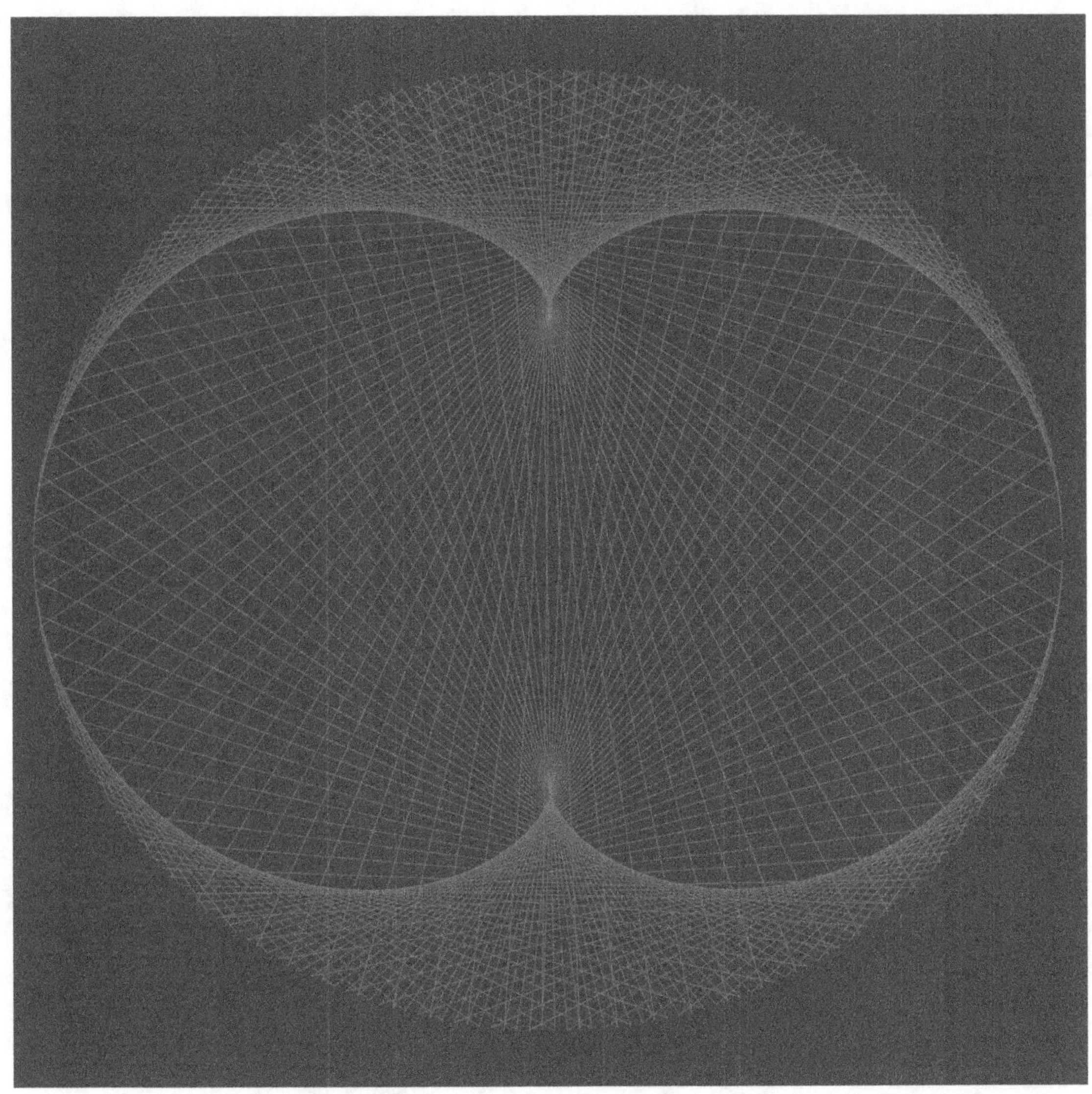

Figure 1: Vertices of a regular 257-gon connected as in $r \mapsto \exp\left[\frac{2\pi i \cdot 3^r}{257}\right]$

86

and finally checks that the values obtained for $S[3, 7, d]$ with $d = 0, 1, \ldots, 127$ agree with equation (II.0.42)

$$2 \cos \frac{2\pi \cdot 3^l}{F_n} = S(n, 2^n - 1, l) \ \text{ for all } 0 \leq l < h$$

by directly calculating $2 \cos\left[\frac{2\pi i \cdot 3^r}{257}\right]$ for $r = 0, 1, \ldots, 127 = h - 1$ and comparing.

```
In[1]:= n = 3; Fermat = Function[n, 2^(2^n) + 1]; Fermat[n]
Out[1]= 257

In[2]:= proot = PrimitiveRootList[Fermat[n]][[1]]
Out[2]= 3

In[3]:= Gauss = Function[{p, proot, d, s},
   e = (p - 1)/d;
   If[IntegerQ[e] && PrimeQ[p],
    ars = PowerMod[proot, Array[s + (# - 1)*d &, e ], p];
    eulersp = Function[t,
      exponent = 2*Pi *I*t/p;
      Exp[exponent]] ;
    Total[Map[eulersp, ars]],
    {"Nonprime or Noninteger", p, e} ]];

In[4]:= Gaussums = Function[{p, proot, d},
   Array[Gauss[ p, proot, d, #] &, d, 0]] ;

In[5]:= vertices = Gaussums[Fermat[n], proot, 2^(2^n)];

In[6]:= ListLinePlot[(Tooltip[{Re[#1], Im[#1]}] &)
                                /@ vertices,
 PlotStyle -> Directive[Hue[0.67, 0.6, 0.6],
 AbsoluteThickness[0.1]], AspectRatio -> 1,
                                Axes -> False]
In[7]:= Clear[sigma]; sigma[q_, l_]
   := sigma[q, l] =
  Sign[Re[N[Gauss[Fermat[n], proot, 2^(q + 1), l] -
      Gauss[Fermat[n], proot, 2^(q + 1), l + 2^q]]]]
```

```
In[8]:= baum = Array[Function[q,
          Array[{q, #} &, 2^q, 0]], 2^n, 0];

In[9]:= Clear[ex0]; ex0[q_, d_] := ex0[q, d] = With[{
    nq = 2^(2^n - q - 1),
    aq = PowerMod[3, 2^q, Fermat[n]]},
   count = 0; For[y = 0, y < nq, y++,
    try = 1 + PowerMod[aq, 2*y + 1, Fermat[n]] -
      PowerMod[3, d, Fermat[n]];
    If[Divisible[try, Fermat[n]], count++]]; count]

In[10]:= Clear[exsn]; exsn[q_, d_] := exsn[q, d] =
  Which[q == 0 , 2^(2^n - 2),
   q == 1 , 2^(2^n - 4),
   q >= 2 && q <= 2^n - 2,
   With[{nq = 2^(2^n - q - 1)},
    Sum[ex0[q, d + x*2^(q + 1)], {x, 0, nq - 1}]],
   True, Indeterminate]

In[11]:= Map[Function[pair, exsn[pair[[1]], pair[[2]]]],
 Most[baum], {2}]

Out[11]= {{64}, {16, 16}, {2, 5, 4, 5}, {2, 0, 2, 0, 1, 2, 1, 0},
 {1,  1, 1, 0, 0, 1, 0, 0, 0, 0, 0, 0, 0, 0, 0, 0},
 {0, 1, 0, 0, 0, 0, 0,  0, 0, 0, 0, 0, 0, 0, 0, 0, 0, 0, 0,
    0, 0, 0, 0, 1, 0, 0, 0, 0, 0, 0, 0, 0},
  {0, 0, 0, 0, 0, 0, 0, 0, 0, 0, 0, 0, 0, 0, 0, 0, 0, 0, 0, 0,
   0, 0, 0, 0, 0, 0, 0, 0, 0, 0, 0, 0,0, 0, 0, 0, 0, 0, 0, 0,0, 0,
   0, 0, 0, 0, 0, 0, 0, 0, 0, 0, 0, 0,1,0,0, 0, 0, 0, 0, 0}}

In[12]:= exsn[6, 56]
Out[12]= 1

In[13]:= Array[Function[q, Sum[exsn[q, d],
   {d, 0, 2^q - 1, 1}]],
 2^n - 1, 0]

Out[13]= {64, 32, 16, 8, 4, 2, 1}
```

88

```mathematica
In[14]:= Array[Function[q, 4 Sum[exsn[q, zweid],
    {zweid, 0, 2^q - 1, 2}]], 2^n - 1, 0]
Out[14]= {256, 64, 24, 24, 8, 0, 4}

In[15]:= Array[
 Function[q,
  2^(2^n - q - 1) +
   Sum[JacobiSymbol[1
             + PowerMod[3, (2 y + 1)*2^q, Fermat[n]],
     Fermat[n]],
    {y, 0, 2^(2^n - q - 1) - 1}]], 2^n - 1, 0]

Out[15]= {128, 64, 24, 24, 8, 0, 4}

In[16]:= Clear[nextS0];
   nextS0 = Function[{q, l, S, mysigma},
  P = S[q, l];
Q = N[Sum[exsn[q, d]*S[q, Mod[l+d,2^q]], {d,0,2^q-1}]];
   solutions = x /. NSolve[x^2 - P*x + Q == 0, x, Reals];
   If[mysigma == -1, solutions[[1]], solutions[[2]] ]];

In[17]:= nextS = Function[{q, l, S}, Which[
    q < 2^n && l >= 0 && l < 2^q,
                nextS0[q, l, S, sigma[q, l]],
    q < 2^n && l >= 2^q && l < 2^(q + 1),
    nextS0[q, l - 2^q, S, -1*sigma[q, l - 2^q]],
    True, Indeterminate]];

In[18]:= ClearAll[S]; S[q_, l_] := S[q, l] =
   Which[q == 0 && l == 0, -1,
    q > 0 && q < 2^n, nextS[q - 1, l, S],
    True, Indeterminate]

In[19]:= Map[Function[pair,sigma[pair[[1]],pair[[2]]]],
 Most[baum], {2}]

Out[19]= {{1}, {1, 1}, {1, -1, 1, -1},{1, 1, 1, 1, 1, 1, -1, 1},
```

```
{1, 1, 1, 1, 1, 1, -1, -1, -1, -1, 1, -1, 1, -1, -1, 1},
 {1, 1, -1, 1, 1, 1, 1, -1, -1, -1, -1, -1, 1, -1, -1, 1,
  -1, -1, -1, -1, -1, -1,1, 1, -1, -1, 1, 1, 1, -1, 1, -1},
  {1, 1, 1, 1, -1, 1,1, -1, -1, -1, -1, 1, 1, 1, 1, -1, 1, 1, 1, 1,
  -1, -1, -1, 1, 1, -1,1, -1, 1, 1, -1, -1, -1, -1, -1, 1, 1,
   -1, -1, -1, -1, -1, -1, -1,1, 1, -1, -1, 1, 1, 1, 1, -1,
    1, -1, 1, 1, 1,-1, -1, 1, -1, -1, -1}}

In[20]:=Map[Function[pair, S[pair[[1]], pair[[2]]]],
    Take[baum, 4], {2}];

In[21]:= Map[Function[pair, S[pair[[1]], pair[[2]]]],
  Take[baum, 6], {2}];

In[22]:= NSbaum =
  Map[Function[pair, S[pair[[1]], pair[[2]]]], baum, {2}];

In[23]:= zweicos = NSbaum[[8]];

In[24]:= Ndef =
  Function[cs2, vert = Fermat[n]*ArcCos[cs2/2]/(2*Pi);
   vert - Round[vert]];

In[25]:= Max[Map[Ndef, zweicos]] ;;
            Min[Map[Ndef, zweicos]]
Out[25]= 1.9061*10^-10 ;; -3.91367*10^-10

In[26]:= Nvert =
  Function[cs2, vert = Fermat[n]*ArcCos[cs2/2]/(2 Pi);
                        Round[vert]];

In[27]:= waslos = Map[Nvert, zweicos]
Out[27]= {1, 3,9,27,81,14,42,126,121,106,61,74,35,105,\
58,83,8,24,72,41,123,112,79,20,60,77,26,78,23,69,50,\
107,64,65,62,71,44,125,118,97,34,102,49,110,73,38,114,\
85,2,6,18,54,95,28,84,5,15,45,122,109,70,47,116,91,\
16,48,113,82,11,33,99,40,120,103,52,101,46,119,100,43,\
128,127,124,115,88,7,21,63,68,53,98,37,111,76,29,87,\
```

```
4,12,36,108,67,56,89,10,30,90,13,39,117,94,25,75,32,\
96,31,93,22,66,59,80,17,51,104,55,92,19,57,86}


In[28]:= twoxvertices = Gausssums[Fermat[n], proot, 2^(2^n - 1)];


In[29]:=  easy =
 Map[Fermat[n]*ExpToTrig[ArcCos[#/2]]/(2 Pi) &, twoxvertices]
Out[29]= {1, 3, 9, 27, 81, 14, 42, 126,121,106, 61, 74, 35, 105, \
58, 83, 8, 24, 72, 41, 123, 112, 79, 20, 60, 77, 26, 78, 23,69,50, \
107, 64, 65, 62, 71, 44, 125, 118, 97, 34,102,49,110,73,38,114, \
85, 2, 6, 18, 54, 95, 28, 84, 5,15,45,122, 109, 70, 47, 116, 91, \
16, 48, 113, 82, 11, 33, 99, 40, 120,103,52,101,46,119,100,43, \
128, 127, 124, 115, 88, 7, 21, 63, 68, 53,98,37,111, 76, 29, 87, \
4, 12, 36,108,67,56,89,10,30,90,13,39,117,94,25,75, 32, \
96, 31, 93, 22, 66, 59, 80, 17, 51,104, 55, 92, 19, 57, 86}


In[30]:= waslos == easy
Out[30]= True


In[31]:= GN = Array[Map[Re[N[#, 20]] &,
     Gausssums[Fermat[n], proot, 2^#]] &, 2^n,0];


In[32]:= Max[GN - NSbaum]  ;; Min[GN - NSbaum]
Out[32]= 9.57412*10^-12  ;;  -9.77995*10^-12


In[33]:= Map[Ndef, NSbaum[[8]]];
```

What about $F_4 = 65\,537$?

II.0.4   Computations for the $65\,537$-gon

The first programming attempt with DrRacket did not succeed because of memory
problems. One needs indeed produce the coefficients $e[4, q, d]$ with *one counting procedure* derived from equation (II.0.32), and has to avoid keeping any intermediate
results. In this manner the amount of computer time needed can be managed. Above
I have given only the program written according to this requirement. Now I finally

am going to write down the nonzero coefficients:

$$e[4, 0, 0] = 16384,$$
$$e[4, 1, 0] = 4096, e[4, 1, 1] = 4096,$$
$$e[4, 2, 0] = 992, e[4, 2, 1] = 1040, e[4, 2, 2] = 1024, e[4, 2, 3] = 1040,$$
$$e[4, 3, 0] = 284, e[4, 3, 1] = 237, e[4, 3, 2] = 272, e[4, 3, 3] = 237,$$
$$e[4, 3, 4] = 256, e[4, 3, 5] = 269, e[4, 3, 6] = 256, e[4, 3, 7] = 237,$$

$$e[4, 4, 0] = 80, e[4, 4, 1] = 62, e[4, 4, 2] = 60, e[4, 4, 3] = 64, e[4, 4, 4] = 57,$$
$$e[4, 4, 5] = 60, e[4, 4, 6] = 61, e[4, 4, 7] = 60, e[4, 4, 8] = 68, e[4, 4, 9] = 64,$$
$$e[4, 4, 10] = 64, e[4, 4, 11] = 58, e[4, 4, 12] = 65, e[4, 4, 13] = 70,$$
$$e[4, 4, 14] = 61, e[4, 4, 15] = 70,$$
$$e[4, 5, 0] = 4, e[4, 5, 1] = 12, e[4, 5, 2] = 20, e[4, 5, 3] = 13, e[4, 5, 4] = 20,$$
$$e[4, 5, 5] = 18, e[4, 5, 6] = 16, e[4, 5, 7] = 19, e[4, 5, 8] = 19, e[4, 5, 9] = 22,$$
$$e[4, 5, 10] = 12, e[4, 5, 11] = 22, e[4, 5, 12] = 13, e[4, 5, 13] = 13, e[4, 5, 14] = 11,$$
$$e[4, 5, 15] = 22, e[4, 5, 16] = 20, e[4, 5, 17] = 15, e[4, 5, 18] = 25, e[4, 5, 19] = 12,$$
$$e[4, 5, 20] = 16, e[4, 5, 21] = 12, e[4, 5, 22] = 16, e[4, 5, 23] = 17, e[4, 5, 24] = 29,$$
$$e[4, 5, 25] = 16, e[4, 5, 26] = 7, e[4, 5, 27] = 17, e[4, 5, 28] = 13, e[4, 5, 29] = 17,$$
$$e[4, 5, 30] = 13, e[4, 5, 31] = 11,$$

$$e[4, 6, 1] = 3, e[4, 6, 2] = 2, e[4, 6, 3] = 5, e[4, 6, 4] = 5, e[4, 6, 5] = 5,$$
$$e[4, 6, 6] = 2, e[4, 6, 7] = 5, e[4, 6, 8] = 2, e[4, 6, 9] = 6, e[4, 6, 10] = 5,$$
$$e[4, 6, 11] = 6, e[4, 6, 12] = 1, e[4, 6, 13] = 6, e[4, 6, 14] = 3, e[4, 6, 15] = 5,$$
$$e[4, 6, 16] = 6, e[4, 6, 17] = 8, e[4, 6, 18] = 3, e[4, 6, 19] = 5, e[4, 6, 20] = 2,$$
$$e[4, 6, 21] = 5, e[4, 6, 22] = 5, e[4, 6, 23] = 1, e[4, 6, 24] = 3, e[4, 6, 25] = 10,$$
$$e[4, 6, 26] = 3, e[4, 6, 27] = 3, e[4, 6, 28] = 4, e[4, 6, 29] = 5, e[4, 6, 30] = 1,$$
$$e[4, 6, 31] = 4, e[4, 6, 32] = 1, e[4, 6, 33] = 6, e[4, 6, 34] = 3, e[4, 6, 35] = 2,$$
$$e[4, 6, 36] = 1, e[4, 6, 37] = 1, e[4, 6, 38] = 7, e[4, 6, 39] = 3, e[4, 6, 40] = 3,$$
$$e[4, 6, 41] = 3, e[4, 6, 42] = 4, e[4, 6, 43] = 6, e[4, 6, 44] = 3, e[4, 6, 45] = 2,$$
$$e[4, 6, 46] = 3, e[4, 6, 47] = 4, e[4, 6, 48] = 7, e[4, 6, 49] = 7, e[4, 6, 50] = 4,$$
$$e[4, 6, 51] = 4, e[4, 6, 52] = 5, e[4, 6, 53] = 6, e[4, 6, 54] = 8, e[4, 6, 55] = 2,$$
$$e[4, 6, 56] = 2, e[4, 6, 57] = 4, e[4, 6, 58] = 3, e[4, 6, 59] = 4, e[4, 6, 60] = 3,$$
$$e[4, 6, 61] = 5, e[4, 6, 62] = 8, e[4, 6, 63] = 3,$$

$$e[4,7,0] = 2, e[4,7,1] = 1, e[4,7,3] = 1, e[4,7,4] = 1, e[4,7,6] = 1,$$
$$e[4,7,7] = 1, e[4,7,8] = 1, e[4,7,10] = 3, e[4,7,14] = 1, e[4,7,15] = 1,$$
$$e[4,7,16] = 1, e[4,7,20] = 1, e[4,7,22] = 1, e[4,7,25] = 1, e[4,7,26] = 1,$$
$$e[4,7,29] = 3, e[4,7,30] = 1, e[4,7,33] = 2, e[4,7,34] = 2, e[4,7,35] = 1,$$
$$e[4,7,36] = 2, e[4,7,37] = 1, e[4,7,38] = 2, e[4,7,40] = 2, e[4,7,41] = 2,$$
$$e[4,7,42] = 2, e[4,7,45] = 1, e[4,7,47] = 1, e[4,7,48] = 1, e[4,7,51] = 1,$$
$$e[4,7,56] = 1, e[4,7,57] = 5, e[4,7,58] = 1, e[4,7,59] = 2, e[4,7,60] = 1,$$
$$e[4,7,61] = 1, e[4,7,63] = 2, e[4,7,65] = 4, e[4,7,67] = 2, e[4,7,70] = 2,$$
$$e[4,7,73] = 2, e[4,7,74] = 2, e[4,7,76] = 2, e[4,7,79] = 2, e[4,7,80] = 1,$$
$$e[4,7,82] = 1, e[4,7,83] = 2, e[4,7,86] = 1, e[4,7,87] = 2, e[4,7,88] = 2,$$
$$e[4,7,89] = 1, e[4,7,90] = 1, e[4,7,94] = 2, e[4,7,95] = 1, e[4,7,97] = 2,$$
$$e[4,7,98] = 1, e[4,7,99] = 1, e[4,7,100] = 3, e[4,7,101] = 2, e[4,7,103] = 2,$$
$$e[4,7,104] = 2, e[4,7,105] = 3, e[4,7,106] = 4, e[4,7,108] = 2, e[4,7,109] = 2,$$
$$e[4,7,110] = 1, e[4,7,111] = 1, e[4,7,114] = 4, e[4,7,115] = 1, e[4,7,116] = 1,$$
$$e[4,7,117] = 2, e[4,7,118] = 1, e[4,7,119] = 1, e[4,7,121] = 2, e[4,7,123] = 1,$$
$$e[4,7,124] = 1, e[4,7,125] = 1, e[4,7,127] = 2,$$

$$e[4,8,3] = 2, e[4,8,4] = 1, e[4,8,7] = 1, e[4,8,14] = 1, e[4,8,19] = 1,$$
$$e[4,8,28] = 2, e[4,8,29] = 1, e[4,8,30] = 1, e[4,8,33] = 1, e[4,8,37] = 1,$$
$$e[4,8,39] = 1, e[4,8,41] = 1, e[4,8,43] = 1, e[4,8,44] = 1, e[4,8,50] = 1,$$
$$e[4,8,51] = 1, e[4,8,53] = 1, e[4,8,56] = 1, e[4,8,59] = 1, e[4,8,61] = 1,$$
$$e[4,8,65] = 1, e[4,8,68] = 1, e[4,8,70] = 1, e[4,8,78] = 1, e[4,8,79] = 1,$$
$$e[4,8,81] = 1, e[4,8,82] = 3, e[4,8,84] = 1, e[4,8,88] = 1, e[4,8,89] = 1,$$
$$e[4,8,106] = 1, e[4,8,109] = 1, e[4,8,112] = 1, e[4,8,117] = 1, e[4,8,124] = 1,$$
$$e[4,8,128] = 1, e[4,8,135] = 1, e[4,8,142] = 1, e[4,8,146] = 1, e[4,8,156] = 2,$$
$$e[4,8,173] = 1, e[4,8,175] = 1, e[4,8,185] = 2, e[4,8,186] = 1, e[4,8,187] = 1,$$
$$e[4,8,188] = 1, e[4,8,195] = 1, e[4,8,200] = 1, e[4,8,212] = 1, e[4,8,219] = 1,$$
$$e[4,8,231] = 1, e[4,8,233] = 1, e[4,8,243] = 1, e[4,8,245] = 2,$$
$$e[4,8,250] = 1,$$
$$e[4,8,252] = 1, e[4,8,253] = 1,$$

$$e[4,9,40] = 1, e[4,9,48] = 1, e[4,9,67] = 2, e[4,9,80] = 1, e[4,9,85] = 2,$$
$$e[4,9,87] = 2, e[4,9,91] = 1, e[4,9,105] = 1, e[4,9,108] = 1, e[4,9,113] = 1,$$
$$e[4,9,134] = 2, e[4,9,174] = 2, e[4,9,210] = 1, e[4,9,225] = 1, e[4,9,232] = 1,$$
$$e[4,9,268] = 1, e[4,9,274] = 1, e[4,9,277] = 1, e[4,9,280] = 1, e[4,9,302] = 1,$$
$$e[4,9,348] = 1, e[4,9,378] = 1, e[4,9,389] = 1, e[4,9,430] = 1, e[4,9,450] = 2,$$
$$e[4,9,464] = 1,$$

$$e[4,10,0] = 2, e[4,10,23] = 1, e[4,10,154] = 2, e[4,10,184] = 1,$$
$$e[4,10,308] = 1, e[4,10,359] = 1, e[4,10,530] = 1,$$
$$e[4,10,666] = 1, e[4,10,718] = 1, e[4,10,733] = 1,$$
$$e[4,10,777] = 2, e[4,10,840] = 1, e[4,10,945] = 1,$$

$$e[4,11,0] = 1, e[4,11,1] = 1, e[4,11,2] = 1,$$
$$e[4,11,777] = 1, e[4,11,800] = 1, e[4,11,1099] = 1,$$
$$e[4,11,1178] = 1, e[4,11,1263] = 1,$$
$$e[4,12,1] = 1, e[4,12,1265] = 1, e[4,12,1899] = 1, e[4,12,4003] = 1,$$
$$e[4,13,3164] = 1, e[4,13,8100] = 1,$$
$$e[4,14,3072] = 1,$$

I include in these recapitulation only the first few sign coefficients σ since I do not think they are really instructive.

$$\sigma(4,0,0) = 1 \quad \sigma(4,1,0) = -1 \quad \sigma(4,1,1) = -1$$
$$\sigma(4,2,0) = -1 \quad \sigma(4,2,1) = 1 \quad \sigma(4,2,2) = 1 \quad \sigma(4,2,3) = -1$$

Thus I may deduct that the construction of the $65\,537$-gon by straightedge and compass is in principle possible.

But one is still missing a serious check of all the above calculations! To this end, I can document more success using mathematica. Same as already for 257-gon, a complete calculation of the symbolic solutions for $2\cos\frac{2\pi l}{65\,537}$ is out of question.

Problem 35. *Estimate roughly how much memory and how much paper would be required.*

But I could at least achieve again a *numeric calculation*. With the standard 10 digit accuracy, the program dies at $q = 13$ because of the accumulation of rounding errors. I did use 35 digits, and I had to avoid the difference occuring in one of the quadratic formula (II.0.40) respectively (II.0.41) and use

$$S(n, q+1, l+2^q) = \frac{S(n, q+1, l) \cdot S(n, q+1, l+2^q)}{S(n, q+1, l)}$$
$$= \frac{\sum_{0 \leq d < 2^q} e[n, q, d] S(n, q, l+d)}{S(n, q+1, l)}$$

With those tricks, the calculation went through to level $q = 2^n - 1 = 15$. Finally I could check the result by recalculation of the counting number l from the numerical value for $2 \cos \frac{2\pi l}{65\,537}$:

```
N129 =  Map[Function[pair, NS[pair[[1]], pair[[2]]]],
                 teilbaum[16][[16]]];
Ndef = Function[cs2, vert
              = 65537 *ArcCos[cs2/2]/(2 Pi);
   def = vert - Round[vert]; N[def]];

Max[Map[Ndef, N129]]
0.000719844

Min[Map[Ndef, N129]]
-0.000725486
```

Hence we have obtained still 4 *accurate digits* for this monster calculation. These computations took about 6 hours. Now I am convinced that my construction of the regular 65 537-gon is correct.

Part III

About Polynomials and Fields

III.1 Polynomials and Fields

In this section, we need a lot of material from modern algebra. My main sources are Michael Artin's book [2] on *Algebra* and the relevant chapter in Robin Hartshorne's book [10].

III.1.1 A Reminder about Polynomials

The integers are denoted by $\mathbf{Z}$. For an *indeterminant* or *variable* x, the expressions $1, x, x^2, x^3, \ldots$ are called monomials. An integer combination of monomials $a_n x^n + a_{n-1} x^{n-1} + \cdots + a_1 x + a_0$ with integer coefficients $a_n, a_{n-1}, \ldots, a_0 \in \mathbf{Z}$ is called an *integer polynomial* or *polynomial over the integers*. The set of all integer polynomials is denoted by $\mathbf{Z}[x]$. One has to put the *indeterminant into square brackets*.

The set of integers $\mathbf{Z}$ and the set $\mathbf{Z}[x]$ are the first examples for a *ring*. The reader should recall the definition of a ring.

Definition 20 (ring). A *ring* is a collection of elements $a, b, \ldots$ which contains the neutral element 0 and allows two binary operations $+$ and $\cdot$ such that for all a, b in the ring, the sum $a + b$ and the product $a \cdot b = ab$ are elements of the ring, too. Moreover for all a, b, c in the ring hold the following arithmetic rules:

neutral element $0 + a = a$

negative element for all a exists the negative element $-a$ such that $a + (-a) = 0$.

commutativity $a + b = b + a$

associativity $(a + b) + c = a + (b + c)$

left distributivity $(a + b)c = ac + bc$

right distributivity $c(a + b) = ca + cb$

For any ring R, we can define the polynomial ring $R[x]$ by taking elements [4] $a_n, a_{n-1}, \ldots, a_0 \in R$ as coefficients in the respective polynomials. Again, provided that R is a ring, the set of polynomials $R[x]$ is a ring, too. It is called the *polynomial ring over the ring R*.

A polynomial with the leading monomial $a_n x^n$ and $a_n \neq 0$ is defined to have the *degree* n. A polynomial with the leading monomial x^n is called *monic*.

---

[4] Many authors use the notation $R[X]$ with capital X. But capitals denote sets, too. Therefore I prefer the notation $R[x]$.

The polynomials of *positive degree* are those with $n \geq 1$ and hence those in which the indeterminant x really appears.

The polynomials with degree 0 are the elements $a_0 \neq 0$ of the ring R. The zero polynomial has the degree undefined or $-\infty$. [5]

Remark. Anyway holds $\deg 0 \neq 0$. Some authors even write $\deg P \geq 0$ to indicate that the polynomial $P \neq 0$.

Definition 21 (field). A *field* is a collection of elements $a, b, c \ldots$ which contains the neutral element 0, the unit element 1, and allows two binary operations $+$ and $\cdot$ such that for all a, b in the field, the sum $a + b$ and the product $a \cdot b = ab$ are elements of the field, too. Moreover for all a, b, c hold the following arithmetic rules:

neutral element $0 + a = a$

negative element for all a exists the negative element $-a$ such that $a + (-a) = 0$.

commutativity $a + b = b + a$

associativity $(a + b) + c = a + (b + c)$

distributivity $(a + b)c = ac + bc$

unit element $1 \cdot a = a$

inverse element for all $a \neq 0$ exists the inverse element a^{-1} such that $aa^{-1} = 1$.

commutativity $ab = ba$

associativity $(ab)c = a(bc)$

A ring is called to have no null divisors if and only if $a \cdot b = 0$ implies that either $a = 0$ or $b = 0$. A ring with no null divisors can be embedded into a quotient field. For any field $\mathbf{F}$, one can define the polynomial ring $\mathbf{F}[x]$. This is not a field, but a ring with important nice properties. Especially the following algorithms work:

- division with remainder;

- the Euclidean algorithm;

- the extended Euclidean algorithm.

---

[5] To put $\deg 0 = -1$ seems to me no good choice neither.

- There exists a greatest common divisor for any two nonzero polynomials.

Division with remainder is possible for polynomials over a field. For any $p, q \in \mathbf{F}[x]$ with $q \neq 0$, we say

$$p : q \text{ is } a \text{ with remainder } r$$

as a short-hand expressing the polynomial equation

$$p = aq + r \quad \text{with} \quad \deg r < \deg q$$

The polynomials a and r are unique up to nonzero factors in the field $\mathbf{F}$. Of course, the remainder r may turn out to be zero.

Furthermore, the Euclidean algorithm and even the *extended Euclidean algorithm* can be performed in the ring $\mathbf{F}[x]$, and always stops after finitely many steps. Hence there exists a greatest common divisor $\gcd(p, q)$ for any two not both zero polynomials $p, q \in \mathbf{F}[x]$.

Proposition 21. *Any two polynomials $p, q \in \mathbf{F}[x]$ not both of which are zero have a greatest common divisor $\gcd(p, q) \in \mathbf{F}[x]$. Moreover, there exist polynomials $r, s \in \mathbf{F}[x]$ such that*

$$rp + sq = \gcd(p, q)$$

The polynomials $\gcd(p, q) \neq 0, r, s$ are only unique up to nonzero factors in the field $\mathbf{F}$.

III.1.2   Lagrange's Theorem

Theorem 7 (Lagrange's Theorem). *Let R be an integral domain or F be any field. A polynomial $p(x) \in R[x]$ or $p(x) \in F[x]$ has at most as many roots as its degree d.*

Especially a polynomial $f(x) \in \mathbf{Z}_p[x]$, with any prime p, has at most as many roots as its degree.

Corollary 14. *Let p be a prime number. Assume that for the monic polynomial $P \in \mathbf{Z}_p[x]$ there are known as many zeros $a_1, \ldots, a_d$ as the degree $d = \deg P$ allows. Then*

$$(III.1.1) \qquad P(x) \equiv \prod_{1 \leq i \leq d} (x - a_i) \quad (\text{mod } \mathbf{Z}_p)$$

Proof. Consider the difference

$$P(x) - \prod_{1 \leq i \leq d} (x - a_i) \in \mathbf{Z}_p[x]$$

Its degree is at most $d - 1$ but it has d zeros. This is impossible for a nonzero polynomial. Hence the difference is zero. $\square$

Problem 36. *Take the ring $\mathbf{Z}_9$ and the polynomial $x(x - 3)(x^2 - 1)(x^2 - 4)$ as an instructive counterexample. Which important assumption about the ring is missing ? What is the degree. How many zeros does the polynomial have.*

Answer. This is an example of a polynomial of degree six with nine zeros. The ring $\mathbf{Z}_9$ is not an integral domain. The "too many" zeros $4, 5, 6$ occur because of two factors divisible by 3, whereas no factor is divisible by nine. $\square$

Proposition 22. *The polynomial ring $\mathbf{Z}_m$ is an integral domain if and only if m is a prime number.*

III.1.3   Algebraic Numbers

Algebraic numbers, algebraic integers, and their conjugates can be defined over any base field $\mathbf{F}$. In case no base field is mentioned, it is assumed, in these notes and in many books by many authors, the base field are the rational numbers.

By a *field extension* is meant a pair of fields $F \subseteq K$. It is customary to write K/F for a field extension. The most common field extensions are $\mathbb{R}/\mathbf{Q}$, $\mathbb{C}/\mathbf{Q}$ and $\mathbb{C}/\mathbb{R}$.

A number α is called *algebraic* or, more accurately, *algebraic over a field* $\mathbf{F}$, if there exists a nonzero polynomial $p \in \mathbf{F}[x]$ such that $p(\alpha) = 0$. In that case α is called a *root of the polynomial p*. The lowest degree of polynomials with the root α is called the *degree of the algebraic number* α. The corresponding polynomial is called the *minimal polynomial of the algebraic number* α. Using division with remainder, it is easy to see that the minimal polynomial is unique up to a factor in $\mathbf{F}$.

A number α is called an *algebraic integer* if it is the root of a nonzero <u>monic</u> minimal polynomial $p \in \mathbf{Z}[x]$.

The numbers $a \in \mathbf{F}$ in the base field, including zero [6], have the degree one. Indeed, $p(x) = x - a$ is a polynomial of degree one with the root a. All other numbers $a \notin \mathbf{F}$ have either degree at least two, or they are never roots for any finite dimensional field extension K/F. The numbers, for which there does not exist any finite dimensional field extension in which they become a root are called *transcendental*.

The set of all real or complex algebraic numbers is denoted by $\mathbf{A}$. This is a <u>countable</u> set.

---

[6] $\deg 0 = 1$ for the number zero, but $\deg 0 = -\infty$ for the polynomial zero.—That is the only contradiction I have ever found in modern algebra!

III.1.4   The Dimension of a Field Extension

For any field extension K/F, the upper field K is a vector space over the base field F. Hence this vector space has a dimension, which can be finite or infinite. We denote the dimension of a field extension by $\dim K/F$ or $[K : F]$. An extension is called finite-dimensional or simply *finite* if the dimension $[K : F]$ is finite. Obviously $[K : F] = 1$ if and only if $K = F$.

Proposition 23 (Tower Theorem). *The dimensions of a chain of two field extensions $F \subseteq L \subseteq K$ are multiplied:*

$$[K : F] = [K : L] \cdot [L : F]$$

Indication of reason. Let $n = [L : F]$ and $l_1, \ldots, l_n$ be a basis of the extension L/F. Let $m = [K : L]$ and $k_1, \ldots, k_m$ be a basis of the extension K/L. Then one can check that the nm products $k_i l_j$ are linearly independent over F and generate the largest field K. Hence the extension K/F has the dimension nm. $\qquad\square$

A polynomial $p \in R[x]$ over any ring is called *irreducible* if any factorization $p = rs$ has one factor which is a unit of the base ring R. A polynomial $p \in \mathbf{F}[x]$ over any field $\mathbf{F}$ is called *irreducible* if any factorization $p = rs$ has one factor of degree zero. Equivalently, we can require that p cannot be factored into any polynomials of lower degree. By these definitions, the zero polynomial is not irreducible.

Every irreducible polynomial $p \in \mathbf{F}[x]$ has degree at least one. Every irreducible polynomial $p \in \mathbf{Z}[x]$ is either a prime number, or has degree at least one.[7]

Given is an extension K/F and one number $\alpha \in K$ which is algebraic over the base field F. The *minimal polynomial* of α is the polynomial in $F[x]$ of lowest degree which has the zero α. The one-element extension $F(\alpha)$ is defined to be the smallest field containing both α and all elements of F.

Proposition 24. *The minimal polynomial is irreducible and unique up to nonzero factors in F. Conversely, let α be a root of an irreducible polynomial. This polynomial is the minimal polynomial of the one-element extension $F(\alpha)$ generated by α.*

Proposition 25 (Extension generated by one element). *The dimension of the extension $F(\alpha)/F$ generated by one algebraic number α is the degree of the minimal polynomial of the algebraic number α. The powers $1, \alpha, \alpha^2, \ldots \alpha^{n-1}$ are a basis for the extension $F(\alpha)/F$.*

---

[7]These two statement are not a contradiction since $\mathbf{Z}$ is not a field.

All further roots which the minimal polynomial may have in K are called the *algebraic conjugates* of α. We see that they have the same minimal polynomial.

Proposition 26. *The extension field $F(\alpha)$ generated by one algebraic number α over the base field F is isomorphic to the ring quotient $F[x]/(p)$ modulo the principal ideal (p) generated by the minimal polynomial p of α.*

The fact that the quotient ring $F[x]/(p)$ is a field guaranties the existence of a field extension corresponding to any irreducible polynomial p. Moreover, any two such extensions are isomorphic, via an isomorphism which fixes all elements of F.

For any algebraic conjugate β of α, the extensions $F(\alpha)$ and $F(\beta)$ are isomorphic, via an isomorphism which fixes all elements of F.

Further and larger extensions K/F can be generated by adjoining any set of numbers $\alpha, \beta, \cdots \in K$. The extension $F(\alpha, \beta)$ is defined to be the smallest field containing both α, β and all elements of F. This definition implies immediately that $F(\alpha, \beta) = F(\alpha)(\beta) = F(\beta)(\alpha) = F(\beta, \alpha)$.

An extension obtained by adjoining finitely many elements is called *finitely generated*. For a finitely generated extension we obtain a spanning set $\alpha^{i-1}\beta^{k-1}$ with $i = 1 \ldots \deg \alpha$ and $k = 1 \ldots \deg \beta$. A similar idea is used in the tower theorem, too. But in the present context, these elements may turn out to be linearly dependent over F. We obtain only an upper bound for the dimension of multiple extensions:

$$[F(\alpha, \beta) : F] = [F(\alpha, \beta) : F(\alpha)][F(\alpha) : F] \le [F(\beta) : F][F(\alpha) : F] = \deg \alpha \cdot \deg \beta$$
$$[F(\alpha, \beta, \gamma) : F] \le \deg \alpha \cdot \deg \beta \cdot \deg \gamma$$

Remark. We may have inequality in the estimate $[F(\alpha, \beta) : F(\alpha)] \le [F(\beta) : F]$. This happens if adjoining α splits the irreducible polynomial $q \in F[x]$ for β into two polynomials $q, r \in F(\alpha)[x]$ of lower degree. In that case $q = rs$ and hence β is either a root of r or s and

$$[F(\alpha, \beta) : F(\alpha)] \le \max(\deg r, \deg s) < \deg q = [F(\beta) : F]$$

A field extension K/F is called *algebraic* if all numbers in K are algebraic over the base field F. Otherwise, the field extension is called transcendental.

Proposition 27. *An extension K/F is finite if and only if it is algebraic and generated by adjoining finitely many elements to the base field.*

There exist finite as well as infinite dimensional algebraic extensions. For example:

- The extensions $\mathbb{C}/\mathbb{R}$ and $\mathbf{Q}(\sqrt{-1})/\mathbf{Q}$ are two dimensional.

- The extensions $\Omega/\mathbf{Q}$ and $K/\mathbf{Q}$ of the Hilbert field and the constructible field over the rational numbers are infinite dimensional.

- The extension of all algebraic numbers $\mathbf{A}/\mathbf{Q}$ over the rational numbers is infinite dimensional.

- The extensions $(\mathbf{A} \cap \mathbb{R})/\mathbf{Q}$ and $\mathbf{A}/\mathbf{Q}(\sqrt{-1})$ are infinite dimensional.

Since $F[x]$ has no null divisors (is an integral domain), the set of quotients p/q of polynomials $p, q \in F[x]$ is a field. The quotient field is also called the field of <u>rational functions</u>. Any one-element transcendental extension is isomorphic to the field of rational functions. Hence every transcendental extension is infinite dimensional.

Proposition 28. *Suppose K/F is a field extension and the polynomial $p \in F[x]$ of degree n has n roots $\alpha_1 \ldots \alpha_n$ in K. Then the polynomial p factors into*

$$p(x) = \prod_{i=1}^{n} (x - \alpha_i)$$

already in the algebraic extension $F(\alpha_1, \ldots, \alpha_n)/F$. The dimension d of this extension is at most the factorial $n!$:

$$d = [F(\alpha_1, \ldots, \alpha_n) : F] \leq 1 \cdot 2 \cdot 3 \cdots (n-1) \cdot n$$

All fields $L \supseteq F$ of dimension d in which the polynomial p splits, are isomorphic with an isomorphism fixing the elements of F.

If the polynomial p is irreducible, then n is a divisor of dimension d.

III.1.5   The Field of Algebraic Numbers

Proposition 29. *The reciprocal of an algebraic number $a \neq 0$ over any field F is algebraic, and contained in the simple extension $F(a)$. Indeed, the reciprocal $1/a$ can be expressed as evaluation of a polynomial $s \in F[x]$ of degree $\deg s < \deg a$:*

$$\frac{1}{a} = s(a)$$

Especially, the reciprocal of an algebraic number over the rational numbers can be expressed as evaluation of an integer polynomial $s \in \mathbf{Z}[x]$ with $\deg s < \deg a$, divided by a positive integer $0 < n \in \mathbf{Z}$:

$$\frac{1}{a} = \frac{s(a)}{n}$$

Proof. We want to express the reciprocal of an algebraic number. Let $a \neq 0$ have the minimal polynomial $p \in F[x]$. We need to obtain the reciprocal $1/a$. To handle this case, we choose $q(x) = x$.

As easily, we can obtain the reciprocal $1/q(a)$ with $q(a) \neq 0$ for any polynomial $q \in F[x]$ with $\deg q < \deg p$. We use the extended Euclidean algorithm for p and q. Their greatest common divisor $g = \gcd(p, q)$ is a divisor of p with degree $\deg g \leq \deg q < \deg p$. Since p is the minimal polynomial, it is irreducible. Hence the divisor g of p is either a multiple of p or a multiple of 1. Since $\deg g < \deg p$ and $g \neq 0$, we see that $\deg g = 0$. In other words, $g \in F$ is a constant polynomial. Since $0 \neq g \in F$, we may even choose $g = 1$. By the extended Euclidean algorithm, the greatest common divisor $g = \gcd(p, q)$ can be expressed as

$$rp + sq = g$$

with polynomials $r, s \in \mathbf{F}[x]$. Evaluation at $x = a$ yields

$$1 = g(a) = r(a)p(a) + s(a)q(a) = s(a)q(a)$$

Hence

$$\frac{1}{q(a)} = s(a)$$

is a representation of the reciprocal of an algebraic number by evaluation of a polynomial over the base field. $\qquad\square$

Remark. Moreover, in the quotient ring $F[x]/(p)$ we get the identity

$$\frac{1}{q} \equiv s \quad \mod p$$

We have proved once more:

Corollary 15. *The quotient ring $F[x]/(p)$ of a polynomial ring of a field over a principal ideal (p) of an* irreducible *polynomial p is a field.*

Question. Write
$$\frac{5 + 4\sqrt{2}}{4 + 2\sqrt{2}}$$
with an integer denominator and cancel.

Answer.

$$\frac{5 + 4\sqrt{2}}{4 + 2\sqrt{2}} = \frac{(5 + 4\sqrt{2})(4 - 2\sqrt{2})}{(4 + 2\sqrt{2})(4 - 2\sqrt{2})} = \frac{20 - 16 + 6\sqrt{2}}{8} = \frac{2 + 3\sqrt{2}}{4}$$

Question. Use the extended Euclidean algorithm for the polynomials $x^3 - 2$ and $x^2 + 1$.

Answer. We need two divisions with remainder until 5 turns out to be the last nonzero remainder.

$$(x^3 - 2) : (x^2 + 1) = x \qquad \text{remainder} \quad -x - 2$$
$$(x^2 + 1) : (-x - 2) = -x + 2 \quad \text{remainder} \quad 5$$

Then go backwards to express the last remainder as a combination of $x^3 - 2$ and $x^2 + 1$.

$$\begin{aligned}
5 &= (x^2 + 1) - (x + 2)(x - 2) \\
&= (x^2 + 1) + [(x^3 - 2) - x(x^2 + 1)](x - 2) \\
&= (x^3 - 2)(x - 2) + (x^2 + 1)(-x^2 + 2x + 1)
\end{aligned}$$

Question. Find the reciprocal

$$\frac{1}{x^2 + 1}$$

in the quotient field $\mathbf{Q}[x]/(x^3 - 2)$.

Answer.

$$5 \equiv (x^2 + 1)(-x^2 + 2x + 1) \quad \mod (x^3 - 2)$$
$$\frac{1}{x^2 + 1} \equiv \frac{-x^2 + 2x + 1}{5} \quad \mod (x^3 - 2)$$

gives the answer in the quotient field $\mathbf{Q}[x]/(x^3 - 2)$.

Question. Write

$$\frac{1}{2^{2/3} + 1}$$

with an integer denominator.

Answer. We use the last answer with $x = \sqrt[3]{2}$ and get

$$\frac{1}{2^{2/3} + 1} = \frac{-2^{2/3} + 2 \cdot 2^{1/3} + 1}{5}$$

As an alternative, we can go back to the identity

$$5 = (x^3 - 2)(x - 2) + (x^2 + 1)(-x^2 + 2x + 1)$$

and plug in $x = \sqrt[3]{2}$. We get the same answer.

Proposition 30. *The sum $a + b$, the product ab, and the quotient a/b with $b \neq 0$ of any two algebraic numbers are algebraic. The sum $a + b$ and the product ab of two algebraic integers is an algebraic integer.*

Indication of reason. Because of the existence of a splitting field, we may assume that the minimal polynomials $p \in F[x]$ with $p(a) = 0$ and $q \in F[x]$ with $q(b) = 0$ split in the field K. Hence

$$p(x) = \prod_{i=1}^{m}(x - a_i) \quad \text{and} \quad q(x) = \prod_{k=1}^{n}(x - b_k)$$

The polynomials

$$S(x) = \prod_{i=1}^{m}\prod_{k=1}^{n}(x - a_i - b_k) \quad \text{and} \quad P(x) = \prod_{i=1}^{m}\prod_{k=1}^{n}(x - a_i b_k)$$

have coefficients which are totally symmetric in the roots a_i as well as b_k. By the Symmetric Functions Theorem, [8] the coefficients of P and Q are F-polynomial functions of the elementary symmetric functions of a_i and b_k. The elementary symmetric functions of a_i are by Viëta's Theorem the coefficients of p, and the elementary symmetric functions of b_k are the coefficients of q. These elementary symmetric function, and hence the coefficients of P and Q are elements of the base field F. Hence $S, P \in F[x]$. This means that the sums $a_i + b_k$ and the products $a_i b_k$ are algebraic over F. $\square$

Theorem 8 (The field of algebraic numbers). *The set $\mathbf{A}$ of all (real or complex) algebraic numbers is a <u>field</u> (Eisenstein 1850). The set $\mathbf{A}$ is <u>countable</u> (Cantor 1874).*

III.2 Gauss' Lemma

Proposition 31. *Assume x is a real or complex zero of any polynomial*

$$a_r x^r + a_{r-1} x^{r-1} + \cdots + a_1 x + a_0 = 0$$

where $a^r \neq 0, a_{r-1}, \ldots a_1, a_0$ are integers. If the root $x = \frac{m}{n}$ is rational with m and n relatively prime, then m is a divisor of a_0 and n is a divisor of a_r.

---

[8]Theorem 16.1.6 in Michael Artin's book [2]

Proof.

$$a_r m^r + a_{r-1} m^{r-1} n + \cdots + a_1 m n^{r-1} + a_0 n^r = 0$$
$$a_r m^r = -\left(a_{r-1} m^{r-1} + \cdots + a_1 m n^{r-2} + a_0 n^{r-1}\right) n$$
$$-\left(a_r m^{r-1} + a_{r-1} m^{r-2} n + \cdots + a_1 n^{r-1}\right) m = a_0 n^r$$

The second line implies that n is a divisor of $a_r m^r$. Since $\gcd(n, m) = 1$, the Euclidean Property stated in Proposition 5 implies that n is a divisor of a_r.

The last line implies that m is a divisor of $a_0 n^r$. Since $\gcd(n, m) = 1$, the Euclidean Property stated in Proposition 5 implies that m is a divisor of a_0. $\qquad\square$

Rule: The rational solutions of an integer polynomial $a_r x^r + \cdots + a_0 = 0$, are among the positive or negatives of the rational numbers top and bottom of which are divisors of top and bottom of the solution of $a_r X + a_0 = 0$.

Proposition 32 (Linear Gauss Lemma). *Suppose that an integer polynomial has a rational zero $\frac{m}{n}$, with n, m are relatively prime. Then it can be factored into* integer *polynomials*

$$(\text{III.2.1}) \qquad a_r x^r + \cdots + a_0 = (nx - m)(b_{r-1} x^{r-1} + \cdots + b_0)$$

If additionally $\gcd(a_r, \ldots, a_0) = 1$, then $\gcd(b_{r-1}, \ldots, b_0) = 1$, too.

III.2.1 Factoring Integer Polynomials

Theorem 9 (Gauss' Lemma). *An integer polynomial which factors over the rational numbers into factors of <u>lower degree</u>, already factored into integer polynomials of lower degree. The latter factoring is obtained from the former one by adjusting integer factors.*

Especially, a monic integer polynomial that factors over the rational numbers, even factors over the integers into monic integer polynomials.

Proof. Suppose that an integer polynomial of degree $r \geq 2$ can be factored into rational polynomials of degree $s \geq 1$ and $t \geq 1$. Multiplying with the least common denominators, we get the integer formula

$$(\text{III.2.2}) \qquad B\left(a_r x^r + \cdots + a_0\right) = A\left(b_s x^s + \cdots + b_0\right)\left(c_t x^t + \cdots + c_0\right)$$

where $a_r x^r + \cdots + a_0$ is obtained from the given integer polynomial by dividing through with the greatest common divisor of the coefficients. Because of using least common denominators everywhere, and cancellation, we get

$$\gcd(A, B) = 1$$

(III.2.3)
$$\gcd(a_r, \ldots, a_0) = 1$$
$$\gcd(b_s, \ldots, b_0) = 1$$
$$\gcd(c_t, \ldots, c_0) = 1$$

Question. Show that $A = 1$.

Answer. Assume any prime number p divides A, and derive a contradiction. By formula (III.2.2) this would imply that p divides all numbers $Ba_r, \ldots Ba_0$. Since $\gcd(a_r, \ldots, a_0) = 1$, we conclude that p divides B. But this contradicts the assumption $\gcd(A, B) = 1$.

Lemma 28. *If a prime p divides all coefficients of the product of two polynomials with integer coefficients*

(III.2.4)
$$(b_s x^s + \cdots + b_0)\left(c_t x^t + \cdots + c_0\right),$$

then p divides either $\gcd(b_s, \ldots, b_0)$ or $\gcd(c_t, \ldots, c_0)$.

Reason for the Lemma. Multiplying the terms of the product (III.2.4), we see that p divides all the integers

$$b_0 c_0$$
$$b_0 c_1 + b_1 c_0$$
$$b_0 c_2 + b_1 c_1 + b_2 c_0$$
$$b_0 c_k + b_1 c_{k-1} + \cdots + b_k c_0$$

for all $k \geq 0$. [9] We see that p divides either b_0 or c_0. Now take the case that p divides b_0. The case that p divides c_0 can be deals with similarly. Because of the assumption $\gcd(b_s, \ldots, b_0) = 1$, there exists an index $0 \leq \sigma < s$ such that p divides $b_0, b_1, \ldots b_\sigma$ but not $b_{\sigma+1}$. But we have seen that p divides all the sums from above.

Successively, we put $k = \sigma + 1, \sigma + 2 \ldots$ and cancel the terms already known to be divisible by p. Hence we see that p divides the sums

$$b_{\sigma+1} c_0$$
$$b_{\sigma+1} c_1 + b_{\sigma+2} c_0$$
$$b_{\sigma+1} c_2 + b_{\sigma+2} c_1 + b_{\sigma+3} c_0$$
$$b_{\sigma+1} c_j + b_{\sigma+2} c_{j-1} + \cdots + b_{\sigma+j+1} c_0$$

---

[9] For simplicity, we put the coefficients of a polynomial equal zero if we exceed its degree.

and so on for all $j \geq 0$. Hence we conclude that p divides $c_0, c_1, c_2, \ldots c_j$ for all $j \geq 0$. $\qquad \square$

Question. Use the Lemma to show that $B = 1$.

Answer. Assume any prime number p divides B, and derive a contradiction. By formula (III.2.2), and since $A = 1$, this would imply that p divides all coefficients of the product (III.2.4). Hence by the Lemma, either p divides all coefficients $b_s, \ldots, b_0$ or all coefficients $c_t, \ldots, c_0$. This has been ruled out by assumption (III.2.3). Hence we conclude that $B = 1$.

Now, since $A = B = 1$, the factoring (III.2.2) is indeed a factoring of the integer polynomial $a_r x^r + \cdots + a_0$ into integer polynomials.

Since the integer polynomial $a_r x^r + \cdots + a_0$ was obtained from the given integer polynomial by dividing through with the greatest common divisor of the coefficients, we have factored the originally given polynomial, too. $\qquad \square$

Problem 37. *Find all symmetric monic polynomials $x^4 + ax^2 + bx^2 + ax + 1$, that factor over the rationals and have*

(i) *Four rational zeros.*

(ii) *Exactly two rational zeros.*

(iii) *No rational zeros.*

Do the cases with one or three zeros occur? Why not?

Solution. The only possible rational zeros are $+1$ and -1. If $x = 1$ is a zero, it is a double zero since

$$P(x) = x^4 P(x^{-1})$$
$$P'(x) = 4x^3 P(x^{-1}) - x^2 P'(x^{-1})$$

Hence $P(x) = 0$ and $x^2 = 1$ imply $P'(1) = -P'(1) = 0$.

(i) Four rational zeros. Hence there are five choices with all zeros in the set $\{+1, -1\}$. Three of them turn out to be symmetric:

$$(x + 1)^4 \, , \; (x^2 - 1)^2 \, , \; (x - 1)^4$$

(ii) Exactly two rational zeros. The cases with a double zero 1 are

$$(x-1)^2(x^2+(a+2)x+1) = x^4 + ax^3 - 2(a+1)x^2 + ax + 1$$

where a is any integer $a \neq 0, a \neq -4$. The cases with a double zero -1 are

$$(x+1)^2(x^2+(a-2)x+1) = x^4 + ax^3 + 2(a-1)x^2 + ax + 1$$

where a is any integer $a \neq 0, a \neq +4$. Too, we see that no solutions with a zero of multiplicity three exist.

(iii) No rational zeros. The polynomial factors into two irreducible monic quadratics: Q and R. The symmetry implies

$$P(x) = Q(x)R(x) = x^2 Q(x^{-1}) x^2 R(x^{-1}) = x^4 P(x^{-1})$$

Uniqueness of factorization into irreducible factors leads to four possibilities:

(a) $x^2 Q(x^{-1}) = Q(x)$ and $x^2 R(x^{-1}) = R(x)$.
(b) $x^2 Q(x^{-1}) = -Q(x)$ and $x^2 R(x^{-1}) = -R(x)$.
(c) $x^2 Q(x^{-1}) = R(x)$ and $x^2 R(x^{-1}) = Q(x)$.
(d) $x^2 Q(x^{-1}) = -R(x)$ and $x^2 R(x^{-1}) = -Q(x)$.

Here are the results:

(a) From $x^2 Q(x^{-1}) = Q(x) = x^2 + cx + 1$ and $x^2 R(x^{-1}) = R(x) = x^2 + dx + 1$, we get

$$Q(x)R(x) = (x^2+cx+1)(x^2+dx+1) = x^4+(c+d)x^3+(cd+2)x^2+(c+d)x+1$$

with any integer $c, d \neq \pm 2$.

(b) $x^2 Q(x^{-1}) = -Q(x)$ and $x^2 R(x^{-1}) = -R(x)$ gives only the solution $(x^2-1)^2$ that we already have.

(c) $x^2 Q(x^{-1}) = R(x)$ and $x^2 R(x^{-1}) = Q(x)$ gives only the special case of (a) with $c = d$, but nothing new.

(d) $-x^2 Q(x^{-1}) = R(x) = x^2 - cx - 1$ and $-x^2 R(x^{-1}) = Q(x) = x^2 + cx - 1$ gives the extra solutions

$$Q(x)R(x) = (x^2 + cx - 1)(x^2 - cx - 1) = x^4 - (c^2 + 2)x^2 + 1$$

These are new solution for any $c \neq 0$.

$\square$

Part IV

Special Polynomials

IV.1 Cyclotomic Polynomials

Lemma 29. *Let $\mathcal{G}$ be any commutative groupoid or ring.* [10] *Let $n \in \mathbf{N} \mapsto A_n \in \mathcal{G}$ and $n \in \mathbf{N} \mapsto B_n \in \mathcal{G}$ be any functions. Let $n \geq 1$ be any integer.*

$$\text{If for all divisors } d \mid n \text{ holds } \prod_{e \mid d} A_e = \prod_{e \mid d} B_e$$

$$\text{then for all divisors } d \mid n \text{ holds } \quad A_d = B_d$$

Proof. Towards a contradiction, assume that the assertion is wrong for some integer n. Hence the assumption holds but nevertheless the conclusion is not true. For all divisors $d \mid n$ holds $\prod_{e \mid d} A_e = \prod_{e \mid d} B_e$. Among the divisors $d \mid n$ there exists a smallest one for which $A_d^* \neq B_d^*$.

Hence for all <u>proper</u> divisors $e \mid d^*$ holds $A_e = B_e$. Too $\prod_{e \mid d^*} A_e = \prod_{e \mid d^*} B_e$. Hence $A_d^* = B_d^*$, a contradiction. $\qquad\square$

IV.1.1   Basic Properties of Cyclotomic Polynomials

Definition 22 (Cyclotomic polynomials). The *cyclotomic polynomials* $\Phi_n(x)$ are monic polynomials which are uniquely determined in either one of the following ways:

(a) The greatest common divisor formula

$$(\text{IV.1.1}) \qquad \Phi_m = \gcd\left\{ \frac{x^m - 1}{x^{m/p} - 1} \; : \; p \mid m, \; p \text{ prime} \right\} \quad \text{for all } m \geq 2.$$

Here the greatest common divisor has to be understood as a monic polynomial in $\mathbf{Z}[x]$. One has to agree that $\Phi_1(x) = x - 1$.

(b) The primitive roots of unity are its zeros.   One has to agree that $\Phi_1(x) = x - 1$, For all $m \geq 2$, the cyclotomic polynomials Φ_m are the monic polynomials in $\mathbb{C}[x]$ the zeros of which are the primitive roots of unity modulo m. These are the $\phi(m)$ complex numbers

$$\exp \frac{2\pi i a}{m} \quad \text{with } \gcd(a, m) = 1 \text{ and } 1 \leq a < m.$$

In other words the zeros of Φ_m are the roots of $z^m - 1 = 0$ which are not roots of $z^d - 1 = 0$ for any proper divisor $d \mid m, d < m$.

---

[10]One only need a commutative and associative multiplication

114

(c) product over divisors The polynomials Φ_n satisfy for all $n \geq 1$

$$\text{(IV.1.2)} \qquad\qquad x^n - 1 = \prod_{d|n} \Phi_d(x)$$

(d) Moebius inversion

$$\text{(IV.1.3)} \qquad\qquad \Phi_n(x) = \prod_{d|n} \left(x^{n/d} - 1\right)^{\mu(d)}$$

Here μ denotes the Moebius function.

Remark. The notion "m-th primitive root of unity" refers to the field $\mathbb{C}$, whereas the notion "primitive root modulo m" refers to the Euler group $G_m^* \subset \mathbf{Z}_m$.

The equivalencies $(a) \Rightarrow (b) \Rightarrow (c) \Rightarrow (a)$ are a "happy merry go round" left to the reader. Clearly property either (a) or (b) determine the cyclotomic polynomials uniquely. By lemma 29 we see that property (c) determines the cyclotomic polynomials uniquely. The equivalence $(c) \Leftrightarrow (d)$ is a consequence of Lemma 1.

Problem 38. *Convince yourself that for all $n \geq 1$*

$$n = \sum_{d|n} \phi(d)$$

$$\phi(n) = \sum_{d|n} \frac{n}{d}\, \mu(d)$$

Proposition 33. *For any m, the cyclotomic polynomial Φ_m has integer coefficients. Its degree is* $\deg \Phi_d(x) = \phi(d)$ *the Euler totient function. For $m \geq 2$, the polynomial is mirror-symmetric:*

$$\Phi_m(x) = x^{\phi(m)} \Phi_m(x^{-1})$$

Proof of symmetry. For $n \geq 2$, the polynomial is mirror-symmetric:

$$x^n - 1 = -x^n\left((x^{-1})^n - 1\right)$$

$$\prod_{d|n} \Phi_d(x) = -x^{\sum_{d|n} \phi(d)} \prod_{d|n} \Phi_d(x^{-1})$$

$$\prod_{d|n} \Phi_d(x) = -\prod_{d|n} x^{\phi(d)} \Phi_d(x^{-1})$$

$$(x-1)\prod_{1<d|n} \Phi_d(x) = -x\left((x^{-1}-1)\right)\prod_{1<d|n} x^{\phi(d)} \Phi_d(x^{-1})$$

$$\prod_{1<d|n} \Phi_d(x) = \prod_{1<d|n} x^{\phi(d)} \Phi_d(x^{-1})$$

By the lemma 29 we conclude that

$$\Phi_d(x) = x^{\phi(d)}\Phi_d(x^{-1}) \quad \text{holds for all divisors } 1 < d \mid n, \text{ and especially for } d = n.$$

$\square$

Proposition 34. *For any $m \neq n$, the cyclotomic polynomials Φ_m and Φ_n are relatively prime.*

Proof. Let $n \neq m \geq 1$ and $g = \gcd(m,n)$. From the little proposition 5, and its proof, one gets that

$$(\text{IV.1.4}) \qquad \gcd(x^m - 1, x^n - 1) = x^g - 1$$

holds in the ring $\mathbf{Z}[x]$ of polynomials. Now the definition (b) of cyclotomic polynomials implies

$$\gcd(\Phi_m(x), \Phi_n(x)) \text{ divides } \gcd\left(\frac{x^m - 1}{x^g - 1}, \frac{x^n - 1}{x^g - 1}\right) = 1$$

Hence Φ_m and Φ_n are relatively prime. $\square$

Lemma 30. *Let $n \geq 1, s \geq 1$, and $p \geq 2$ be a prime that does not divide n.*

$$(\text{IV.1.5}) \qquad \Phi_n(x^{p^s}) = \prod_{0 \leq t \leq s} \Phi_{np^t}(x)$$

Proof. Since $p \nmid n$, one sees that $e \mid np^s$ holds if and only if $d \mid n$, $0 \leq t \leq s$ and $e = dp^t$.

$$(x^{p^s})^n - 1 = x^{np^s} - 1$$

$$\prod_{d \mid n} \Phi_d(x^{p^s}) = \prod_{e \mid np^s} \Phi_e(x)$$

$$\prod_{d \mid n} \Phi_d(x^{p^s}) = \prod_{d \mid n} \left[\prod_{0 \leq t \leq s} \Phi_{dp^t}(x) \right]$$

For all divisors $d \mid n$ holds $p \nmid d$. Hence we get similarly for all divisors $d \mid n$

$$\prod_{e \mid d} \Phi_e(x^{p^s}) = \prod_{e \mid d} \left[\prod_{0 \leq t \leq s} \Phi_{ep^t}(x) \right]$$

By the lemma 29 we conclude that for all divisors $d \mid n$ including $d = n$ holds

$$\Phi_d(x^{p^s}) = \prod_{0 \leq t \leq s} \Phi_{dp^t}(x)$$

$\square$

Proposition 35 (Properties of the cyclotomic polynomials). *Relatively easy calculations can be based on the following formulas. Let $n \geq 1$, prime $p \geq 2$ and $s \geq 1$.*

(a) *If prime p does not divide n holds*

$$(\text{IV.1.6}) \qquad \Phi_{np}(x) = \frac{\Phi_n(x^p)}{\Phi_n(x)}$$

$$(\text{IV.1.7}) \qquad \Phi_{np^s}(x) = \frac{\Phi_n(x^{p^s})}{\Phi_n(x^{p^{s-1}})}$$

$$(\text{IV.1.8}) \qquad \Phi_{np^s}(x) = \Phi_{np^{s-1}}(x^p) = \Phi_{np^{s-2}}(x^{p^2}) = \cdots = \Phi_{np}(x^{p^{s-1}})$$

(b) *If either prime p divides n holds*

$$(\text{IV.1.9}) \qquad \Phi_{np}(x) = \Phi_n(x^p)$$

$$(\text{IV.1.10}) \qquad \Phi_{np^s}(x) = \Phi_{np^{s-1}}(x^p) = \Phi_{np^{s-2}}(x^{p^2}) = \cdots = \Phi_{np}(x^{p^{s-1}}) = \Phi_n(x^{p^s})$$

(c) *The values for prime powers are*

$$(\text{IV.1.11}) \qquad \Phi_p(x) = \frac{x^p - 1}{x - 1} = \sum_{0 \leq k \leq p-1} x^k$$

$$(\text{IV.1.12}) \qquad \Phi_{p^s}(x) = \Phi_p(x^{p^{s-1}}) = \frac{x^{p^s} - 1}{x^{p^{s-1}} - 1} = \sum_{0 \leq k \leq p-1} x^{k \cdot p^{s-1}}$$

$$(\text{IV.1.13}) \qquad \Phi_{2^s}(x) = (x^{2^{s-1}}) + 1$$

(d) $\Phi_1(x) = x - 1$ *and* $\Phi_2(x) = x + 1$. *If $n \geq 3$ is odd and $p \geq 3$ is prime hold*

$$(\text{IV.1.14}) \qquad \Phi_{2n}(x) = \Phi_n(-x)$$

$$(\text{IV.1.15}) \qquad \Phi_{2p^s}(x) = \Phi_{p^s}(-x) = \sum_{0 \leq k \leq p-1} (-x)^{k \cdot p^{s-1}}$$

$$(\text{IV.1.16}) \qquad \Phi_{4n}(x) = \Phi_n(-x^2) = \Phi_n(x) \cdot \Phi_n(-x)$$

$$(\text{IV.1.17}) \qquad \Phi_{2^s n}(x) = \Phi_n(-x^{2^{s-1}})$$

Problem 39. *Use formula (IV.1.5) to proof part (a) and (b) of proposition 35*

Check of part (a). Assume that prime p does not divide n. Formula (IV.1.5) with $s = 1$ yields $\Phi_n(x^p) = \Phi_n(x)\Phi_{np}(x)$ and hence claim (IV.1.6). The quotient of formula (IV.1.5) with s and $s - 1$ yields claim

(IV.1.7)
$$\Phi_{np^s}(x) = \frac{\Phi_n(x^{p^s})}{\Phi_n(x^{p^{s-1}})}$$

In the last formula, one replaces s by $s - 1$ and x by x^p to obtain

$$\Phi_{np^{s-1}}(x^p) = \frac{\Phi_n(x^{p^s})}{\Phi_n(x^{p^{s-1}})}$$

which together imply claim (IV.1.8). I may leave part (b) to the reader. $\qquad\square$

Proof of item (d) *from proposition 35* . Assume that $n \geq 3$ is odd.

$$x^{2n} - 1 = \prod_{d|n} \Phi_d(x) \cdot \prod_{d|n} \Phi_{2d}(x) = (x^n - 1) \cdot \prod_{d|n} \Phi_{2d}(x)$$

$$\frac{x^{2n} - 1}{x^n - 1} = x^n + 1 = \prod_{d|n} \Phi_{2d}(x)$$

$$x^n + 1 = -((-x)^n - 1) = -\prod_{d|n} \Phi_d(-x)$$

$$\prod_{d|n} \Phi_{2d}(x) = -\prod_{d|n} \Phi_d(-x)$$

$$(x + 1) \prod_{1<d|n} \Phi_{2d}(x) = -(-x - 1) \prod_{1<d|n} \Phi_d(-x)$$

$$\prod_{1<d|n} \Phi_{2d}(x) = \prod_{1<d|n} \Phi_d(-x)$$

The last line holds for all odd $n \geq 3$. By the lemma 29 we conclude that

$$\Phi_{2n}(x) = \Phi_n(-x) \text{ holds for all odd } n \geq 3.$$

$$(-x^2)^n - 1 = (x^n - 1) \cdot ((-x)^n - 1)$$

$$\prod_{d|n} \Phi_d(-x^2) = \prod_{d|n} \Phi_d(x) \cdot \prod_{d|n} \Phi_d(-x)$$

$$(-x^2 - 1) \prod_{1<d|n} \Phi_d(-x^2) = (x - 1)(-x - 1) \prod_{1<d|n} \Phi_d(x) \cdot \Phi_d(-x)$$

$$\prod_{1<d|n} \Phi_d(-x^2) = \prod_{1<d|n} \Phi_d(x) \cdot \Phi_d(-x)$$

The last line holds for all odd $n \geq 3$. By the lemma 29 we conclude that

$$\Phi_n(-x^2) = \Phi_n(x) \cdot \Phi_n(-x) \ \text{ holds for all odd } n \geq 3.$$

Item (a) for $p = 2$ and n odd implies now

$$\Phi_{n2^s}(x) = \Phi_{2n}(x^{2^{s-1}}) = \Phi_n(-x^{2^{s-1}})$$
$$\Phi_{4n}(x) = \Phi_n(-x^2) = \Phi_n(x) \cdot \Phi_n(-x)$$

$\square$

Problem 40. *Prove that the function $\Phi_n(x)$ is even if and only if n is divisible by 4.*

Solution. Formula

(IV.1.16)
$$\Phi_{4n}(x) = \Phi_n(-x^2) = \Phi_n(x) \cdot \Phi_n(-x)$$

shows that $\Phi_n(x)$ is an even function if n is divisible by 4.

Conversely, assume that Φ_n is even. Hence $\Phi_n'(0) = 0$. Let

$$n = \prod_i 2^s p_i^{r_i}$$

be the prime factorization of n. The open cases are $s = 0$ or 1. We reduce to the case that n is odd by means of formula

(IV.1.14)
$$\Phi_{2n}(x) = \Phi_n(-x)$$

We reduce to the case that n is square free by means of formula

(IV.1.8)
$$\Phi_{np^s}(x) = \Phi_{np^{s-1}}(x^p) = \Phi_{np^{s-2}}(x^{p^2}) = \cdots = \Phi_{np}(x^{p^{s-1}})$$

Now formula

(IV.1.21)
$$\Phi_n'(0) = \begin{cases} 1 & \text{if } n = 1; \\ -\mu(n) & \text{if } n \geq 2. \end{cases}$$

shows that $\Phi_n'(0) = 1$ if n has an odd number of prime factors, and $\Phi_n'(0) = -1$ if n has an even number of prime factors. Hence the function $\Phi_n(x)$ cannot be even if $4 \nmid n$. $\square$

Problem 41. *Prove that the function $\Phi_n(x)$ is even if and only if n is divisible by 4. Use definition 22 item* (b) *to give a simple convincing solution.*

Proof. Assume that $\Phi_n(x)$ is an even function. Its roots occur in pairs $\pm z$. There exists zeros, and $\Phi_n(z) = 0$ implies both $z^n - 1 = 0$ and $(-z)^n - 1 = 0$. For odd n this leads to a contradiction. Hence n is even. The primitive n-th roots of unity are its zeros of Φ_n. These are the $\phi(n)$ complex numbers

$$\exp\frac{2\pi i a}{n} \quad \text{with } \gcd(a, n) = 1 \text{ and } 1 \le a < n.$$

Since the roots of Φ_n occur in pairs $\pm z$ we see that $\gcd(a, n) = 1$ implies $\gcd(a + \frac{n}{2}, n) = 1$. Since n is even, both a and $a + \frac{n}{2}$ are odd. Hence $\frac{n}{2}$ is even and $4 \mid n$.

The converse is now easy enough to follow, and left to the reader. $\qquad\square$

Lemma 31.

$$(\text{IV.1.18}) \qquad \Phi_n(0) = \begin{cases} -1 & \text{if } n = 1; \\ +1 & \text{if } n \ge 2. \end{cases}$$

$$(\text{IV.1.19}) \qquad \Phi_n(1) = \begin{cases} 0 & \text{if } n = 1; \\ p & \text{if } n = p^r \text{ is a prime power;} \\ 1 & \text{if } n \text{ has at least two prime divisors.} \end{cases}$$

$$(\text{IV.1.20}) \qquad \Phi_n(-1) = \begin{cases} -2 & \text{if } n = 1; \\ 0 & \text{if } n = 2; \\ 2 & \text{if } n = 2^r \text{ and } r \ge 2; \\ p & \text{if } n = 2p^r \text{ and } p \text{ odd prime;} \\ 1 & \text{otherwise.} \end{cases}$$

Proof of formula (IV.1.19). At first note that formula (IV.1.19) holds for sure in the case of n a prime-power. Suppose there is a smallest counterexample. This n is not a prime power.

$$\frac{x^n - 1}{x - 1} = \prod_{1 < d \mid n} \Phi_d(x)$$

$$\lim_{x \to 1} \frac{x^n - 1}{x - 1} = n = \prod_{1 < d \mid n} \Phi_d(1)$$

$$\prod_{1 < d \mid n} \lambda(d) = \prod_{1 < d \mid n} \Phi_d(1)$$

$$\lambda(d) = \Phi_d(1) \quad \text{holds for all } d \mid n \text{ such that } 1 < d < n$$

The last two lines imply $\lambda(n) = \Phi_n(1)$. No smallest counterexample does exists, the assertion holds for all n $\qquad\qquad\square$

Proof of formula (IV.1.20). Assume that $n \geq 3$ is odd. Hence $2n$ is not a prime-power. One puts $x = 1$ into formula

$$(\text{IV.1.14}) \qquad\qquad \Phi_{2n}(x) = \Phi_n(-x)$$
$$1 = \Phi_{2n}(1) = \Phi_n(-1)$$

One puts $x = -1$ into the same formula (IV.1.14) and get

$$\Phi_{2n}(-1) = \Phi_n(1)$$

Assume $s \geq 2$. One puts $x = -1$ into formula

$$(\text{IV.1.17}) \qquad\qquad \Phi_{2^s n}(x) = \Phi_n(-x^{2^{s-1}})$$
$$\Phi_{2^s n}(-1) = \Phi_n(-(-1)^{2^{s-1}}) = \Phi_n(-1) = 1$$

For $s \geq 2$ one gets $\Phi_n(-(-1)^{2^{s-1}}) = \Phi_n(-1) = 1$. $\qquad\qquad\square$

Lemma 32.

$$(\text{IV.1.21}) \qquad\qquad \Phi_n'(0) = \begin{cases} 1 & \text{if } n = 1; \\ -\mu(n) & \text{if } n \geq 2. \end{cases}$$

Proof. The Moebius function satisfies $\mu(1) = 1$ and

$$\sum_{d \mid n} \mu(d) = 0 \ \text{ for all } n \geq 2.$$

At first note that formula (IV.1.21) holds in the cases of $n = 1, 2, 3$ as well as n a prime-power. Suppose there is a smallest counterexample $n \geq 2$. Logarithmic derivative at $x = 0$ yields

$$x^n - 1 = (x - 1) \prod_{1 < d \mid n} \Phi_d(x)$$

$$\ln(x^n - 1) = \ln(x - 1) + \sum_{1 < d \mid n} \ln \Phi_d(x)$$

$$\frac{nx^{n-1}}{x^n - 1} = \frac{1}{x - 1} + \sum_{1 < d \mid n} \frac{\Phi_d'(x)}{\Phi_d(x)}$$

$$0 = -1 + \sum_{1 < d \mid n} \Phi_d'(0)$$

$$0 = -1 - \sum_{1 < d \mid n} \mu(d)$$

$$\sum_{1 < d \mid n} \Phi_d'(0) = - \sum_{1 < d \mid n} \mu(d)$$

By the minimality of n holds

$$\Phi_d'(0) = -\mu(d) \ \text{ for all } d \text{ such that } d \mid n \text{ and } 1 < d < n.$$

The last two lines imply $\Phi_n'(0) = -\mu(n)$. No smallest counterexample does exists, the assertion holds for all n $\qquad\qquad\square$

n	$\Phi_n(x)$	$\phi(n)$
1	$x - 1$	1
2	$x + 1$	1
3	$x^2 + x + 1$	2
4	$x^2 + 1$	2
5	$x^4 + x^3 + x^2 + x + 1$	4
6	$x^2 - x + 1$	2
7	$x^6 + x^5 + x^4 + x^3 + x^2 + x + 1$	6
8	$x^4 + 1$	4
9	$x^6 + x^3 + 1$	6
10	$x^4 - x^3 + x^2 - x + 1$	4
11	$x^{10} + x^9 + x^8 + x^7 + x^6 + x^5 + x^4 + x^3 + x^2 + x + 1$	10
12	$x^4 - x^2 + 1$	4

Problem 42. *Calculate Φ_{15}.*

Solution. Since $15 = 3 \cdot 5$ and $\Phi_3(x) = x^2 + x + 1$

$$\Phi_{15}(x) = \frac{\Phi_3(x^5)}{\Phi_3(x)} = \frac{x^{10} + x^5 + 1}{x^2 + x + 1} = x^8 + \ldots$$

One needs to do the polynomials division. In lack of a symbolic manipulator, one may proceed as follows. Put $x = 10$ and calculate $\frac{10^{10} + 10^5 + 1}{111}$. Add $\frac{10^9 - 1}{9}$ and read off the result

$$\Phi_{15}(10) + 111\,111\,111 = 201\,202\,102$$
$$\Phi_{15}(10) = 10^8 - 10^7 + 10^5 - 10^4 + 10^3 - 10^1 + 1$$
$$\Phi_{15}(x) = x^8 - x^7 + x^5 - x^4 + x^3 - x + 1$$

$\square$

Problem 43. *Calculate Φ_{105}. This is the smallest case where not all coefficients are $-1, 0, +1$.*

$$\begin{aligned}
\Phi_{105}(x) &= \frac{\Phi_{15}(x^7)}{\Phi_{15}(x)} \\
&= x^{48} + x^{47} + x^{46} - x^{43} - x^{42} - 2\,x^{41} - x^{40} - x^{39} \\
&\quad + x^{36} + x^{35} + x^{34} + x^{33} + x^{32} + x^{31} - x^{28} - x^{26} \\
&\quad - x^{24} \\
&\quad - x^{22} - x^{20} + x^{17} + x^{16} + x^{15} + x^{14} + x^{13} + x^{12} \\
&\quad - x^9 - x^8 - 2\,x^7 - x^6 - x^5 + x^2 + x + 1
\end{aligned}$$

Problem 44. *Check the elementary estimates*

(IV.1.22)
$$|\Phi_n(z)| \geq (|z| - 1)^{\phi(n)} \quad \text{for } |z| > 1;$$
$$|\Phi_n(z)| \geq (1 - |z|)^{\phi(n)} \quad \text{for } |z| < 1.$$

Equality is only possible in the cases $n = 1$ and $n = 2$.

IV.1.2   Infinitely many Primes in Arithmetic Sequences

The following Theorem 10 is taken from the newest book *Lehrbuch dr Algebra* by Gerd Fischer [8].

Theorem 10. *Let $n \geq 2$. The arithmetic sequence $\{1 + kn : k \geq 1\}$ contains infinitely many primes.*

Proof. Assume that the primes $p_1, \ldots, p_i$ have been found in the sequence $\{1 + kn : k \geq 1\}$. We define

$$x := n \cdot p_1 \cdots p_i \quad \text{if } i \geq 1$$

and put $x := n$ in the special case $i = 0$. From estimate (IV.1.22) we get

$$|\Phi_n(x)| \geq (x - 1)^{\phi(n)} \geq 1$$

and $\Phi_n(x) = 1$ implies $x = 2$ hence $n = 2$, $i = 0$. But in that case $\Phi_2(2) = 3$ again gives the inequality $\Phi_n(x) > 1$. Hence there exists a prime number

$$p \mid \Phi_n(x)$$

Because of $p \mid \Phi_n(x) \mid x^n - 1$ we conclude that

$$p \nmid x; \quad \text{hence we define } p_{i+1} := p$$

to be the new prime.

    We still need to show that p lies in the arithmetic sequence $\{1 + kn : k \geq 1\}$. Let $m \mid n$ be the smallest natural number such that $p \mid x^m - 1$, in other words m is the order of x modulo p. From the little Fermat Theorem we know that $m \mid p - 1$. I claim that $m = n$, in other words x is a primitive root modulo p.

    We assume towards a contradiction that $m < n$. Hence $n = d \cdot m$ and $d \geq 2$. By formula (IV.1.1)

$$p \mid \Phi_n(x) \mid \frac{x^n - 1}{x^m - 1} = \sum_{0 \leq l < d} x^{ml}$$

But on the other hand

$$\sum_{0 \leq l < d} x^{ml} \equiv d \pmod{p}$$

Since $p \nmid n$ and $p \nmid d$ we conclude

$$p \nmid \sum_{0 \leq l < d} x^{ml}$$

which is a contradiction.

Hence $m = n$ and $n \mid p-1$. Thus p lies in the arithmetic sequence $\{1+kn \, : \, k \geq 1\}$, as to be shown. $\qquad\square$

Corollary 16. *We assume that $n \geq 3$ and $k \in \mathbf{Z}$. All <u>odd</u> prime factors of $\Phi_n(nk)$ lie in the arithmetic sequence $\{1 + ln \, : \, l \geq 1\}$.*

Proof. In the special case $k = 0$ holds $\Phi_n(nk) = 1$ and we are ready. I now assume $k \neq 0$ It does not matter if $k < 0$. Take any prime number

$$p \mid \Phi_n(nk)$$

Because of $p \mid \Phi_n(nk) \mid (nk)^n - 1$ we conclude that

$$p \nmid nk$$

We need to show that p lies in the arithmetic sequence $\{1 + ln \, : \, l \geq 1\}$. It does not matter if $k < 0$. For both signs of k we define $x := nk \bmod p$. Let m be the smallest natural number such that $p \mid x^m - 1$, in other words m is the order of x modulo p. Since $p \mid x^n - 1$, we know that $m \mid n$. From the little Fermat Theorem we know that $m \mid p - 1$. I claim that $m = n$, in other words x is a primitive root modulo p.

We assume towards a contradiction that $1 \leq m < n$. Hence $n = d \cdot m$ and $d \geq 2$. By formula (IV.1.1)

$$p \mid \Phi_n(x) \mid \frac{x^n - 1}{x^m - 1} = \sum_{0 \leq l < d} x^{ml}$$

But on the other hand

$$\sum_{0 \leq l < d} x^{ml} \equiv \sum_{0 \leq l < d} 1^l \equiv d \quad (\bmod \ p)$$

Since $p \nmid n$ and $p \nmid d$ we conclude

$$p \nmid \sum_{0 \leq l < d} x^{ml}$$

which is a contradiction.

Hence $m = n$ and $n \mid p - 1$. Thus p and all odd prime divisors of $\Phi_n(nk)$ lie in the arithmetic sequence $\{1 + ln \, : \, l \geq 1\}$, as to be shown. $\qquad\square$

Problem 45. *How many factors 2^s may $\Phi_n(nk)$ have?*

The following problem asks about a conjecture that, according to a few mathematica calculations, seems to be true.

Problem 46. *Let $n \geq 2$ and $k \in \mathbf{Z}$. Does the product*

$$\prod_{|k| \leq N} \Phi_n(nk)$$

contain any prime from the arithmetic sequence $\{1 + ln : l \geq 1\}$. for large enough N.

Proposition 36. *For all m, the cyclotomic polynomials Φ_m are irreducible both in $\mathbf{Q}[x]$ as well as $\mathbf{Z}[x]$*

At first, I give a more direct proof of proposition 36 in the cases that m is a prime or a square of a prime, based on the Eisenstein criterium.

IV.1.3   Eisenstein's Irreducibility Criterium

Proposition 37 (The Eisenstein criterium). *For a polynomial P to be irreducible in the ring $\mathbf{Z}[z]$, it is <u>sufficient</u> that there exists a prime number p such that*

1. *the leading coefficient of the polynomial is not divisible by p;*

2. *all other coefficients except the leading one are divisible by p;*

3. *the constant coefficient is not divisible by p^2.*

Proof. Assume toward a contradiction that a reducible polynomial would satisfy the criterium. We would get the integer factorization

$$(IV.1.23) \qquad a_0 + \cdots + a_r x^r = \left(b_0 + \cdots + b_s x^s\right)\left(c_0 + \cdots + c_t x^t\right),$$

where $r = s + t$ and $s, t, \geq 1$. Multiplying the terms of the product yields

$$\begin{aligned}
a_0 &= b_0 c_0 \\
a_1 &= b_0 c_1 + b_1 c_0 \\
a_2 &= b_0 c_2 + b_1 c_1 + b_2 c_0 \\
a_k &= b_0 c_k + b_1 c_{k-1} + \cdots + b_k c_0
\end{aligned}$$

for all $k \geq 0$. It is assumed that the prime p, but not p^2 divides $a_0 = b_0 c_0$. Hence p divides exactly one of the two numbers b_0 and c_0. We may assume that p divides

c_0, but does not divide b_0. Furthermore, by assumption, p divides $a_0, a_1, \ldots a_{r-1}$, but not a_r. Recursively, we conclude that p divides $b_0 c_1 = a_1 - b_1 c_0$ and hence c_1, next $b_0 c_2 = a_2 - b_1 c_1 - b_2 c_0$ and hence $c_2, \ldots, c_{r-1}$.

Hence p divides <u>all</u> coefficients of the factor $c_t x^t + \cdots + c_0$, since $t \le r - 1$. This implies that p divides all coefficients of the original polynomial $a_r x^r + \cdots + a_0$, contradicting the assumption that the prime p does not divide the leading coefficient a_r.

From this contradiction, we see that no factoring of the polynomial into polynomials of lower degree is possible, and hence it is irreducible. $\qquad\square$

Problem 47. *Find the irreducible polynomial with zero $z_1 = \sqrt{\frac{5-\sqrt{5}}{2}}$. Find all its algebraic conjugates, and show that the number is a totally real algebraic integer.*

Solution of Problem 47. Simple arithmetic shows that $(2z^2 - 5)^2 - 5 = 0$ and hence

$$P(z) := z^4 - 5z^2 + 5 = 0$$

is the monic polynomial in the ring $\mathbf{Z}[z]$ with zero z_1. Hence z_1 is an algebraic integer. The Eisenstein criterium applies with $p = 5$. Hence its zeros are exactly the algebraic conjugates of z. Obviously, these zeros are

$$\sqrt{\frac{5 + \sqrt{5}}{2}} \,, \quad -\sqrt{\frac{5 + \sqrt{5}}{2}} \,, \quad \sqrt{\frac{5 - \sqrt{5}}{2}} \,, \quad -\sqrt{\frac{5 - \sqrt{5}}{2}}$$

which all four turn out to be real. $\qquad\square$

Lemma 33. *For any prime p, the binomial coefficients*

$$\binom{p}{k} \quad \text{for } k = 1 \ldots p - 1$$

are divisible by p, but not by p^2.

Proposition 38. *For any prime p, the polynomial*

$$\Phi_p(z) = \frac{z^p - 1}{z - 1}$$

is irreducible over the integers.

Increasing powers. The assertion is true for $p = 2$. We assume now that p is an odd prime. We substitute $z = 1 + x$ and use the Eisenstein criterium to show that the resulting polynomial $P(x) = \Phi_p(1 + x)$ is irreducible. The binomial formula implies

$$P(x) = \frac{(1+x)^p - 1}{x} = \sum_{k=1}^{p} \binom{p}{k} x^{k-1}$$

$$= p + \binom{p}{2} x + \binom{p}{3} x^2 + \cdots + \binom{p}{p-2} x^{p-3} + p x^{p-2} + x^{p-1}$$

Because of Lemma 33, we see that all three assumptions for the Eisenstein criterium (Proposition 37) are satisfied. Hence the polynomial P and hence Φ_p are irreducible over the integers. $\qquad\square$

Lemma 34. *For any odd prime p, the binomial coefficients*

$$\binom{p^2}{k} \quad \text{for } k = 1 \ldots p - 1$$

are divisible by p.

Proposition 39. *For any prime p, the polynomial*

$$\Phi_{p^2}(z) = \frac{z^{p^2} - 1}{z^p - 1}$$

is irreducible over the integers.

Proof. In the case $p = 2$, we get $\Phi_4 = 1 + z^2$. which we can check to be irreducible. We assume now that p is an odd prime. We substitute $z = 1 + x$ and use the Eisenstein criterium to show that the resulting polynomial

$$\Phi_{p^2}(1 + x) = \sum_{i=0}^{p(p-1)} b_i x^i = p + \cdots + x^{p(p-1)}$$

is irreducible. The definition and the binomial formula imply

$$\frac{z^{p^2} - 1}{z - 1} = \Phi_{p^2}(z) \cdot \frac{z^p - 1}{z - 1}$$

$$\frac{(1+x)^{p^2} - 1}{x} = \Phi_{p^2}(1 + x) \cdot \frac{(1+x)^p - 1}{x}$$

$$\sum_{l=1}^{p^2} \binom{p^2}{l} x^{l-1} = \sum_{i=0}^{p(p-1)} b_i x^i \cdot \sum_{k=1}^{p} \binom{p}{k} x^{k-1}$$

We have obtained the integer factorization

$$a_0 + \cdots + a_r x^r = \left(b_0 + \cdots + b_s x^s\right)\left(c_0 + \cdots + c_t x^t\right),$$

which is really possible in this context since it has been constructed above. The degrees are now $r = p^2 - 1$, $s = \deg \Phi_{p^2} = p(p-1)$ and $t = \deg \Phi_p = p - 1$. Multiplying the polynomials and comparing the coefficients yields once more

$$
\begin{aligned}
a_0 &= b_0 c_0 \\
a_1 &= b_0 c_1 + b_1 c_0 \\
a_2 &= b_0 c_2 + b_1 c_1 + b_2 c_0 \\
a_k &= b_0 c_k + b_1 c_{k-1} + \cdots + b_k c_0
\end{aligned}
$$

for all $k \geq 0$. We know from Lemma 33 and 34 that the prime p divides all the coefficients $a_0, \ldots, a_{p^2-1}$, but $a_{p^2} = 1$. That is enough to check inductively that p divides all coefficients $b_0, \ldots, b_{p^2-p-1}$. But the leading coefficient of $\Phi_{p^2}(1 + x)$ is $b_{p^2-p} = 1$.

We see that the polynomial $\Phi_{p^2}(1+x)$ satisfies all three assumptions for the Eisenstein criterium (Proposition 37). Hence $\Phi_{p^2}(1 + x)$ and hence $\Phi_{p^2}(x)$ are irreducible over the integers. $\qquad\square$

Example $p = 3, p^2 = 9$. We claim that

$$\Phi_9(1 + x) = \sum_{i=0}^{6} b_i x^i$$

is irreducible. The definition and the binomial formula imply

$$
\begin{aligned}
\frac{z^9 - 1}{z - 1} &= \Phi_9(z) \cdot \frac{z^3 - 1}{z - 1} \\
\frac{(1 + x)^9 - 1}{x} &= \Phi_9(1 + x) \cdot \frac{(1 + x)^3 - 1}{x} \\
\sum_{l=1}^{9} \binom{9}{l} x^{l-1} &= \sum_{i=0}^{6} b_i x^i \cdot \left[3 + 3x + x^2\right]
\end{aligned}
$$

We have obtained an integer factorization. Multiplying the polynomials and comparing the coefficients yields once more

$$\binom{9}{1} = b_0 \cdot 3 , \quad \binom{9}{2} = b_0 \cdot 3 + b_1 \cdot 3 , \quad \binom{9}{3} = b_0 \cdot 1 + b_1 \cdot 3 + b_2 \cdot 3$$

$$\binom{9}{4} = b_1 \cdot 1 + b_2 \cdot 3 + b_3 \cdot 3 , \quad \binom{9}{5} = b_2 \cdot 1 + b_3 \cdot 3 + b_4 \cdot 3 , \quad \binom{9}{6} = b_3 \cdot 1 + b_4 \cdot 3 + b_5 \cdot 3$$

$$\binom{9}{7} = b_4 \cdot 1 + b_5 \cdot 3 + b_6 \cdot 3 , \quad \binom{9}{8} = b_5 \cdot 1 + b_6 \cdot 3 , \quad \binom{9}{9} = b_6 \cdot 1$$

We get the constant coefficients $a_0 = 9$, $b_0 = 3$ and $c_0 = 3$ and the leading coefficient $b_6 = 1$. As in the general case, we to use the inductive argument to conclude that 3 is a divisor of the coefficients $b_0, \ldots, b_5$. That is all we need. All assumptions in the Eisenstein criterium are satisfied and hence $\Phi_9(1 + x)$ and $\Phi_9(x)$ are irreducible over the integers. $\qquad\square$

Remark. The explicit result $\Phi_9(1 + x) = 3 + 9x + 18x^2 + 21x^3 + 15x^4 + 6x^5 + x^6$ is not needed.

IV.1.4   Irreducibility of Cyclotomic Polynomials

Lemma 35. *Let p be a prime. One uses the natural homomorphism*

$$h_p : \mathbf{Z} \mapsto \mathbf{Z}_p \quad and \quad h_p : \mathbf{Z}[x] \mapsto \mathbf{Z}_p[x]$$

As an easy consequence of the Little Fermat Theorem

$$h_p(R(x^p)) = (h_p(R(x)))^p$$

holds for any integer polynomial $R \in \mathbf{Z}[x]$.

Lemma 36. *Let p be a prime not dividing n. Assume that the cyclotomic polynomial Φ_n becomes reducible after mapping by the homomorphism h_p and let*

$$h_p(\Phi_n) = \overline{Q} \cdot \overline{R}$$

where the polynomials $\overline{Q}, \overline{R} \in \mathbf{Z}_p[x]$ are proper factors. Then $\overline{Q}$ and $\overline{R}$ are relatively prime. Especially $h_p(\Phi_n)$ has only simple zeros.

Proof. Assume towards a contradiction that the polynomials $\overline{Q}$ and $\overline{R}$ have a common proper factor $T \in \mathbf{Z}_p[x]$. Since the polynomial $h_p(\Phi_n)$ is monic, the degree of T is at least one. Hence T^2 divides $h_p(\Phi_n) = \overline{Q} \cdot \overline{R}$ and hence $Y = h_p(x^n - 1)$, too. Let $Y = F \cdot T^2$.

Now we use that elementary calculus holds for polynomials in $\mathbf{Z}_p[x]$. Hence $Y' = F'T^2 + 2FTT' = (F'T + 2FT')T$. We see that T is a divisor of both Y and Y'. Hence T is a divisor of $xY' - nY = n$, a constant polynomial.

On the other hand we know that $T \in \mathbf{Z}_p[x]$ is a nonconstant polynomial. Hence $n = 0 \in \mathbf{Z}_p[x]$, in other words the prime p is a divisor of n, contradicting the assumption $p \nmid n$. $\qquad\square$

General proof of proposition 36. This is an adaptation of the proof from Grillet's book [13] *Abstarct Algebra*, the section on Cyclotomy p. 211.

Towards a contradiction, we suppose that the cyclotomic polynomial Φ_n can be factored in $\mathbf{Q}[x]$. By Gauss' lemma we get even a factoring

$$\Phi_n(x) = Q(x)R(x)$$

with integer polynomials $Q, R \in \mathbf{Z}[x]$. Too, we may assume one factor Q to be irreducible. Any root a of Q is a primitive root of unity. Hence the powers $a, a^2, \ldots, a^n$ are a complete set of roots of $x^n - 1$. Hence there exists an integer l such that $R(a^l) = 0$. Let k be the minimal integer

$$k := \min\{l \geq 1 : \text{ there exists } a \text{ such that } Q(a) = 0 \text{ and } R(a^l) = 0\}$$

Since Φ_n and indeed $x^n - 1$ have only simple zeros, we know that $k \geq 2$. Too, $\gcd(k, n) = 1$ since a^k is a primitive root of unity.

Lemma 37. *The minimal k is a prime $p = k$ not dividing n. Moreover there exists $a \in \mathbb{C}$ such that $Q(a) = 0$ and $R(a^p) = 0$. If Q is irreducible, there exists a polynomial $S \in \mathbf{Z}[x]$ such that $R(x^p) = Q(x)S(x)$.*

Proof. Indeed, assume towards a contradiction that $p \mid k$ is a proper prime divisor. We put $b = a^p$. Now b is a primitive root of unity and hence either $Q(b) = 0$ or $R(b) = 0$. In the first case $Q(b) = 0, R(b^{k/p}) = 0$ would contradict the minimality of k. In the second case $Q(a) = 0, R(a^p) = 0$ would contradict the minimality of k.

Hence $k = p$ is a prime. Since $\gcd(k, n) = \gcd(p, n) = 1$, the prime p does not divide n. If Q is irreducible, the greatest common divisor $\gcd(Q, R(x^p)) \in \mathbf{Z}[x]$ is either (i) a constant nonzero polynomial or (ii) the polynomial Q. There exists $a \in \mathbb{C}$ such that $Q(a) = 0$ and $R(a^p) = 0$. This excludes the first possibility (i). Hence the greatest common divisor $\gcd(Q, R(x^p)) = Q$ and hence there exists a polynomial $S \in \mathbf{Z}[x]$ such that $R(x^p) = Q(x)S(x)$. $\qquad\square$

We now use lemma 35, and argue in the ring $\mathbf{Z}_p[x]$. Little Fermat yields $h_p(R(x^p)) = (h_p(R(x)))^p$. Hence

$$h_p(R)^p = h_p(R(x^p)) = h_p(Q)h_p(S) = \overline{Q} \cdot h_p(S).$$

With polynomials $\overline{Q} = h_p(Q)$ and $\overline{R} = h_p(R)$ in $\mathbf{Z}_p[x]$ we get factoring

$$h_p(\Phi_n) = \overline{Q} \cdot \overline{R}$$

We distinguish two cases

(a) $\overline{Q} \neq 0$. We know that $\overline{Q}(a) = 0$ and hence $\overline{Q}$ has degree at least one;

(b) $\overline{Q} = 0$.

Take case (a). Since the polynomials $\overline{Q}$ and $\overline{R}^p$ have the common factor $\overline{Q}$ of degree at least one, we conclude that the polynomials $\overline{Q}$ and $\overline{R}$ have a common factor of degree at least one. [11] Since $p \nmid n$, this is impossible by lemma 36.

Take case (b) that $h_p(Q) = 0$. Now $\Phi_n = QR$ implies $h_p(\Phi_n) = h_p(Q) \cdot h_p(R) = 0$ and hence of $Y = h_p(x^n - 1) = 0$, too, which is impossible.

The contradiction occurs in both cases (a) and (b) showing the cyclotomic polynomial Φ_n cannot be factored in $\mathbf{Q}[x]$. $\qquad\square$

IV.1.5   Constructible Roots of Unity

This essay is inspired by Ptolemy's ancient work, which amounts in modern language to setting up a table for the trigonometric functions. His goal was to get indeed numeric values. I am not interested in those numbers, but in exact formulas in terms of rather simple root expressions. As one bisects the angle further and further, more and more boxed roots appear. I stop somewhere after it gets too complicated. My goal is to set up a list of angles, for which the sinus and cosinus functions can be given by rational combinations of the roots $1, \sqrt{2}, \sqrt{3}, \sqrt{5}$ only,—but nolens volens I needed more boxed roots nevertheless.

Problem 48. *Find all different powers—exactly, expressed using square roots:*

(a) $\quad \omega = \dfrac{-1 + i\sqrt{3}}{2}.$

---

[11] Since $\overline{Q}$ may be reducible, we cannot conclude that $\overline{Q}$ is a divisor of $\overline{R}$. But indeed each irreducible factor of $\overline{Q}$ is a divisor of $\overline{R}$. Similarly, 9 and 36 have the common factor 9, but 9 and 6 have the common factor 3.

132

(b) $z = \dfrac{\sqrt{5} - 1 + i\sqrt{10 + 2\sqrt{5}}}{4}$.

(c) $w = \dfrac{1 + i}{\sqrt{2}}$.

You need not show all calculations. Simplify your answers by using the conjugate complex. What is remarkable about these expressions?

Answer. It is remarkable that the different powers can be obtained simply by some sign changes of some square roots:

(a)
$$\omega = \frac{-1 + i\sqrt{3}}{2} \, , \quad \omega^2 = \frac{-1 - i\sqrt{3}}{2} = \overline{\omega} \, , \quad \omega^3 = 1$$

(b)
$$z = \frac{\sqrt{5} - 1 + i\sqrt{10 + 2\sqrt{5}}}{4} \, , \quad z^2 = \frac{-\sqrt{5} - 1 + i\sqrt{10 - 2\sqrt{5}}}{4} \, ,$$

$$z^3 = \overline{z^2} \, , \quad z^4 = \overline{z}$$

(c)
$$w = \frac{1 + i}{\sqrt{2}} \, , \quad w^2 = i \, , \quad w^3 = \frac{-1 + i}{\sqrt{2}} \, , \quad w^4 = -1 \, ,$$

$$w^5 = \overline{w^3} \, , \quad w^6 = -i \, , \quad w^7 = \overline{w}$$

An exact trigonometric table of $\cos 3°k$ and $\sin 3°k$ for all $k = 1, \ldots, 15$ can be obtained using the three roots of unity ω, z and w. This trick goes in principle back to Ptolemy!

(a) From the primitive third and eighth root of unity ω and w from Problem 48, we get a product of argument $15°$.

$$\omega^{-1} w^3 = e^{2\pi i \left[-\frac{1}{3} + \frac{3}{8}\right]} = e^{\frac{2\pi i}{24}} = \cos 15° + i \sin 15°$$

(b) We can calculate this product in terms of the exact square roots, obtained in Problem 48.

$$\omega^{-1} w^3 = \frac{-1 - i\sqrt{3}}{2} \cdot \frac{-1 + i}{\sqrt{2}} = \frac{1 + \sqrt{3} + i(\sqrt{3} - 1)}{2\sqrt{2}}$$

(c) Finally, we compare the real- and imaginary parts obtained form (a) and (b), and use the common denominator 4.

$$\cos 15° = \frac{\sqrt{2} + \sqrt{6}}{4}$$

$$\sin 15° = \frac{\sqrt{6} - \sqrt{2}}{4}$$

$k\,15°$	$\cos k\,15°$	$\sin k\,15°$
$15°$	$\dfrac{\sqrt{2}+\sqrt{6}}{4}$	$\dfrac{\sqrt{6}-\sqrt{2}}{4}$
$30°$	$\dfrac{\sqrt{3}}{2}$	$\dfrac{1}{2}$
$45°$	$\dfrac{\sqrt{2}}{2}$	$\dfrac{\sqrt{2}}{2}$
$60°$	$\dfrac{1}{2}$	$\dfrac{\sqrt{3}}{2}$
$75°$	$\dfrac{\sqrt{6}-\sqrt{2}}{4}$	$\dfrac{\sqrt{6}+\sqrt{2}}{4}$
$90°$	0	1

(d) In this case, we get the table immediately.

Problem 49. *Calculate* $\cos 15° + i \sin 15°$ *directly by formulas*

$$\cos \frac{\theta}{2} = \sqrt{\frac{1 + \cos \theta}{2}} \ , \quad \sin \frac{\theta}{2} = \sqrt{\frac{1 - \cos \theta}{2}}$$

and confirm that you get the same values.

Answer.

$$\cos \frac{30°}{2} = \sqrt{\frac{1 + \cos 30°}{2}} = \sqrt{\frac{2 + \sqrt{3}}{4}} = \frac{\sqrt{2 + \sqrt{3}}}{2}$$

Similarly, one gets $\sin 15° = \frac{\sqrt{2-\sqrt{3}}}{2}$. By squaring both sides, one checks that indeed

$$\frac{\sqrt{2 + \sqrt{3}}}{2} = \frac{\sqrt{6} + \sqrt{2}}{4} \ , \quad \frac{\sqrt{2 - \sqrt{3}}}{2} = \frac{\sqrt{6} - \sqrt{2}}{4}$$

Problem 50. *The following steps can be elaborated to obtain an exact trigonometric table of* $\cos 9°k$ *and* $\sin 9°k$ *for all* $k = 1, \ldots, 40$.

(a) *Find natural numbers* p, q *such that* $-\frac{p}{5} + \frac{q}{8} = \frac{1}{40}$.

(b) *With* z, w *from Problem 48, calculate the argument* θ *in*

$$z^{-p} w^{q} = e^{2\pi i \theta}$$

(c) *Hint: calculate*

$$\frac{\sqrt{5} + 1 - i\sqrt{10 - 2\sqrt{5}}}{4} \cdot \frac{1 + i}{\sqrt{2}}$$

(d) *Calculate* $z^{-p} w^{q}$ *with the exact square roots, obtained in Problem 48.*

(e) *Use Euler's*

$$\text{(IV.1.24)} \qquad e^{\frac{2\pi i}{40}} = \cos\frac{2\pi}{40} + i\sin\frac{2\pi}{40}$$

and get exact expressions for $\cos 9°$ *and* $\sin 9°$. *Use the common denominator* 8 *in your results.*

Answer. **(a)** The natural numbers p, q have to satisfy $-8p + 5q = 1$. The smallest solution is $p = 3, q = 5$.

(b) With z, w from Problem 48,

$$z^{-p}w^q = e^{2\pi i\left[-\frac{3}{5}+\frac{5}{8}\right]} = e^{\frac{2\pi i}{40}}$$

(d) With the same z, w from Problem 48, using $z^5 = 1$ and $w^2 = i$, we can simplify and get

$$z^{-3}w^5 = -z^2w = \frac{\sqrt{5}+1 - i\sqrt{10-2\sqrt{5}}}{4} \cdot \frac{1+i}{\sqrt{2}}$$

$$= \frac{\sqrt{10}+\sqrt{2}+2\sqrt{5-\sqrt{5}}}{8} + i\,\frac{\sqrt{10}+\sqrt{2}-2\sqrt{5-\sqrt{5}}}{8}$$

(e) Comparing real- and imaginary part from parts (b) and (c) yields

$$\text{(IV.1.25)} \qquad \cos 9° = \frac{\sqrt{10}+\sqrt{2}+2\sqrt{5-\sqrt{5}}}{8}$$

$$\text{(IV.1.26)} \qquad \sin 9° = \frac{\sqrt{10}+\sqrt{2}-2\sqrt{5-\sqrt{5}}}{8}$$

Problem 51. *Give the exact expression for the perimeter of a regular 20-gon, inscribed into the unit circle. Compare with* 2π. *How large is the deviation?*

Answer. The circumference of a regular 20-gon is

$$40\sin 9° = 5\left[\sqrt{10}+\sqrt{2}-2\sqrt{5-\sqrt{5}}\right] \approx 6.257378602$$

Bur $2\pi \approx 6.283185307$ so only one decimal point. But the relative error of less than a percent is a beginning.

In a similar way, we can use the twelfth and fifth root of unity, and get sin and cos for the multiples of 6°.

(a) The natural numbers p, q have to satisfy $12p - 5q = \pm 1$. The smallest solution is $p = 3, q = 7$. Dividing by 60 yields

$$\frac{3}{5} - \frac{7}{12} = \frac{1}{60}$$

(b) With the twelfth and fifth roots of unity τ and z we get

$$z^3 \tau^{-7} = e^{2\pi i \left[\frac{3}{5} - \frac{7}{12}\right]} = e^{\frac{2\pi i}{60}} = \cos 6° + i \sin 6°$$

(d) We can calculate this product with the exact values of the square roots, simplify and get

$$z^3 \tau^{-7} = -z^3 \tau^{-1} = \frac{\sqrt{5} + 1 + i\sqrt{10 - 2\sqrt{5}}}{4} \cdot \frac{\sqrt{3} - i}{2}$$
$$= \frac{\sqrt{15} + \sqrt{3} + \sqrt{10 - 2\sqrt{5}}}{8} + i \frac{-\sqrt{5} - 1 + \sqrt{3}\sqrt{10 - 2\sqrt{5}}}{8}$$

(e) Comparing real- and imaginary part from parts (b) and (c) yields

$$(\text{IV.1.27}) \qquad \cos 6° = \frac{\sqrt{15} + \sqrt{3} + \sqrt{10 - 2\sqrt{5}}}{8}$$

$$(\text{IV.1.28}) \qquad \sin 6° = \frac{-\sqrt{5} - 1 + \sqrt{3}\sqrt{10 - 2\sqrt{5}}}{8}$$

Problem 52. *Give the exact expression for the perimeter of a regular 30-gon, inscribed into the unit circle. Compare with 2π. How large is the deviation?*

Answer. The circumference of a regular 30-gon is

$$60 \sin 6° = \frac{15}{2}\left[-\sqrt{5} - 1 + \sqrt{3}\sqrt{10 - 2\sqrt{5}}\right] \approx 6.271707796$$

But $2\pi \approx 6.283185307$—so just one decimal point coincides.

Problem 53. *Prove that*

$$(\text{IV.1.29}) \qquad \cos 24° = \frac{\sqrt{5} + 1 + \sqrt{3}\sqrt{10 - 2\sqrt{5}}}{8}$$

$$(\text{IV.1.30}) \qquad \sin 24° = \frac{\sqrt{15} + \sqrt{3} - \sqrt{10 - 2\sqrt{5}}}{8}$$

Here is a table assembling the information we have gathered so far.

θ	$\cos\theta$	$\sin\theta$
$6°$	$\dfrac{\sqrt{15}+\sqrt{3}+\sqrt{10-2\sqrt{5}}}{8}$	$\dfrac{-\sqrt{5}-1+\sqrt{3}\sqrt{10-2\sqrt{5}}}{8}$
$9°$	$\dfrac{\sqrt{10}+\sqrt{2}+2\sqrt{5-\sqrt{5}}}{8}$	$\dfrac{\sqrt{10}+\sqrt{2}-2\sqrt{5-\sqrt{5}}}{8}$
$15°$	$\dfrac{\sqrt{2}+\sqrt{6}}{4}$	$\dfrac{\sqrt{6}-\sqrt{2}}{4}$
$18°$	$\sqrt{\dfrac{5+\sqrt{5}}{8}}$	$\dfrac{\sqrt{5}-1}{4}$
$24°$	$\dfrac{\sqrt{5}+1+\sqrt{3}\sqrt{10-2\sqrt{5}}}{8}$	$\dfrac{\sqrt{15}+\sqrt{3}-\sqrt{10-2\sqrt{5}}}{8}$
$30°$	$\dfrac{\sqrt{3}}{2}$	$\dfrac{1}{2}$
$36°$	$\dfrac{\sqrt{5}+1}{4}$	$\sqrt{\dfrac{5-\sqrt{5}}{8}}$
$45°$	$\dfrac{\sqrt{2}}{2}$	$\dfrac{\sqrt{2}}{2}$

Problem 54. *An exact trigonometric table of $\cos 3°k$ and $\sin 3°k$ for all $k = 1, \ldots, 15$ can be obtained with the same trick—which goes in principle back to Ptolemy!*

(a) *Calculate $\dfrac{1}{3} - \dfrac{1}{5} - \dfrac{1}{8}$. With ω, z and w from Problem 48, get the argument θ in*

$$\omega z^{-1} w^{-1} = e^{2\pi i\theta} = \cos 2\pi\theta + i\sin 2\pi\theta$$

(b) *Calculate $\omega z^{-1} w^{-1}$ with the exact square roots, obtained in Problem 48. Get a readable result, do not distribute every term.*

(c) *Get exact expressions for $\cos 3°$ and $\sin 3°$.*

Solution. **(a)** Since $\dfrac{1}{3} - \dfrac{1}{5} - \dfrac{1}{8} = \dfrac{1}{120}$, we get the argument

$$\omega z^{-1} w^{-1} = e^{2\pi i\left[\frac{1}{3}-\frac{1}{5}-\frac{1}{8}\right]} = \cos\frac{2\pi}{120} + i\sin\frac{2\pi}{120}$$

(b) With the exact square roots from Problem 48,

$$\omega z^{-1} w^{-1} = \omega \overline{w} \overline{z} = \left[\frac{-1+i\sqrt{3}}{2}\right]\left[\frac{1-i}{\sqrt{2}}\right]\frac{\sqrt{5}-1-i\sqrt{10+2\sqrt{5}}}{4}$$

$$= \frac{\sqrt{3}-1+i(\sqrt{3}+1)}{2\sqrt{2}}\,\frac{\sqrt{5}-1-i\sqrt{10+2\sqrt{5}}}{4}$$

$$= \frac{(\sqrt{3}-1)(\sqrt{5}-1)+(\sqrt{3}+1)\sqrt{10+2\sqrt{5}}}{8\sqrt{2}} + i\,\frac{(\sqrt{3}+1)(\sqrt{5}-1)-(\sqrt{3}-1)\sqrt{10+2\sqrt{5}}}{8\sqrt{2}}$$

(e) We get exact expressions

$$\cos 3^\circ = \frac{\sqrt{2}(\sqrt{3}-1)(\sqrt{5}-1)+2(\sqrt{3}+1)\sqrt{5+\sqrt{5}}}{16}$$

$$\sin 3^\circ = \frac{\sqrt{2}(\sqrt{3}+1)(\sqrt{5}-1)-2(\sqrt{3}-1)\sqrt{5+\sqrt{5}}}{16}$$

$\square$

Problem 55. *Calculate the circumference of a regular 60-gon. Which approximation of 2π do we get? How many decimal points are correct?*

Answer. The circumference of an 60-gon is

$$120\sin 3^\circ = \frac{15}{2}\left[\sqrt{2}(\sqrt{3}+1)(\sqrt{5}-1)-2(\sqrt{3}-1)\sqrt{5+\sqrt{5}}\right] \approx 6.280314749$$

$$2\pi \approx 6.283185307$$

One gets two correct decimal points.

Problem 56. *Gather the entire table with $\cos k3^\circ$ and $\sin k3^\circ$ for all $k = 1\ldots 30$. Go on as far as you get.*

$$\omega^k z^{-k} w^{-k} = e^{\frac{2\pi i k}{120}} = \left[\frac{\sqrt{3}-1+i(\sqrt{3}+1)}{2\sqrt{2}}\right]^k \left[\frac{\sqrt{5}-1-i\sqrt{10+2\sqrt{5}}}{4}\right]^k$$

$$\omega^2 z^{-2} w^{-2} = e^{\frac{4\pi i}{120}} = \left[\frac{-\sqrt{3}+i}{2}\right]\left[\frac{-\sqrt{5}-1-i\sqrt{10-2\sqrt{5}}}{4}\right]$$

$$= \frac{\sqrt{15}+\sqrt{3}+\sqrt{10-2\sqrt{5}}-i\sqrt{5}-i+i\sqrt{3}\sqrt{10-2\sqrt{5}}}{8}$$

$$\omega^4 z^{-4} w^{-4} = e^{\frac{8\pi i}{120}} = -\omega z = \left[\frac{1-i\sqrt{3}}{2}\right]\left[\frac{\sqrt{5}-1+i\sqrt{10+2\sqrt{5}}}{4}\right]$$

$$= \frac{\sqrt{5}-1+\sqrt{3}\sqrt{10+2\sqrt{5}}-i\sqrt{15}+i\sqrt{3}+i\sqrt{10+2\sqrt{5}}}{8}$$

$$\omega^7 z^{-7} w^{-7} = e^{\frac{14\pi i}{120}} = \omega w z^{-2} = \left[\frac{-\sqrt{3}-1+i(\sqrt{3}-1)}{2\sqrt{2}}\right]\left[\frac{-\sqrt{5}-1-i\sqrt{10-2\sqrt{5}}}{4}\right]$$

$$= \frac{\sqrt{2}(\sqrt{3}+1)(\sqrt{5}+1)+2(\sqrt{3}-1)\sqrt{5-\sqrt{5}}}{16} + i\,\frac{\sqrt{2}(-\sqrt{3}+1)(\sqrt{5}+1)+2(\sqrt{3}+1)\sqrt{5-\sqrt{5}}}{16}$$

Further values were obtained by trial and error for the signs of the roots.

θ	$\cos\theta$	$\sin\theta$
$3°$	$\dfrac{\sqrt{2}(\sqrt{3}-1)(\sqrt{5}-1)+2(\sqrt{3}+1)\sqrt{5+\sqrt{5}}}{16}$	$\dfrac{\sqrt{2}(\sqrt{3}+1)(\sqrt{5}-1)-2(\sqrt{3}-1)\sqrt{5+\sqrt{5}}}{16}$
$6°$	$\dfrac{\sqrt{15}+\sqrt{3}+\sqrt{10-2\sqrt{5}}}{8}$	$\dfrac{-\sqrt{5}-1+\sqrt{3}\sqrt{10-2\sqrt{5}}}{8}$
$9°$	$\dfrac{\sqrt{10}+\sqrt{2}+2\sqrt{5-\sqrt{5}}}{8}$	$\dfrac{\sqrt{10}+\sqrt{2}-2\sqrt{5-\sqrt{5}}}{8}$
$12°$	$\dfrac{\sqrt{5}-1+\sqrt{3}\sqrt{10+2\sqrt{5}}}{8}$	$\dfrac{-\sqrt{15}+\sqrt{3}+\sqrt{10+2\sqrt{5}}}{8}$
$15°$	$\dfrac{\sqrt{2}+\sqrt{6}}{4}$	$\dfrac{\sqrt{6}-\sqrt{2}}{4}$
$18°$	$\sqrt{\dfrac{5+\sqrt{5}}{8}}$	$\dfrac{\sqrt{5}-1}{4}$
$21°$	$\dfrac{\sqrt{2}(\sqrt{3}+1)(\sqrt{5}+1)+2(\sqrt{3}-1)\sqrt{5-\sqrt{5}}}{16}$	$\dfrac{\sqrt{2}(-\sqrt{3}+1)(\sqrt{5}+1)+2(\sqrt{3}+1)\sqrt{5-\sqrt{5}}}{16}$
$24°$	$\dfrac{\sqrt{5}+1+\sqrt{3}\sqrt{10-2\sqrt{5}}}{8}$	$\dfrac{\sqrt{15}+\sqrt{3}-\sqrt{10-2\sqrt{5}}}{8}$
$27°$	$\dfrac{\sqrt{10}-\sqrt{2}+2\sqrt{5+\sqrt{5}}}{8}$	$\dfrac{-\sqrt{10}+\sqrt{2}+2\sqrt{5+\sqrt{5}}}{8}$
$30°$	$\dfrac{\sqrt{3}}{2}$	$\dfrac{1}{2}$
$33°$	$\dfrac{-\sqrt{2}(\sqrt{3}-1)(\sqrt{5}-1)+2(\sqrt{3}+1)\sqrt{5+\sqrt{5}}}{16}$	$\dfrac{\sqrt{2}(\sqrt{3}+1)(\sqrt{5}-1)+2(\sqrt{3}-1)\sqrt{5+\sqrt{5}}}{16}$
$36°$	$\dfrac{\sqrt{5}+1}{4}$	$\sqrt{\dfrac{5-\sqrt{5}}{8}}$
$39°$	$\dfrac{-\sqrt{2}(-\sqrt{3}+1)(\sqrt{5}+1)+2(\sqrt{3}+1)\sqrt{5-\sqrt{5}}}{16}$	$\dfrac{\sqrt{2}(\sqrt{3}+1)(\sqrt{5}+1)-2(\sqrt{3}-1)\sqrt{5-\sqrt{5}}}{16}$
$42°$	$\dfrac{\sqrt{15}-\sqrt{3}+\sqrt{10+2\sqrt{5}}}{8}$	$\dfrac{-\sqrt{5}+1+\sqrt{3}\sqrt{10+2\sqrt{5}}}{8}$
$45°$	$\dfrac{\sqrt{2}}{2}$	$\dfrac{\sqrt{2}}{2}$
$48°$	$\dfrac{-\sqrt{5}+1+\sqrt{3}\sqrt{10+2\sqrt{5}}}{8}$	$\dfrac{\sqrt{15}-\sqrt{3}+\sqrt{10+2\sqrt{5}}}{8}$
$51°$	$\dfrac{\sqrt{2}(\sqrt{3}+1)(\sqrt{5}+1)-2(\sqrt{3}-1)\sqrt{5-\sqrt{5}}}{16}$	$\dfrac{-\sqrt{2}(-\sqrt{3}+1)(\sqrt{5}+1)+2(\sqrt{3}+1)\sqrt{5-\sqrt{5}}}{16}$
$54°$	$\sqrt{\dfrac{5-\sqrt{5}}{8}}$	$\dfrac{\sqrt{5}+1}{4}$
$57°$	$\dfrac{\sqrt{2}(\sqrt{3}+1)(\sqrt{5}-1)+2(\sqrt{3}-1)\sqrt{5+\sqrt{5}}}{16}$	$\dfrac{-\sqrt{2}(\sqrt{3}-1)(\sqrt{5}-1)+2(\sqrt{3}+1)\sqrt{5+\sqrt{5}}}{16}$
$60°$	$\dfrac{1}{2}$	$\dfrac{\sqrt{3}}{2}$
$63°$	$\dfrac{-\sqrt{10}+\sqrt{2}+2\sqrt{5+\sqrt{5}}}{8}$	$\dfrac{\sqrt{10}-\sqrt{2}+2\sqrt{5+\sqrt{5}}}{8}$
$66°$	$\dfrac{\sqrt{15}+\sqrt{3}-\sqrt{10-2\sqrt{5}}}{8}$	$\dfrac{\sqrt{5}+1+\sqrt{3}\sqrt{10-2\sqrt{5}}}{8}$
$69°$	$\dfrac{\sqrt{2}(-\sqrt{3}+1)(\sqrt{5}+1)+2(\sqrt{3}+1)\sqrt{5-\sqrt{5}}}{16}$	$\dfrac{\sqrt{2}(\sqrt{3}+1)(\sqrt{5}+1)+2(\sqrt{3}-1)\sqrt{5-\sqrt{5}}}{16}$
$72°$	$\dfrac{\sqrt{5}-1}{4}$	$\sqrt{\dfrac{5+\sqrt{5}}{8}}$
$75°$	$\dfrac{\sqrt{6}-\sqrt{2}}{4}$	$\dfrac{\sqrt{6}+\sqrt{2}}{4}$
$78°$	$\dfrac{-\sqrt{15}+\sqrt{3}+\sqrt{10+2\sqrt{5}}}{8}$	$\dfrac{\sqrt{5}-1+\sqrt{3}\sqrt{10+2\sqrt{5}}}{8}$
$81°$	$\dfrac{\sqrt{10}+\sqrt{2}-2\sqrt{5-\sqrt{5}}}{8}$	$\dfrac{\sqrt{10}+\sqrt{2}+2\sqrt{5-\sqrt{5}}}{8}$
$84°$	$\dfrac{-\sqrt{5}-1+\sqrt{3}\sqrt{10-2\sqrt{5}}}{8}$	$\dfrac{\sqrt{15}+\sqrt{3}+\sqrt{10-2\sqrt{5}}}{8}$
$87°$	$\dfrac{\sqrt{2}(\sqrt{3}+1)(\sqrt{5}-1)-2(\sqrt{3}-1)\sqrt{5+\sqrt{5}}}{16}$	$\dfrac{\sqrt{2}(\sqrt{3}-1)(\sqrt{5}-1)+2(\sqrt{3}+1)\sqrt{5+\sqrt{5}}}{16}$
$90°$	0	1

Decomposition of the corresponding cyclotomic polynomials

The exponentials $\exp(i\phi_k) = \cos\phi_k + i\sin\phi_k$ with $\phi_k = \frac{2\pi k}{120}$ for $-60 < k \le 60$ are the vertices of a regular 120-gon. They are the zeros of the polynomial $z^{120} - 1$. In the ring $\mathbf{Z}[z]$ of polynomials with integer coefficients, it has the factorization into the irreducible cyclotomic polynomials

$$z^{120} - 1 = \prod_{d \mid 120} \Phi_d(z)$$

Let me order the above ϕ_k as zeros of the cyclotomic polynomials. Plus or minus of the real and imaginary parts of the ϕ_k appear in the above list, unless they are the units $\pm 1, \pm i$. For any one divisor $d \mid 120$ they are the zeros of the same cyclotomic polynomial Φ_d. Since $\Phi_d \in \mathbf{Z}[z]$ is irreducible, these zeros are all algebraic conjugates. The zeros lie all in the extension $K + iK$ of the constructible fields K, since the regular 120-gon is constructible. This was already known in antiquity, and follows of course from the Gauss-Wantzel Theorem, too. For the members of the field $K + iK$, the algebraic conjugates are obtained from the same root expression by change of the appropriate plus and minus signs in front of *some* of the square roots. The sign changes of several square roots may be linked during this a bid tricky process.

It is worth while to confirm these general remarks with the explicit formulas. I begin with case where d is divisible by 4. The mathematica code is contained in the file Ptolemy1.nb.

In this case the cyclotomic polynomial Φ_d is symmetric. For simplicity, I have put the four zeros at the conjugate complexes $\phi_{\pm k}$, and at their supplementary angles $\phi_{\pm(60-k)}$ together.

$d = 120$

$$(\text{IV.1.31}) \qquad \Phi_{120}(z) = 1 + z^4 - z^{12} - z^{16} - z^{20} + z^{28} + z^{32}$$

$$= \prod_{s_1=\pm 1, s_2=\pm 1, s_3=\pm 1} \left(z^2 + 1\right)^2 - \frac{z^2}{32} \times$$

$$\left(s_1 + \sqrt{3}s_2 + \sqrt{5}s_3 + \sqrt{15}s_1 s_2 s_3 - s_1 s_2\sqrt{2\left(5 - \sqrt{5}s_1 s_3\right)} + \sqrt{6\left(5 - \sqrt{5}s_1 s_3\right)}\right)^2$$

It is worth while to check to which one of the angles ϕ_k these 4×8 zeros belong. The numbers $\{\pm k, 60 \pm k\}$ have to cover all integers modulo 120 which are relatively prime to 120.

s_1	s_2	s_3	$\{\mp(60-k), \mp k\}$	
1	1	1	$\{-53, 53, -7, 7\}$	$\cos 21° = \dfrac{\sqrt{2}+\sqrt{6}+\sqrt{10}+\sqrt{30}-2\sqrt{5-\sqrt{5}}+2\sqrt{3(5-\sqrt{5})}}{16}$
1	1	-1	$\{-31, 31, -29, 29\}$	$\cos 87° = \dfrac{-\sqrt{2}-\sqrt{6}+\sqrt{10}+\sqrt{30}+2\sqrt{\sqrt{5}+5}-2\sqrt{3(\sqrt{5}+5)}}{16}$
1	-1	1	$\{-37, 37, -23, 23\}$	$\cos 69° = \dfrac{\sqrt{2}-\sqrt{6}+\sqrt{10}-\sqrt{30}+2\sqrt{5-\sqrt{5}}+2\sqrt{3(5-\sqrt{5})}}{16}$
1	-1	-1	$\{-59, 59, -1, 1\}$	$\cos 3° = \dfrac{\sqrt{2}-\sqrt{6}-\sqrt{10}+\sqrt{30}+2\sqrt{\sqrt{5}+5}+2\sqrt{3(\sqrt{5}+5)}}{16}$
-1	1	1	$\{-49, 49, -11, 11\}$	$\cos 33° = \dfrac{-\sqrt{2}+\sqrt{6}+\sqrt{10}-\sqrt{30}+2\sqrt{\sqrt{5}+5}+2\sqrt{3(\sqrt{5}+5)}}{16}$
-1	1	-1	$\{-47, 47, -13, 13\}$	$\cos 39° = \dfrac{-\sqrt{2}+\sqrt{6}-\sqrt{10}+\sqrt{30}+2\sqrt{5-\sqrt{5}}+2\sqrt{3(5-\sqrt{5})}}{16}$
-1	-1	1	$\{-41, 41, -19, 19\}$	$\cos 57° = \dfrac{-\sqrt{2}-\sqrt{6}+\sqrt{10}+\sqrt{30}-2\sqrt{\sqrt{5}+5}+2\sqrt{3(\sqrt{5}+5)}}{16}$
-1	-1	-1	$\{-43, 43, -17, 17\}$	$\cos 51° = \dfrac{\sqrt{2}+\sqrt{6}+\sqrt{10}+\sqrt{30}+2\sqrt{5-\sqrt{5}}-2\sqrt{3(5-\sqrt{5})}}{16}$

Lemma 38. *Assume that $d \mid 120$ is a divisor of 120. The zeros of Φ_d are those roots of unity ϕ_k for which*

$$\frac{120}{d} \mid k \quad and \quad \gcd\left(d, \frac{d \cdot k}{120}\right) = 1$$

Proof. The zeros of Φ_d are those $\exp(\frac{2\pi i \cdot l}{d})$ for which $-d < l \le d$ and $\gcd(d, l) = 1$. The above roots of unity are those ϕ_k with $k = \frac{120 l}{d}$ and $\gcd(d, l) = \gcd(d, \frac{d \cdot k}{120}) = 1$. This is equivalent to the conditions for k given above. $\qquad\square$

Remark. With $e = 120/d$ holds

$$e \mid k \quad and \quad \gcd\left(\frac{k}{e}, \frac{120}{e}\right) = 1$$

$d = 60$

$$(\text{IV.1.32})$$

$$\Phi_{60}(z) = 1 + z^2 - z^6 - z^8 - z^{10} + z^{14} + z^{16}$$

$$= \prod_{s_1=\pm 1, s_2=\pm 1} (z^2 + 1)^2 - \frac{1}{16} z^2 \left(\sqrt{2\left(5 - \sqrt{5}s_1 s_2\right)} + \sqrt{3}s_1 + \sqrt{15}s_2\right)^2$$

The numbers $\{\pm k, 60 \pm k\}$ which satisfy the conditions given in the lemma are

$$\{\{-58, 58, -2, 2\}, \{-34, 34, -26, 26\}, \{-46, 46, -14, 14\}, \{-38, 38, -22, 22\}\}$$

and we get these ϕ_k once more as zeros of the product of formula (IV.1.32).

s_1	s_2	$\{\mp(60-k), \mp k\}$	$\cos 3k$
1	1	$\{-58, 58, -2, 2\}$	$\cos 6° = \dfrac{\sqrt{3}+\sqrt{15}+\sqrt{2\left(5-\sqrt{5}\right)}}{8}$
1	−1	$\{-34, 34, -26, 26\}$	$\cos 78° = \dfrac{\sqrt{3}-\sqrt{15}+\sqrt{2\left(\sqrt{5}+5\right)}}{8}$
−1	1	$\{-46, 46, -14, 14\}$	$\cos 42° = \dfrac{-\sqrt{3}+\sqrt{15}+\sqrt{2\left(\sqrt{5}+5\right)}}{8}$
−1	−1	$\{-38, 38, -22, 22\}$	$\cos 66° = \dfrac{\sqrt{3}+\sqrt{15}-\sqrt{2\left(5-\sqrt{5}\right)}}{8}$

$d = 40$

(IV.1.33)

$$\Phi_{40}(z) = 1 - z^4 + z^8 - z^{12} + z^{16}$$

$$= \prod_{s_1=\pm 1, s_2=\pm 1} (z^2+1)^2 - \frac{1}{16}z^2 \left(2\sqrt{5-\sqrt{5}}s_1 s_2 + \sqrt{2}s_1 + \sqrt{10}s_2\right)^2$$

s_1	s_2	$\{\mp(60-k), \mp k\}$	$\cos 3k$
1	1	$\{-57, 57, -3, 3\}$	$\cos 9° = \dfrac{\sqrt{2}+\sqrt{10}+2\sqrt{5-\sqrt{5}}}{8}$
1	−1	$\{-39, 39, -21, 21\}$	$\cos 63° = \dfrac{\sqrt{2}-\sqrt{10}+2\sqrt{\sqrt{5}+5}}{8}$
−1	1	$\{-51, 51, -9, 9\}$	$\cos 27° = \dfrac{\sqrt{2}-\sqrt{10}+2\sqrt{\sqrt{5}+5}}{8}$
−1	−1	$\{-33, 33, -27, 27\}$	$\cos 81° = \dfrac{\sqrt{2}+\sqrt{10}-2\sqrt{5-\sqrt{5}}}{8}$

$d = 24$

(IV.1.34)
$$\Phi_{24}(z) = 1 - z^4 + z^8$$

$$= \prod_{s_1=\pm 1} (z^2+1)^2 - \frac{1}{2}\left(\sqrt{3}s_1+1\right)^2 z^2$$

s_1	$\{\mp(60-k),\mp k\}$	$\cos 3k$
1	$\{-55,55,-5,5\}$	$\cos 15° = \frac{\sqrt{2}+\sqrt{6}}{4}$
-1	$\{-35,35,-25,25\}$	$\cos 75° = \frac{-\sqrt{2}+\sqrt{6}}{4}$

$d = 20$

$$(IV.1.35) \qquad \Phi_{20}(z) = 1 - z^2 + z^4 - z^6 + z^8$$

$$= \prod_{s_1=\pm 1} \left(z^2+1\right)^2 - \frac{1}{2}\left(\sqrt{5}s_1 + 5\right) z^2$$

s_1	$\{\mp(60-k),\mp k\}$	$\cos 3k$
1	$\{-54,54,-6,6\}$	$\cos 18° = \frac{10+2\sqrt{5}}{4}$
-1	$\{-42,42,-18,18\}$	$\cos 54° = \frac{10-2\sqrt{5}}{4}$

$d = 12$

$$(IV.1.36) \qquad \Phi_{12}(z) = 1 - z^2 + z^4 = -3z^2 + (1+z^2)^2$$

$\{\mp(60-k),\mp k\}$	$\cos 3k$
$\{-50,50,-10,10\}$	$\cos 30° = \frac{\sqrt{3}}{2}$

$d = 8$

$$(IV.1.37) \qquad \Phi_8(z) = 1 + z^4 = -2z^2 + (1+z^2)^2$$

$\{\mp(60-k),\mp k\}$	$\cos 3k$
$\{-45,45,-15,15\}$	$\cos 45° = \frac{\sqrt{2}}{2}$

Next, here are the cases where d is not divisible by 4. Here the cyclotomic polynomial $|Phi_d$ is not symmetric. One can one put the conjugate zeros $\phi_{\pm k}$ together. The supplementary angle $\phi_{30\pm k}$ does not give a zero of the same cyclotomic polynomial.

$d = 30$

(IV.1.38)

$$\Phi_{30}(z) = 1 + z - z^3 - z^4 - z^5 + z^7 + z^8$$

$$= \prod_{s_1 = \pm 1, s_2 = \pm 1} -\frac{1}{4}z\left(\sqrt{6\left(\sqrt{5}s_1 + 5\right)}s_2 + \sqrt{5}s_1 - 1\right) + z^2 + 1$$

s_1	s_2	$\{\mp k\}$	$\cos 3k$
1	1	$\{-4, 4\}$	$\cos 12° = \dfrac{-1+\sqrt{5}+\sqrt{6\left(\sqrt{5}+5\right)}}{8}$
1	-1	$\{-44, 44\}$	$\cos 132° = \dfrac{-1+\sqrt{5}-\sqrt{6\left(\sqrt{5}+5\right)}}{8}$
-1	1	$\{-28, 28\}$	$\cos 84° = \dfrac{-1-\sqrt{5}+\sqrt{6\left(5-\sqrt{5}\right)}}{8}$
-1	-1	$\{-52, 52\}$	$\cos 156° = \dfrac{-1-\sqrt{5}-\sqrt{6\left(5-\sqrt{5}\right)}}{8}$

$d = 15$

(IV.1.39)

$$\Phi_{15}(z) = 1 - z + z^3 - z^4 + z^5 - z^7 + z^8$$

$$= \prod_{s_1 = \pm 1, s_2 = \pm 1} -\frac{1}{4}z\left(\sqrt{6\left(5 - \sqrt{5}s_1\right)}s_2 + \sqrt{5}s_1 + 1\right) + z^2 + 1$$

s_1	s_2	$\{\mp k\}$	$\cos 3k$
1	1	$\{-8, 8\}$	$\cos 24° = \dfrac{1+\sqrt{5}+\sqrt{6\left(\sqrt{5}-5\right)}}{8}$
1	-1	$\{-32, 32\}$	$\cos 96° = \dfrac{1+\sqrt{5}-\sqrt{6\left(5-\sqrt{5}\right)}}{8}$
-1	1	$\{-16, 16\}$	$\cos 48° = \dfrac{1-\sqrt{5}+\sqrt{6\left(5+\sqrt{5}\right)}}{8}$
-1	-1	$\{-56, 56\}$	$\cos 168° = \dfrac{1-\sqrt{5}-\sqrt{6\left(5+\sqrt{5}\right)}}{8}$

146

$d = 10$

(IV.1.40)
$$\Phi_{10}(z) = 1 - z + z^2 - z^3 + z^4$$
$$= \prod_{s_1 = \pm 1} \left[-\frac{1}{2}\left(\sqrt{5}s_1 + 1\right) z + z^2 + 1 \right]$$

s_1	$\{\mp k\}$	$\cos 3k$
1	$\{-12, 12\}$	$\cos 36° = \frac{1+\sqrt{5}}{4}$
-1	$\{-36, 36\}$	$\cos 108° = \frac{1-\sqrt{5}}{4}$

$d = 6$

(IV.1.41)
$$\Phi_6(z) = 1 - z + z^2$$

$\{\mp k\}$	$\cos 3k$
$\{-20, 20\}$	$\cos 60° = \frac{1}{2}$

$d = 5$

(IV.1.42)
$$\Phi_5(z) = 1 + z + z^2 + z^3 + z^4$$

s_1	$\{\mp k\}$	$\cos 3k$
1	$\{-24, 24\}$	$\cos 72° = \frac{-1+\sqrt{5}}{4}$
-1	$\{-48, 48\}$	$\cos 144° = \frac{-1-\sqrt{5}}{4}$

$d = 3$

(IV.1.43)
$$\Phi_3(z) = 1 + z + z^2$$

$\{\mp k\}$	$\cos 3k$
$\{-40, 40\}$	$\cos 120° = \frac{-1}{2}$

Finally, there were left out the three cases $d = 4, 2, 1$ where all zeros are units.

$$\Phi_4(z) = 1 + z^2$$
$$\Phi_2(z) = 1 + z$$
$$\Phi_1(z) = -1 + z$$

The table given above contains a bid different expressions for the cosin function. Here I have put them side by side for comparison. The reader may check for correctness.

θ	$\cos\theta$	$\cos\theta$
$3°$	$\dfrac{\sqrt{2}(\sqrt{3}-1)(\sqrt{5}-1)+2(\sqrt{3}+1)\sqrt{5+\sqrt{5}}}{16}$	$\dfrac{\sqrt{2}-\sqrt{6}-\sqrt{10}+\sqrt{30}+2\sqrt{\sqrt{5}+5}+2\sqrt{3\left(\sqrt{5}+5\right)}}{16}$
$6°$	$\dfrac{\sqrt{15}+\sqrt{3}+\sqrt{10-2\sqrt{5}}}{8}$	$\dfrac{\sqrt{3}+\sqrt{15}+\sqrt{2\left(5-\sqrt{5}\right)}}{8}$
$9°$	$\dfrac{\sqrt{10}+\sqrt{2}+2\sqrt{5-\sqrt{5}}}{8}$	$\dfrac{\sqrt{2}+\sqrt{10}+2\sqrt{5-\sqrt{5}}}{8}$
$12°$	$\dfrac{\sqrt{5}-1+\sqrt{3}\sqrt{10+2\sqrt{5}}}{8}$	$\dfrac{-1+\sqrt{5}+\sqrt{6\left(\sqrt{5}+5\right)}}{8}$
$15°$	$\dfrac{\sqrt{2}+\sqrt{6}}{4}$	$\dfrac{\sqrt{2}+\sqrt{6}}{4}$
$18°$	$\sqrt{\dfrac{5+\sqrt{5}}{8}}$	$\dfrac{10+2\sqrt{5}}{4}$
$21°$	$\dfrac{\sqrt{2}(\sqrt{3}+1)(\sqrt{5}+1)+2(\sqrt{3}-1)\sqrt{5-\sqrt{5}}}{16}$	$\dfrac{\sqrt{2}+\sqrt{6}+\sqrt{10}+\sqrt{30}-2\sqrt{5-\sqrt{5}}+2\sqrt{3\left(5-\sqrt{5}\right)}}{16}$
$24°$	$\dfrac{\sqrt{5}+1+\sqrt{3}\sqrt{10-2\sqrt{5}}}{8}$	$\dfrac{1+\sqrt{5}+\sqrt{6\left(\sqrt{5}-5\right)}}{8}$
$27°$	$\dfrac{\sqrt{10}-\sqrt{2}+2\sqrt{5+\sqrt{5}}}{8}$	$\dfrac{\sqrt{2}-\sqrt{10}+2\sqrt{\sqrt{5}+5}}{8}$
$30°$	$\dfrac{\sqrt{3}}{2}$	$\dfrac{\sqrt{3}}{2}$
$33°$	$\dfrac{-\sqrt{2}(\sqrt{3}-1)(\sqrt{5}-1)+2(\sqrt{3}+1)\sqrt{5+\sqrt{5}}}{16}$	$\dfrac{-\sqrt{2}+\sqrt{6}+\sqrt{10}-\sqrt{30}+2\sqrt{\sqrt{5}+5}+2\sqrt{3\left(\sqrt{5}+5\right)}}{16}$
$36°$	$\dfrac{\sqrt{5}+1}{4}$	$\dfrac{1+\sqrt{5}}{4}$
$39°$	$\dfrac{-\sqrt{2}(-\sqrt{3}+1)(\sqrt{5}+1)+2(\sqrt{3}+1)\sqrt{5-\sqrt{5}}}{16}$	$\dfrac{-\sqrt{2}+\sqrt{6}-\sqrt{10}+\sqrt{30}+2\sqrt{5-\sqrt{5}}+2\sqrt{3\left(5-\sqrt{5}\right)}}{16}$
$42°$	$\dfrac{\sqrt{15}-\sqrt{3}+\sqrt{10+2\sqrt{5}}}{8}$	$\dfrac{-\sqrt{3}+\sqrt{15}+\sqrt{2\left(\sqrt{5}+5\right)}}{8}$
$45°$	$\dfrac{\sqrt{2}}{2}$	$\dfrac{\sqrt{2}}{2}$
$48°$	$\dfrac{-\sqrt{5}+1+\sqrt{3}\sqrt{10+2\sqrt{5}}}{8}$	$\dfrac{1-\sqrt{5}+\sqrt{6\left(5+\sqrt{5}\right)}}{8}$
$51°$	$\dfrac{\sqrt{2}(\sqrt{3}+1)(\sqrt{5}+1)-2(\sqrt{3}-1)\sqrt{5-\sqrt{5}}}{16}$	$\dfrac{\sqrt{2}+\sqrt{6}+\sqrt{10}+\sqrt{30}+2\sqrt{5-\sqrt{5}}-2\sqrt{3\left(5-\sqrt{5}\right)}}{16}$
$54°$	$\sqrt{\dfrac{5-\sqrt{5}}{8}}$	$\dfrac{10-2\sqrt{5}}{4}$
$57°$	$\dfrac{\sqrt{2}(\sqrt{3}+1)(\sqrt{5}-1)+2(\sqrt{3}-1)\sqrt{5+\sqrt{5}}}{16}$	$\dfrac{-\sqrt{2}-\sqrt{6}+\sqrt{10}+\sqrt{30}-2\sqrt{\sqrt{5}+5}+2\sqrt{3\left(\sqrt{5}+5\right)}}{16}$
$60°$	$\dfrac{1}{2}$	$\dfrac{1}{2}$
$63°$	$\dfrac{-\sqrt{10}+\sqrt{2}+2\sqrt{5+\sqrt{5}}}{8}$	$\dfrac{\sqrt{2}-\sqrt{10}+2\sqrt{\sqrt{5}+5}}{8}$
$66°$	$\dfrac{\sqrt{15}+\sqrt{3}-\sqrt{10-2\sqrt{5}}}{8}$	$\dfrac{\sqrt{3}+\sqrt{15}-\sqrt{2\left(5-\sqrt{5}\right)}}{8}$
$69°$	$\dfrac{\sqrt{2}(-\sqrt{3}+1)(\sqrt{5}+1)+2(\sqrt{3}+1)\sqrt{5-\sqrt{5}}}{16}$	$\dfrac{\sqrt{2}-\sqrt{6}+\sqrt{10}-\sqrt{30}+2\sqrt{5-\sqrt{5}}+2\sqrt{3\left(5-\sqrt{5}\right)}}{16}$
$72°$	$\dfrac{\sqrt{5}-1}{4}$	$\dfrac{-1+\sqrt{5}}{4}$
$75°$	$\dfrac{\sqrt{6}-\sqrt{2}}{4}$	$\dfrac{-\sqrt{2}+\sqrt{6}}{4}$
$78°$	$\dfrac{-\sqrt{15}+\sqrt{3}+\sqrt{10+2\sqrt{5}}}{8}$	$\dfrac{\sqrt{3}-\sqrt{15}+\sqrt{2\left(\sqrt{5}+5\right)}}{8}$
$81°$	$\dfrac{\sqrt{10}+\sqrt{2}-2\sqrt{5-\sqrt{5}}}{8}$	$\dfrac{\sqrt{2}+\sqrt{10}-2\sqrt{5-\sqrt{5}}}{8}$
$84°$	$\dfrac{-\sqrt{5}-1+\sqrt{3}\sqrt{10-2\sqrt{5}}}{8}$	$\dfrac{-1-\sqrt{5}+\sqrt{6\left(5-\sqrt{5}\right)}}{8}$
$87°$	$\dfrac{\sqrt{2}(\sqrt{3}+1)(\sqrt{5}-1)-2(\sqrt{3}-1)\sqrt{5+\sqrt{5}}}{16}$	$\dfrac{-\sqrt{2}-\sqrt{6}+\sqrt{10}+\sqrt{30}+2\sqrt{\sqrt{5}+5}-2\sqrt{3\left(\sqrt{5}+5\right)}}{16}$
$90°$	1	0

IV.2   Nonrectangular Numbers and Polynomials

Take a rectangle $A \times B$ with the finite sides $|A| = m$ and $|B| = n$. Among the nonempty subsets of $A \times B$, some are the Cartesian product of a subset of A with a subset of B, but most are not. The nonempty subsets of the second kind I nickname *nonrectangular subsets.* How many such nonrectangular subsets do exist? We know that a set of n elements has 2^n subsets, of which $2^n - 1$ are nonempty. One gets for the number of nonrectangular subsets of a rectangle of size $m \times n$

$$\mathcal{D}isrec(m, n) = 2^{mn} - 1 - (2^m - 1)(2^n - 1)$$

which I nick name the *nonrectangular number.*

IV.2.1   Nonrectangular Numbers

Problem 57. *Determine the nonrectangular number with $m = 3$ and $n = 4$ and decompose it into prime factors. Primefactor a few more nonrectangular numbers. Report what looks remarkable.*

$$2^{12} - 1 - (2^3 - 1)(2^4 - 1) = 4095 - 7 \cdot 15 = 3990 = 2 \cdot 3 \cdot 5 \cdot 7 \cdot 19$$

Often one finds many small primefactors. For $2 = m \leq n$ the nonrectangular number becomes

$$2^{2n} - 1 - 3 \cdot (2^n - 1) = (2^n - 1)(2^n - 2)$$

which are already two remarkable factors, which can be factored further.

Problem 58. *Prove that for $2 \leq m \leq n$, the nonrectangular numbers are divisible by 6, but not by 4.*

Answer. For all $m, n \geq 1$, the nonrectangular number $\mathcal{D}isrec(m, n)$ is a difference of two odd numbers, and hence even. A calculation modulo 3 shows

$$\mathcal{D}isrec(m, n) \equiv (-1)^{mn} - 1 - ((-1)^m - 1)((-1)^n - 1) \mod 3$$

For even m or even n, the right-hand side is zero. For mn odd, the right-hand side is $-2 - (-2)(-2) = -6 \equiv 0 \mod 3$. Hence in all cases $\mathcal{D}isrec(m, n)$, is divisible by 3. Since the number is divisible by 2 and by 3, it is divisible by 6. A calculation modulo 4 yields for $m, n \geq 2$

$$\mathcal{D}isrec(m, n) \equiv -1 - (-1)(-1) = -2 \mod 4$$

Hence the number is not divisible by 4. $\qquad\square$

150

To get more results I need to recall the little proposition 5.

Proposition 40 (The little proposition). *I use the convention from mathematica that the greatest common divisor is always positive, except for* $\gcd(0,0) = 0$. *Given are integers* $a \neq 0$ *and* $n, m, k, l \geq 0$. *If* k *divides* l, *then* $a^k - 1$ *divides* $a^l - 1$. *Moreover*

(IV.2.1) $\gcd(a^m - 1, a^n - 1) = |a^{\gcd(m,n)} - 1|$ *in all cases;*

(IV.2.2) $\gcd(a^m - 1, a^n - 1) = a^{\gcd(m,n)} - 1$ *if* $a > 0$ *or* n, m *both even;*

(IV.2.3) $\gcd(a^m - 1, a^n - 1) = -a^{\gcd(m,n)} + 1$ *if* $a < 0$, *and either* m *or* n *is odd.*

Proof. The case with $a \neq 0$ and $mn = 0$ can be checked directly. Too, the case $a = 0$ and $m \geq 1$, $n \geq 1$ can be checked directly. Assume now $a \neq 0$ and $m, n \geq 1$. To check the first part, assume that $m = sn$. The geometric series with quotient $q := a^n$ yields

$$a^m - 1 = q^s - 1 = (q - 1)(1 + q + q^2 + \cdots + q^{s-1})$$

Hence $a^m - 1$ is divisible by $a^n - 1 = q - 1$.

We see from this first step that $a^{\gcd(m.n)} - 1$ is a divisor of $\gcd(a^m - 1, a^n - 1)$. To confirm the reversed divisibility, the Euclidean algorithm is needed. I use the notation $m \mid n$ to indicate that m is a divisor of n. We begin with the first division with remainder $m = qn + r$ with $0 \leq r < n$ and see

$$a^n - 1 \mid (a^{qn} - 1)a^r$$
$$\gcd(a^m - 1, a^n - 1) \mid \gcd(a^{qn+r} - 1, a^{qn+r} - a^r) \mid a^r - 1$$
$$\gcd(a^m - 1, a^n - 1) \mid \gcd(a^n - 1, a^r - 1)$$

Let $r_0 = m, r_1 = n, r_2, \ldots, r_M = \gcd(m, n), r_{M+1} = 0$ be the sequence of remainders occurring in the Euclidean algorithm. Successively we get

$$\gcd(a^m - 1, a^n - 1) \mid \gcd(a^n - 1, a^r - 1) \mid \gcd(a^{r_2} - 1, a^{r_3} - 1) \mid \ldots$$
$$\ldots \mid \gcd(a^{r_M} - 1, a^{r_{M+1}} - 1) = \gcd(a^{\gcd(m,n)} - 1, 0)$$
$$= a^{\gcd(m,n)} - 1$$

From both $a^{\gcd(m,n)} - 1 \mid \gcd(a^m - 1, a^n - 1)$ and $\gcd(a^m - 1, a^n - 1) \mid a^{\gcd(m,n)} - 1$ we conclude the equality (IV.2.1), as claimed. I use the convention from mathematica that the greatest common divisor is always positive, except for $\gcd(0,0) = 0$. Note that we still have the awkward absolute value $|a^{\gcd(m,n)} - 1|$. In the case $a < 0$ and $\gcd(m, n)$ odd, one gets $a^{\gcd(m,n)} - 1 \leq -2$, and hence $|a^{\gcd(m,n)} - 1| = -a^{\gcd(m,n)} + 1$. Hence the reversed sign in the third equation. $\qquad \square$

Already from the first easier part of the little proposition, one concludes that

$$2^m - 1 \mid \mathcal{D}isrec(m, n) \ \text{ and } \ 2^n - 1 \mid \mathcal{D}isrec(m, n)$$

But we can indeed get even more divisors. Remember the second factor $2^n - 2$ in $\mathcal{D}isrec(2, n)$. To see that this factor appears always, I do a calculation modulo $2^{n-1} - 1$. One gets $2^n \equiv 2$ and $2^{mn} \equiv 2^m$, and for the nonrectangular number is obtained

$$\mathcal{D}isrec(m, n) = 2^{mn} - 1 - (2^m - 1)(2^n - 1) \equiv 2^m - 1 - (2^m - 1)(2 - 1) = 0 \quad \mathrm{mod}\ 2^{n-1} - 1$$

The corresponding result holds with m and n exchanged. We have obtained

$$2^{m-1} - 1 \mid \mathcal{D}isrec(m, n) \ \text{ and } \ 2^{n-1} - 1 \mid \mathcal{D}isrec(m, n)$$

We may put the result together. Since $\gcd(m - 1, m) = 1$, the little proposition yields $\gcd(2^{m-1} - 1, 2^m - 1) = 1$. The least common multiple of these two numbers is their product. Hence the get

$$(2^{m-1} - 1)(2^m - 1) \mid \mathcal{D}isrec(m, n) \ \text{ and } \ (2^{n-1} - 1)(2^n - 1) \mid \mathcal{D}isrec(m, n)$$

One may even go a step further

Proposition 41. *Let $2 \le m, n$ be integers and put*

$$\mathcal{P}rod_{m,n} = (2^m - 1)(2^{m-1} - 1)(2^n - 1)(2^{n-1} - 1)$$
$$\mathcal{G}cd_{m,n} = \gcd((2^m - 1)(2^{m-1} - 1), (2^n - 1)(2^{n-1} - 1))$$

The nonrectangular number $\mathcal{D}isrec_{m,n}$ has the integer factorization

(IV.2.4)

$$\mathcal{D}isrec_{m,n}$$
$$= 2 \cdot \frac{(2^m - 1)(2^{m-1} - 1)(2^n - 1)(2^{n-1} - 1)}{(2^{\gcd(m,n)} - 1)(2^{\gcd(m,n-1)} - 1)(2^{\gcd(m-1,n)} - 1)(2^{\gcd(m-1,n-1)} - 1)} \cdot \mathcal{D}isrec_{m,n}$$

This nice result explains why the prime factorization of the nonrectangular number has so many small prime factors. The result begs the question. After factoring the above fraction, is one left with a prime factorization of $\mathcal{D}isrec_{m,n}$. The answer is no, as I found out with the help of mathematica.

152

IV.2.2 The Related Polynomials

The idea is obvious: replace 2 by an undeterminate x to obtain from the nonrectungular numbers the nonrectangular polynomials. It turns out a little trick saves the day even better. Let me define

$$\mathcal{N}R_{m,n}(x) = (x^{mn} - 1)(x - 1) - (x^m - 1)(x^n - 1)$$

and nickname this to be the *nonrectangular polynomial.* Corresponding to the above prime factorization of the nonrectangular numbers, we get some factorizations of the nonrectangular polynomials in the ring $\mathbf{Z}[x]$ of polynomials with integer coefficients. The case $m = 2$ is once more a bid too simple, but nevertheless useful to get some clues.

$$\begin{aligned}
\mathcal{N}R_{2,n}(x) &= (x^{2n} - 1)(x - 1) - (x^2 - 1)(x^n - 1) \\
&= (x^n - 1)(x^n + 1)(x - 1) - (x - 1)(x + 1)(x^n - 1) \\
&= (x - 1)(x^n - 1)(x^n + 1 - x - 1) = x(x - 1)(x^n - 1)(x^{n-1} - 1)
\end{aligned}$$

While doing the exercises to follow, it is convenient to get for later use the signs of the nonzero results.

Problem 59. *Assuming* $2 \le m \le n$, *check that* $x = 0$ *is a simple zero of the polynomial* $\mathcal{N}R_{m,n}(x)$.

Proof. One sees immediately that $x = 0$ is a zero of $\mathcal{N}R_{m,n}(x)$. I expand and calculate the derivative at $x = 0$. Monomials are in the following ordered with increasing degree.

$$\begin{aligned}
\mathcal{N}R_{m,n}(x) &= (x^{mn} - 1)(x - 1) - (x^m - 1)(x^n - 1) \\
&= x^{mn+1} - x^{mn} - x + 1 - x^{m+n} + x^m + x^n - 1 \\
&= -x + x^m + x^n - x^{m+n} - x^{mn} + x^{mn+1} \\
\mathcal{N}R'_{m,n}(0) &= -1
\end{aligned}$$

Hence the zero $x = 0$ is simple. $\qquad\square$

Problem 60. *Assuming* $2 \le m \le n$, *check that* $x = -1$ *is a simple zero of the polynomial* $\mathcal{N}R_{m,n}(x)$.

Solution.

$$\mathcal{N}R_{m,n}(x) = -x + x^m + x^n - x^{m+n} - x^{mn} + x^{mn+1}$$

$$\mathcal{N}R'_{m,n}(x) = -1 + mx^{m-1} + nx^{n-1} - (m+n)x^{m+n-1} - mnx^{mn-1} + (mn+1)x^{mn}$$

$$\mathcal{N}R'_{m,n}(-1) = -1 + m(-1)^{m-1} + n(-1)^{n-1} - (m+n)(-1)^{m+n-1}$$
$$- mn(-1)^{mn-1} + (mn+1)(-1)^{mn}$$

$$\mathcal{N}R'_{m,n}(-1) = -1 + m + n - (m+n)(-1) - mn + (mn+1)(-1)$$
$$= -2(m-1)(n-1) < 0 \ \text{ if } mn \text{ odd}$$

$$\mathcal{N}R'_{m,n}(-1) = -1 + m(-1) + n(-1)^{n-1} - (m+n)(-1)^{n-1} - mn(-1) + (mn+1)$$
$$= m(2n - 1 + (-1)^n) > 0 \ \text{ if } m \text{ even}$$

The zero is simple in all cases. $\square$

The next exercise gets still a bid more cumbersome.

Problem 61. *Assuming $2 \le m \le n$, check that $x = 1$ is a threefold zero of the polynomial $\mathcal{N}R_{m,n}(x)$.*

Proof.

$$\mathcal{N}R^{(3)}_{m,n}(x) = m(m-1)(m-2)x^{m-4} + n(n-1)(n-2)x^{n-4}$$
$$- (m+n)(m+n-1)(m+n-2)x^{m+n-3}$$
$$- mn(mn-1)(mn-2)x^{mn-3} + (mn+1)mn(mn-1)x^{mn-2}$$

$$\mathcal{N}R^{(3)}_{m,n}(1) = m(m-1)(m-2) + n(n-1)(n-2) - (m+n)(m+n-1)(m+n-2)$$
$$- mn(mn-1)(mn-2) + (mn+1)mn(mn-1)$$
$$= 3(m-1)m(n-1)n > 0$$

$$\mathcal{N}R^{(2)}_{m,n}(1) = m(m-1) + n(n-1) - (m+n)(m+n-1)$$
$$- mn(mn-1) + (mn+1)m \cdot n = 0$$

$$\mathcal{N}R'_{m,n}(1) = m + n - (m+n) - mn + (mn+1) = 0$$

At $x = 1$, the third derivative is indeed the lowest nonzero derivative. Hence $x = 1$ is a threefold zero. $\square$

Main Theorem 2 (Rule of Descartes). *The rule of Descartes tells, that the number of positive zeros of any real polynomial, counting their multiplicities, is equal or less than the number of sign changes in the list of coefficients. The monomials have to be ordered, say in increasing degree. As Gauss has remarked: the difference of the two counts is even.*

Problem 62. *Assume $2 \leq m \leq n$. Use the rule of Descartes, to confirm that the polynomial $\mathcal{N}R_{m,n}(x)$ does not have any further positive zeros, except the threefold zero at $x = 1$.*

Solution. For the polynomial

$$-x + x^m + x^n - x^{m+n} - x^{mn} + x^{mn+1}$$

the list of coefficients, still using increasing order for the degrees of the monomials, is in the generic case $m > n > 2$

$$-, +, +, -, -, +$$

and in the special case $m = n > 2$

$$-, +, -, -, +$$

and in the special case $m = n = 2$

$$-, +, -, +$$

One gets in all cases three sign changes. Hence three are either one or three zeros. The triple zero at $x = 1$ is hence the only positive zero. $\qquad\square$

Problem 63. *Assume $3 \leq m \leq n$ and mn odd. Use the rule of Descartes, to confirm that the polynomial $\mathcal{N}R_{m,n}(x)$ has a further negative simple zero in the interval $(-1, 0)$, and hence together two negative zeros.*

Solution. One needs to count the number of sign changes in the list of coefficients of the polynomial $\mathcal{N}R_{m,n}(-x)$. This is done by plugging $x = -1$ into the list of monomials of the original polynomial

$$\mathcal{N}R_{m,n}(x) = -x + x^m + x^n - x^{m+n} - x^{mn} + x^{mn+1}$$

still using increasing order for their degrees. In the case with $m < n$ both odd is obtained

$$+, -, -, -, +, +$$

in the case of $m = n$ odd

$$+, -, -, +, +$$

In these cases there are two sign changes, and hence either no or two negative zeros. There exists a simple zero at -1, hence there needs to be a further simple zero. From above we know that $\mathcal{N}R'_{m,n}(-1) = -2(m - 1)(n - 1) < 0$ and $\mathcal{N}R'_{m,n}(0) = -1$. Hence there exists a zero in the interval $(-1, 0)$. From Descartes rule, we conclude additionally that this zero is simple, and no further zeros exist. $\qquad\square$

Problem 64. *Assume $2 \leq m \leq n$ and m and n are both even. Use the rule of Descartes, to confirm that the polynomial $\mathcal{N}R_{m,n}(x)$ no further negative zeros besides $x = -1$.*

Solution. For m and n both even, and $x = -1$, the signs for the list of monomials is

$$+, \, +, \, +, \, -, \, -, \, -$$

There is one sign change, and hence a unique negative zeros. There exists a simple zero at -1. Hence this is the unique negative zero. $\quad\square$

Finally, here is the case where the good luck no longer works completely.

Problem 65. *Assume $3 \leq m \leq n$ and m odd and n is even. Use the rule of Descartes, to confirm that the polynomial $\mathcal{N}R_{m,n}(x)$ either no or two further negative zeros besides $x = -1$.*

Solution. For m odd, n even, and $x = -1$, the signs for the list of monomials is

$$+, \, -, \, +, \, +, \, -, \, -$$

There are three sign changes, and hence either one or three negative zeros. There exists a simple zero at -1. Hence there exists either no or two further negative zero.

To get a sharper result I try my luck, and use what is known about the graph of the function: $\lim_{x \to -\infty} \mathcal{N}R_{m,n}(x) = -\infty$, $\mathcal{N}R'_{m,n}(-1) = n(2m - 1 + (-1)^m) > 0$ and $\mathcal{N}R'_{m,n}(0) = -1$. This does not give any further information. A graph for the case $m = 3, n = 24$ shows a local minimal between -1 and 0, but still a positive value at the minimum. $\quad\square$

In the above problem we have checked that the polynomial $\mathcal{N}R_{m,n} \in \mathbf{Z}[x]$ has the divisor $x(x + 1)(x - 1)^3$. To get a better result about the factorization, we use the following fact, which is the analog to the little proposition 5 for the polynomial ring $\mathbf{Z}[x]$.

Proposition 42 (The little proposition for polynomials). *Given are integers $n, m \geq 1$. If m divides n, then the polynomial $x^m - 1$ divides $x^n - 1$.*

$$(\text{IV.2.5}) \qquad\qquad \gcd(x^m - 1, x^n - 1) = x^{\gcd(m,n)} - 1$$

holds with the greatest common divisor in the ring $\mathbf{Z}[x]$.

It is now not too hard to guess and confirm the result corresponding to proposition 41 above. Let $2 \leq m, n$ be integers and put

$$\mathcal{P}rod_{m,n}(x) = (x^m - 1)(x^{m-1} - 1)(x^n - 1)(x^{n-1} - 1)$$
$$\mathcal{G}cd_{m,n}(x) = (x^{\gcd(m,n)} - 1)(x^{\gcd(m,n-1)} - 1)$$
$$(x^{\gcd(m-1,n)} - 1)(x^{\gcd(m-1,n-1)} - 1)$$

Proposition 43. *Assume $m, n \geq 2$. In the ring $\mathbf{Z}[x]$, the polynomial $(x+1)\mathcal{G}cd_{m,n}$ is a divisor of $\mathcal{P}rod_{m,n}$. The polynomial $\mathcal{N}R_{m,n} \in \mathbf{Z}[x]$ has the factorization*

(IV.2.6) $$\mathcal{N}R_{m,n} = x(x+1)(x-1)^3 \cdot \frac{\mathcal{P}rod_{m,n}}{(x+1)\mathcal{G}cd_{m,n}} \cdot \mathcal{N}RR_{m,n}$$

Proof. Here is a natural, although a bid tedious proof. We start with the definition of the nonrectangular polynomial and divide by $x - 1$. In the first step, a division by $x^m - 1$ prepares an application of the geometric series. In the second step, a division by $x^{m-1} - 1$ prepares a further application of the geometric series. Thus one obtains still a polynomial with integer coefficients.

$$\mathcal{N}R_{m,n}(x) = (x^{m*n} - 1)(x - 1) - (x^m - 1)(x^n - 1)$$
$$\frac{\mathcal{N}R_{m,n}(x)}{(x-1)(x^m-1)} = \frac{x^{m*n} - 1}{x^m - 1} - \frac{x^n - 1}{x - 1} = \sum_{1 \leq k \leq n-1} (x^{mk} - x^k)$$
$$\frac{\mathcal{N}R_{m,n}(x)}{x(x-1)(x^m-1)(x^{m-1}-1)} = \sum_{1 \leq k \leq n-1} x^{k-1} \frac{x^{(m-1)k} - 1}{x^{m-1} - 1} = \sum_{1 \leq k \leq n-1} \sum_{0 \leq l \leq k-1} x^{(m-1)l+k-1}$$
$$\frac{\mathcal{N}R_{m,n}(x)}{x(x-1)(x^m-1)(x^{m-1}-1)} = \sum_{0 \leq k \leq n-2} \sum_{0 \leq l \leq k} x^{(m-1)l+k}$$

Hence $x(x-1)(x^m - 1)(x^{m-1} - 1)$ is a divisor of $\mathcal{N}R_{m,n}(x)$. One now exchanges m and n and takes the greatest common divisor.

$$\frac{\mathcal{N}R_{m,n}(x)}{x(x-1)\mathrm{lcm}((x^m-1)(x^{m-1}-1),(x^n-1)(x^{n-1}-1))}$$
$$= \gcd\left(\sum_{0 \leq k \leq n-2} \sum_{0 \leq l \leq k} x^{(m-1)l+k}, \sum_{0 \leq k \leq m-2} \sum_{0 \leq l \leq k} x^{(n-1)l+k} \right) =: \mathcal{N}RR_{m,n}(x)$$

The new extra factor I have nicknamed the *nonrectangular remainder polynomial.*

One has obtained a factorization of the nonrectangular polynomial.

$$\mathcal{N}R_{m,n}(x) = x(x-1) \cdot \operatorname{lcm}((x^m - 1)(x^{m-1} - 1), (x^n - 1)(x^{n-1} - 1)) \cdot \mathcal{N}RR_{m,n}(x)$$

$$= x(x-1)^3 \cdot \frac{(x^m - 1)(x^{m-1} - 1)(x^n - 1)(x^{n-1} - 1)}{\gcd((x^m - 1)(x^{m-1} - 1), (x^n - 1)(x^{n-1} - 1))} \cdot \mathcal{N}RR_{m,n}(x)$$

Why the extra factor $(x-1)^2$? Only with the assumption $\gcd(a,b) = \gcd(c,d) = 1$ holds

$$\gcd(ab, cd) = \gcd(a,c)\gcd(a,d)\gcd(b,c)\gcd(b,d)$$

Let

$$Q_m(x) := \frac{x^m - 1}{x - 1}$$

One needs to divide the top by $(x-1)^4$, the bottom only by $(x-1)^2$ in order to produce the relatively prime $Q_m.Q_{m-1}$ as well as Q_n, Q_{n-1}.

$$\mathcal{N}R_{m,n}(x) = x(x+1)(x-1)^3 \cdot \mathcal{N}RR_{m,n}(x) \cdot$$

$$\cdot \frac{Q_m Q_{m-1}/(x+1) \cdot Q_n Q_{n-1}/(x+1)}{\gcd(Q_m, Q_n)/(x+1) \cdot \gcd(Q_m, Q_{n-1})\gcd(Q_{m-1}, Q_n)\gcd(Q_{m-1}, Q_{n-1})}$$

$$= x(x+1)(x-1)^3 \cdot \mathcal{N}RR_{m,n}(x) \cdot$$

$$\cdot \frac{(x^m - 1)(x^{m-1} - 1)/(x+1) \cdot (x^n - 1)(x^{n-1} - 1)/(x+1)}{(x^{\gcd(m,n)} - 1)/(x+1) \cdot (x^{\gcd(m,n-1)} - 1)(x^{\gcd(m-1,n)} - 1)(x^{\gcd(m-1,n-1)} - 1)}$$

I have done a few additional steps to get the result announced in the proposition. $x^m - 1$ and $Q_m(x)$ are divisible by $x + 1$ if and only if m is even. Since either m or $m - 1$ is even, the polynomial $Q_m(x)Q_{m-1}(x)/(x+1)$ has integer coefficients. In the denominator, one may divide by $x + 1$ only for the single term where both m, n or $m, n - 1$ or $m - 1, n$ or $m - 1, n - 1$ are even. $\qquad\square$

Problem 66. *Calculate the degree of the polynomial $\mathcal{N}RR_{m,n}(x)$. Check that the constant coefficient is always equal to 1.*

Problem 67. *Calculate $\mathcal{N}RR_{m,n}(1)$.*

Answer.

$$\deg \mathcal{N}RR_{m,n} = (m-2)(n-2) - 5$$
$$+ \gcd(m,n) + \gcd(m, n-1) + \gcd(m-1, n) + \gcd(m-1, n-1)$$
$$\mathcal{N}RR_{m,n}(1) = \gcd\left(\frac{m(m-1)}{2}, \frac{n(n-1)}{2}\right)$$

Together with problem 61 one obtains a nice check. Any analytic function has at any point a in its domain a Taylor series, also called Taylor expansion. We use the expansion of the polynomial $\mathcal{N}R_{m,n}(x)$ at point $a = 1$. From problem 61, we get the leading term

$$\mathcal{N}R_{m,n}(x) = \mathcal{N}R_{m,n}^{(3)}(1)\frac{(x-1)^3}{3!} + O(x-1)^4 = \frac{(m-1)m(n-1)n}{2}(x-1)^3 + O(x-1)^4$$

On the other hand, one can do the same power expansion for the factorization from above. Since we need only the leading term, we may plug $x = 1$ into all factors *which do not vanish* at $x = 1$, and obtain

$$\mathcal{N}R_{m,n}(x) = (x-1)^3 \cdot \mathcal{N}RR_{m,n}(1) \cdot$$
$$\cdot \frac{Q_m Q_{m-1} \cdot Q_n Q_{n-1}}{\gcd(Q_m, Q_n) \cdot \gcd(Q_m, Q_{n-1})\gcd(Q_{m-1}, Q_n)\gcd(Q_{m-1}, Q_{n-1})} + O(x-1)^4$$

$$= (x-1)^3 \cdot \gcd\left(\frac{m(m-1)}{2}, \frac{n(n-1)}{2}\right) \cdot$$
$$\cdot \frac{m(m-1)n(n-1)}{\gcd(m,n)\gcd(m,n-1)\gcd(m-1,n)\gcd(m-1,n-1)}$$

Of course, the two expansions are equal. Hence we have obtained

$$\frac{(m-1)m(n-1)n}{2}(x-1)^3 = (x-1)^3 \cdot \gcd\left(\frac{m(m-1)}{2}, \frac{n(n-1)}{2}\right) \cdot$$
$$\cdot \frac{m(m-1)n(n-1)}{\gcd(m,n)\gcd(m,n-1)\gcd(m-1,n)\gcd(m-1,n-1)}$$

which is indeed correct. $\qquad\qquad\qquad\qquad\qquad\qquad\qquad\qquad\qquad\qquad\qquad\square$

After these exercises have provided some confirmation that our results are correct, I come to the main questions for which the above proposition begs. Are we now able to find the complete factorization of the nonrectangular polynomials into irreducible polynomials? For all, except the last factor, we use the known factorization of $x^n - 1$ into the cyclotomic polymonials, which are known to be irreducible in $\mathbf{Z}[x]$:

$$x^n - 1 = \prod_{d|n} \Phi_d(x)$$

Here is my list of conjectures, that I have checked with mathematica to hold for all $m, n \leq 30$:

Remark (The main conjecture). The polynomials $\mathcal{N}RR_{m,n}(x) \in \mathbf{Z}[x]$ are irreducible for all $m, n \geq 2$.

Remark (Simple zeros). The polynomials $\mathcal{N}RR_{m,n}(x) \in \mathbf{Z}[x]$ have only simple zeros for all $m, n \geq 2$.

Remark (No zeros with $|x| = 1$). The polynomials $\mathcal{N}RR_{m,n}(x) \in \mathbf{Z}[x]$ have no zeros on the unit circle.

The numerical evidence seems to be strong enough for these claims. But the above methods are most probably too weak to prove anyone of these good guesses.

IV.3 Polynomials from the Euler Group

Definition 23 (Euler-type polynomial). Let $m \geq 2$ and let $1 = a_1 < a_2 < \cdots < a_{\phi(m)} = m - 1$ be a list of integers relatively prime to m which represent the Euler group G_m^*. The polynomial

$$(\text{IV.3.1}) \quad Q_m(x) := \{\prod(x - a_i) \,:\, 1 \leq i \leq \phi(m) \text{ and } \gcd(a_i, m) = 1\} = \sum_{k=0}^{\phi(m)} q_k x^k$$

is called *Euler-type polynomial.* In the beginning, this polynomial is literally the integer polynomial $Q_m \in \mathbf{Z}[x]$. Usually appears in the results below the polynomial projected into the ring $\mathbf{Z}_m$ with the coefficients q_k understood modulo m.

Proposition 44. *For any prime p, the Euler-type polynomial is*

$$(\text{IV.3.2}) \quad Q_p(x) = \prod_{1 \leq i \leq p-1} (x - i) \equiv x^{p-1} - 1 \quad (\text{mod } \mathbf{Z}_p)$$

Proof of Proposition 44. Let $P(x) = x^{p-1} - 1 \in \mathbf{Z}_p[x]$. By Fermat's Little Theorem 7 we get $p - 1$ zeros:

$$P(i) \equiv 0 \quad (\text{mod } p) \text{ for } 1 \leq i \leq p - 1.$$

Thus the polynomial P has as many zeros of as its degree allows. Moreover, the polynomial P is monic. By the corollary 14 to Lagrange's Theorem we conclude

$$(\text{IV.3.3}) \quad x^{p-1} - 1 \equiv \prod_{1 \leq i \leq d} (x - i) \quad (\text{mod } \mathbf{Z}_p)$$

as claimed. $\qquad\qquad\qquad\qquad\qquad\qquad\qquad\qquad\qquad\qquad\qquad\square$

IV.3.1 Wilson's Theorem and a Relative

Corollary 17 (Wilson's Theorem). *For any prime p holds*

$$(\text{IV.3.4}) \quad (p - 1)! \equiv -1 \quad (\text{mod } p)$$

whereas for any composite number $m \geq 4$ holds of course $(m - 1)! \equiv 0 \pmod{m}$.

Proof. We compare the constant coefficients on both sides of equation (IV.3.3). $\quad\square$

Proposition 45 ("half-Wilson" Theorem). *Let p be an odd prime and let*

$$(\text{IV.3.5}) \quad J :\equiv \left(\frac{p-1}{2}\right)! \quad (\text{mod } p)$$

(a) *In the case $p \equiv 1 \pmod 4$ holds*

$$(\text{IV.3.6}) \qquad\qquad\qquad J^2 \equiv -1 \pmod p$$

Hence -1 is a quadratic residue.

(b) *In the case $p \equiv 3 \pmod 4$ holds*

$$(\text{IV.3.7}) \qquad either\ J \equiv 1 \pmod p\ or\ J \equiv -1 \pmod p$$

```
(require math/number-theory)
(define largenumber 3000)
(define arithmetic3+4
  (build-list largenumber (lambda (x) (+ 3(* 4 x)))))

(define primes3+4 (filter prime? arithmetic3+4))

(define half-Wilson (lambda(p)
  (let* ( [p-1-over_2 (/(sub1 p)2)]
          [facto (factorial p-1-over_2)]
          [posmod (modulo facto p)])
          (if (positive? (- (* 2 posmod) p))
              (- posmod p) posmod)
  )))

(define (sigmalist anylist)
   (if (null? anylist)
       (lambda(x) 0)
       (let ([jump (lambda(x)
                      (if (>= x (car anylist)) 1 0))])
             (lambda(x)
                (+ ((sigmalist(cdr anylist))x)
                   (jump x))))))

(define analytic (lambda(x) (/ x  (log x))))

(require plot)

(define half-Wilson3+4plus
```

```
  (filter (lambda(item) (= 1 (half-Wilson item))) primes3+4))
(define half-Wilson3+4pluscount
  (lambda(x) ((sigmalist half-Wilson3+4plus)x)))


(define half-Wilson3+4minus
  (filter (lambda(item) (= -1 (half-Wilson item))) primes3+4))
(define half-Wilson3+4minuscount
  (lambda(x) ((sigmalist half-Wilson3+4minus)x)))


;(plot-new-window? #t)
(plot (list ;(axes)
              (function (lambda(x) (half-Wilson3+4pluscount x))
                        #:color "red")
              (function (lambda(x) (half-Wilson3+4minuscount x))
                        #:color "blue")
              (function (lambda(x) (/(analytic x)4)) 1 10000
                        #:color 0 #:style 'dot))
          #:x-min 0   #:x-max 10000
          #:y-min 0   ); #:y-max 14)
```

Remark. One asks whether for primes $p \equiv 3 \pmod 4$ exists some criterium to decide whether $J \equiv 1 \pmod p$ or $J \equiv -1 \pmod p$ holds. I could not find a simple criterium to decide. On the other hand, some numerical experiments show that the two alternatives occur about one as often as the other one, with astonishing accuracy.

In the figure on page 163 are plotted the number of primes $p \equiv 3 \pmod 4$ up to variable $x \le 10\,000$. In the mostly upper curve are counted the primes for which $J \equiv 1 \pmod p$, in the middle one the primes for which $J \equiv -1 \pmod p$. The dotted curve is the function $\frac{x}{4 \log x}$.

IV.3.2   Symmetry of Polynomials

Definition 24 (Mirror symmetry of polynomials). The polynomial $P(x)$ of degree d is called *mirror-symmetric* iff $x^d P(x^{-1}) = P(x)$. The polynomial is called *mirror-antisymmetric* iff $x^d P(x^{-1}) = -P(x)$.

Definition 25 (Usual symmetry of polynomials). The polynomial $P(x)$ is called *symmetric* iff $P(-x) = P(x)$. The polynomial is called *antisymmetric* iff $P(-x) = -P(x)$.

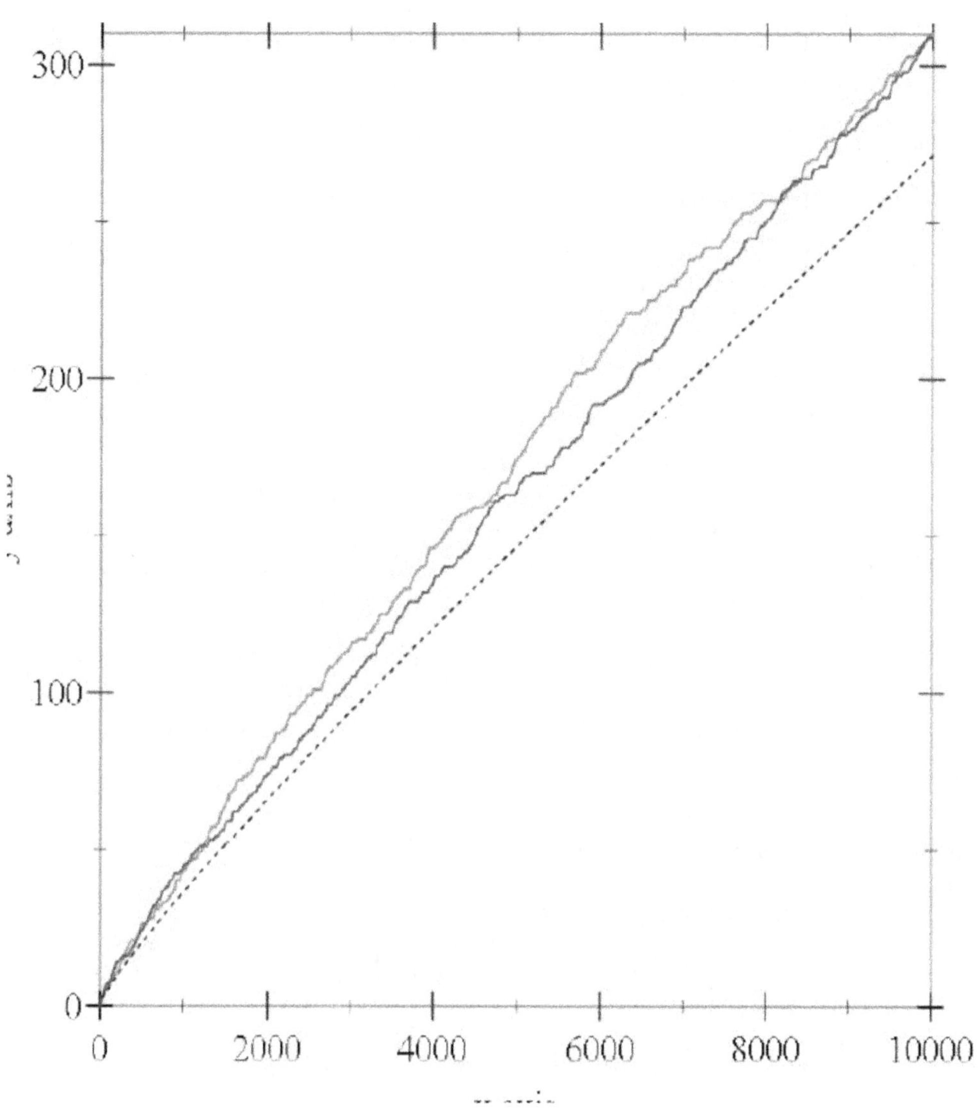

Figure 2: Distribution of primes

Lemma 39. *For any integer $m \geq 3$, the Euler-type polynomial $Q_m(x) \in \mathbf{Z}_m[x]$ is a polynomial of x^2, thus symmetric in the usual sense.*

Proof. Let $m \geq 3$ and hence $\phi(m)$ is even. By lemma 11, the mapping $a \mapsto i(a) = m - a$ is an involution $G_m^* \mapsto G_m^*$ which has no fixed points. The Euler-type polynomial as given by definition (IV.3.1) may now be partitioned as follows

$$Q_m(x) = \{\prod (x - a_i) \, : \, 1 \leq i \leq \phi(m) \text{ and } \gcd(a_i, m) = 1\}$$
$$= \{\prod (x - a_i)(x - m + a_i) \, : \, 1 \leq i \leq \tfrac{\phi(m)}{2} \text{ and } \gcd(a_i, m) = 1\}$$
$$\equiv \{\prod (x^2 - a_i^2) \, : \, 1 \leq i \leq \tfrac{\phi(m)}{2} \text{ and } \gcd(a_i, m) = 1\} \quad (\bmod \, \mathbf{Z}_m)$$

which is in $\mathbf{Z}_m[x]$ a polynomial of x^2. $\qquad\square$

Remark. Of course $Q_3[x] = (x - 1)(x - 2) = x^2 - 3x + 2 \in \mathbf{Z}[x]$ is <u>not</u> a polynomial of x^2. Only the <u>projection</u> $Q_3[x] \equiv x^2 - 1 \in \mathbf{Z}_3[x]$ is a polynomial of x^2. The same situation holds for all m.

Proposition 46. *Let $m \geq 2$. The Euler-type polynomial $Q_m(x) \in \mathbf{Z}_m[x]$ is mirror-antisymmetric in the cases that there exists a primitive root modulo m. The polynomial $Q_m(x) \in \mathbf{Z}_m[x]$ is mirror-symmetric in the cases that there does not exist a primitive root modulo m.*

Proof. In the case $m = 2$ holds $Q_2(x) = x - 1$ which is mirror-antisymmetric as claimed. Excluding this case let $m \geq 3$. The mapping $a \mapsto a^{-1}$ is an involution $G_m^* \mapsto G_m^*$. One partitions the Euler group into the three disjoint parts $G_m^* = H + H^{-1} + Q$ with

$$Q := \{a \in G_m^* : a^2 \equiv 1 \mod m\}$$

and the Euler-type polynomial from definition (IV.3.1) as follows

$$Q_m(x) = \{\prod (x - a_i) \, : \, a \in G_m^*\}$$
$$\equiv \prod_{a \in H} (x - a)(x - a^{-1}) \cdot \prod_{a \in Q} (x - a) \quad (\bmod \, \mathbf{Z}_m)$$
$$x^{\phi(m)} Q_m(x^{-1}) \equiv \prod_{a \in H} x^2 (x^{-1} - a)(x^{-1} - a^{-1}) \cdot \prod_{a \in Q} x(x^{-1} - a)$$
$$x^{\phi(m)} Q_m(x^{-1}) \equiv \prod_{a \in H} (x - a)(x - a^{-1}) \cdot \prod_{a \in Q} (-a)(x - a)$$
$$x^{\phi(m)} Q_m(x^{-1}) \equiv \left[\prod_{a \in Q} (-a) \right] \cdot Q_m(x) \quad (\bmod \, \mathbf{Z}_m)$$

We now consider the case where a primitive root modulo m exists. By proposition 16 holds $Q = \{\pm 1\}$ and hence

$$\prod_{a \in Q} (-a) \equiv -1 \quad (\text{mod } m)$$

$$x^{\phi(m)} Q_m(x^{-1}) \equiv (-1) Q_m(x) \quad (\text{mod } \mathbf{Z}_m)$$

which confirms that the polynomial $Q_m(x) \in \mathbf{Z}_m[x]$ is mirror-antisymmetric.

We now consider the case of nonexistence of a primitive root modulo m. By proposition 16 and since $|Q|$ is even holds

$$\prod_{a \in Q} (-a) \equiv +1 \quad (\text{mod } m)$$

$$x^{\phi(m)} Q_m(x^{-1}) \equiv Q_m(x) \quad (\text{mod } \mathbf{Z}_m)$$

which confirms that the polynomial $Q_m(x) \in \mathbf{Z}_m[x]$ is indeed mirror-symmetric. $\square$

IV.3.3 Main Results

My goal is the calculation of the polynomial Euler-type $Q_m(x) \in \mathbf{Z}_m[x]$ projected into the ring $\mathbf{Z}_m$. The following theorems facilitate the project at least to some degree.

Theorem 11. *Let $m \geq 2$ and $n \geq 2$ be relatively prime. We assume the Euler-type polynomials $Q_m(x) \in \mathbf{Z}_m$ and $Q_n(x) \in \mathbf{Z}_n$ to be known. Then Q_{mn} is the solution of the Chinese remainder problem*
(IV.3.8)
$$Q_{mn}(x) \equiv (Q_m(x))^{\phi(n)} \quad (\text{mod } \mathbf{Z}_m) \quad \text{and} \quad Q_{mn}(x) \equiv (Q_n(x))^{\phi(m)} \quad (\text{mod } \mathbf{Z}_n)$$

Proof. Let $1 = a_1 < a_2 < \cdots < a_{\phi(m)} = m - 1$ be a list of integers relatively prime to m which represent the Euler group G_m^*. Let $1 = b_1 < b_2 < \cdots < b_{\phi(n)} = n - 1$ be a list of integers relatively prime to n which represent the Euler group G_n^*. For $1 \leq i \leq \phi(m)$ and $1 \leq j \leq \phi(n)$ we solve the Chinese remainder problems

$$c_{ij} \equiv a_i \quad (\text{mod } m) \quad \text{and} \quad c_{ij} \equiv b_j \quad (\text{mod } n)$$

to obtain the set of integers c_{ij} relatively prime to mn which represent the Euler group G_{mn}^*. From definition (IV.3.1) follows

$$Q_{mn}(x) \equiv \{\prod (x - c_{ij}) : 1 \leq i \leq \phi(m) \text{ and } 1 \leq j \leq \phi(n)\} \quad (\text{mod } \mathbf{Z}_{mn})$$

$$Q_{mn}(x) \equiv \{\prod (x - a_i) : 1 \leq i \leq \phi(m) \text{ and } 1 \leq j \leq \phi(n)\} \equiv (Q_m(x))^{\phi(n)} \quad (\text{mod } \mathbf{Z}_m)$$

$$Q_{mn}(x) \equiv \{\prod (x - b_j) : 1 \leq i \leq \phi(m) \text{ and } 1 \leq j \leq \phi(n)\} \equiv (Q_n(x))^{\phi(m)} \quad (\text{mod } \mathbf{Z}_n)$$

$\square$

Problem 68. *Use theorem 11 and proposition 44 to calculate $Q_{65}(x) \in \mathbf{Z}_{65}$.*

Solution. Let $m = 5$ and $n = 13$. According to equation (IV.3.8)

$$Q_{65}(x) \equiv (Q_5(x))^{\phi(13)} \quad (\mathrm{mod}\ \mathbf{Z}_5) \quad \text{and}$$
$$Q_{65}(x) \equiv (Q_{13}(x))^{\phi(5)} \quad (\mathrm{mod}\ \mathbf{Z}_{13})$$

Now equation (IV.3.2) yields

$$Q_{65}(x) \equiv (x^4 - 1)^{12} \quad (\mathrm{mod}\ \mathbf{Z}_5) \quad \text{and}$$
$$Q_{65}(x) \equiv (x^{12} - 1)^4 \quad (\mathrm{mod}\ \mathbf{Z}_{13})$$

Some factoring facilitates the further calculation

$$(x^4 - 1)^{12} = (x^4 - 1)^4(x^4 - 1)^8 \quad \text{and}$$
$$(x^{12} - 1)^4 = (x^4 - 1)^4(x^8 + x^4 + 1)^4$$

We may solve the simpler problem

$$R(x) \equiv (x^4 - 1)^2 \quad (\mathrm{mod}\ \mathbf{Z}_5) \quad \text{and} \quad R(x) \equiv x^8 + x^4 + 1 \quad (\mathrm{mod}\ \mathbf{Z}_{13}) \quad \text{hence}$$
$$Q_{65}(x) \equiv (x^4 - 1)^4 R(x)^4 \quad (\mathrm{mod}\ \mathbf{Z}_{65})$$

This leads to the even simpler Chinese remainder problem

$$c \equiv 2 \quad (\mathrm{mod}\ 5) \quad \text{and} \quad c \equiv 1 \quad (\mathrm{mod}\ 13)$$
$$\text{We get} \quad R(x) \equiv x^8 + cx^4 + 1 \quad (\mathrm{mod}\ \mathbf{Z}_{65})$$

Put together we obtain the solution

$$c \equiv 27 \quad (\mathrm{mod}\ 65)$$
$$R(x) \equiv x^8 + 27x^4 + 1 \quad (\mathrm{mod}\ \mathbf{Z}_{65})$$
$$Q_{65}(x) \equiv (x^4 - 1)^4 R(x)^4 \quad (\mathrm{mod}\ \mathbf{Z}_{65})$$

which gives in the end

$$Q_{65}(x) \equiv (x^4 - 1)^4(x^8 + 27x^4 + 1)^4 \quad (\mathrm{mod}\ \mathbf{Z}_{65})$$

$\square$

Problem 69. *Use theorem 11 and the result of problem 68 to calculate $Q_{130}(x) \in \mathbf{Z}_{130}$.*

Solution. Let $m = 65$ and $n = 2$. According to equation (IV.3.8)

$$Q_{130}(x) \equiv (Q_2(x))^{\phi(65)} \pmod{\mathbf{Z}_2} \text{ and}$$
$$Q_{130}(x) \equiv (Q_{65}(x))^{\phi(2)} \pmod{\mathbf{Z}_{65}}$$

Now the result of problem 68 and some factoring yields

$$Q_{130}(x) \equiv (x-1)^{48} \equiv (x^4-1)^4(x^4-1)^8 \pmod{\mathbf{Z}_2} \text{ and}$$
$$Q_{130}(x) \equiv (x^4-1)^4(x^8+27x^4+1)^4 \pmod{\mathbf{Z}_{65}}$$

We may solve the simpler problem

$$R(x) \equiv (x^4-1)^2 \pmod{\mathbf{Z}_2} \text{ and}$$
$$R(x) \equiv x^8 + 27x^4 + 1 \pmod{\mathbf{Z}_{65}} \text{ to get}$$
$$Q_{130}(x) \equiv (x^4-1)^4 R(x)^4 \pmod{\mathbf{Z}_{130}}$$

This leads to the even simpler Chinese remainder problem

$$c \equiv 0 \pmod 2 \text{ and}$$
$$c \equiv 27 \pmod{65} \text{ to get}$$
$$R(x) \equiv x^8 + cx^4 + 1 \pmod{\mathbf{Z}_{130}}$$

The solution is

$$c \equiv 27 + 65 = 92 \pmod{130}$$
$$R(x) \equiv x^8 + 92x^4 + 1 \pmod{\mathbf{Z}_{130}}$$
$$Q_{130}(x) \equiv (x^4-1)^4 R(x)^4 \pmod{\mathbf{Z}_{130}}$$
$$Q_{130}(x) \equiv (x^4-1)^4(x^8 + 92x^4 + 1)^4 \pmod{\mathbf{Z}_{130}}$$

$$\square$$

Theorem 12 (Main Theorem for Euler-type polynomials). *Let p be prime, $m \geq 1$, and $r \geq 1$. We need to assume that $mp \geq 3$. Let the polynomial $P(x) \in \mathbf{Z}[x]$ be given such that the Euler-type polynomial satisfies*

(IV.3.9) $$Q_{mp}(x) \equiv P(x) \pmod{\mathbf{Z}_{mp}}$$

Then holds for all $r \geq 1$

(IV.3.10) $$Q_{mp^r}(x) \equiv (P(x))^{p^{r-1}} \pmod{\mathbf{Z}_{mp^r}}$$

Corollary 18. *For any odd prime p and $r \geq 1$ holds*

(IV.3.11)
$$Q_{p^r}(x) \equiv (x^{p-1} - 1)^{p^{r-1}} \pmod{\mathbf{Z}_{p^r}}$$

Problem 70. *For which values of r holds*

((?))
$$(x-1)^{2^{r-1}} \equiv (x^2 - 1)^{2^{r-2}} \pmod{\mathbf{Z}_{2^r}}$$

Solution. The equivalence is false for all $r \geq 2$, as one already sees by comparing the coefficient of x on both sides: of cause $2^{r-1}x \not\equiv 0 \pmod{\mathbf{Z}_{2^r}}$. $\square$

Remark. Theorem 12 does not hold in the case $mp = 2$ and $r \geq 2$,—instead one gets with $m = 1, p = 2$ the following result 19.

Corollary 19. *For any $r \geq 2$ holds*

(IV.3.12)
$$Q_{2^r}(x) \equiv (x^2 - 1)^{2^{r-2}} \pmod{\mathbf{Z}_{2^r}}$$

Proof. Use Theorem 12 with $p = 2$ and $m = 2$ and $P(x) = x^2 - 1$. Finally rename $r + 1$ to be the new r. $\square$

Corollary 20. *Let $m \geq 2$. Assume that the polynomial $P(x) \in \mathbf{Z}[x]$ is known to satisfy*
$$Q_{2m}(x) \equiv P(x) \pmod{\mathbf{Z}_{2m}}$$
Then holds for all $r \geq 1$

(IV.3.13)
$$Q_{m2^r}(x) \equiv (P(x))^{2^{r-1}} \pmod{\mathbf{Z}_{m2^r}}$$

Problem 71. *Use corollary 20 and the result of problem 69 to calculate $Q_{260}(x) \in \mathbf{Z}_{260}$.*

Problem 72. *Use the Chinese remainder theorem to check that*
$$Q_{260}(x) \equiv (x^4 - 1)^8 (x^8 + 92x^4 + 1)^8 \pmod{\mathbf{Z}_{260}}$$

Corollary 21. *Let $m \geq 3$ have the square-free part $f \geq 3$. Let the polynomial $P(x) \in \mathbf{Z}[x]$ be given such that*

(IV.3.14)
$$Q_f(x) \equiv P(x) \pmod{\mathbf{Z}_f}$$

Then holds

(IV.3.15)
$$Q_m(x) \equiv (P(x))^{\frac{m}{f}} \pmod{\mathbf{Z}_m}$$

Proof. Let

$$m = \prod_{1 \leq i \leq t} p_i^{r_i} \quad \text{and} \quad f = \prod_{1 \leq i \leq t} p_i$$

be the prime factorizations of m and its square-free part f. Put the primes p_i in increasing order. We define the finite sequence m_s with $0 \leq s \leq t$ by setting

$$m_0 = f \quad \text{and} \quad m_s = \prod_{1 \leq i \leq s} p_i^{r_i} \cdot \prod_{s < i \leq t} p_i \text{ for } 1 \leq s \leq t \text{ ending with } m_t = m.$$

We prove successively for $0 \leq s \leq t$

$$(IV.3.16) \qquad Q_{m_s}(x) \equiv P(x)^{\frac{m_s}{f}} \pmod{\mathbf{Z}_{m_s}} \text{ with } \frac{m_s}{f} = \prod_{1 \leq i \leq s} p_i^{r_i - 1}.$$

Assumed by the corollary is

$$(IV.3.14) \qquad Q_f(x) \equiv P(x) \pmod{\mathbf{Z}_f}$$

which is equation (IV.3.16) above for $s = 0$. For the step $s \to s+1$ one has to put

$$p := p_s, \; r := r_s, \; m := \prod_{1 \leq i < s} p_i^{r_i} \cdot \prod_{s < i \leq t} p_i$$

into an instance of theorem 12. Note that $mp = m_s \geq 3$ and

$$m_{s+1} = m_s \cdot p^{r_s - 1}$$

Thus theorem 12 yields the implication that

$$Q_{m_s}(x) \equiv P(x)^{\frac{m_s}{f}} \pmod{\mathbf{Z}_{m_s}}$$

implies

$$Q_{m_{s+1}}(x) \equiv (P(x))^{\frac{m_s}{f} \cdot p^{r_s - 1}} = (P(x))^{\frac{m_{s+1}}{f}} \pmod{\mathbf{Z}_{m_{s+1}}}$$

the latter formula is equation (IV.3.16) with $s + 1$. After t induction steps the claim (IV.3.15) is obtained. $\qquad \square$

After having seen some of its consequences, we now proceed to the proof of the main Theorem 12 for Euler-type polynomials.

Lemma 40. *Let $m \geq 1$, let p be a prime and $r \geq 1$. Let $1 = a_1 < a_2 < \cdots < a_{\phi(mp^r)} = mp^r - 1$ be a list of integers relatively prime to mp which represent the Euler group $G^*_{mp^r}$. Then*

$$\{a_i + jmp^r : 1 \leq i \leq \phi(mp^r) \text{ and } 0 \leq j \leq p-1\}$$

of integers relatively prime to mp^{r+1} which represent the Euler group $G^*_{mp^{r+1}}$.

Lemma 41. *Let $P \in \mathbf{Z}[x]$ be an integer polynomial and h be an integer. Then for all $k \geq 0$ the derivative $\frac{P^{(k)}(x)}{k!}$ is an integer polynomial. The well-known formula*

$$(\text{IV.3.17}) \qquad P(x+h) = P(x) + P'(x)h + R(x,h)h^2$$

holds not only in the sense of an approximation for small h. Too, the remainder term $R(x,h) \in \mathbf{Z}[x,h]$ is an integer polynomial.

Remark. To indicate that all coefficients of $R(x,h)h^2$ are divisible by h^2, I write $\mathcal{D}(h^2)$ in place of $R(x,h)h^2$.

Proof. For all $k \geq 0$ and $l \geq 0$ holds

$$\frac{1}{k!}\frac{d^k x^l}{dx^k} = \binom{l}{k} x^{l-k} \in \mathbf{Z}[x]$$

Use the Taylor's expansion to see that $R(x,h)$ is an integer polynomial.

$$P(x+h) = P(x) + P'(x)h + \sum_{2 \leq k \leq d} \frac{P^{(k)}(x)}{k!} h^k$$

$$R(x,h) = \sum_{2 \leq k \leq d} \frac{P^{(k)}(x)}{k!} h^{k-2}$$

Here $d = \deg P$ is the degree of the polynomial P. $\qquad \square$

Remark. One may also proceed as follows. Use the binomial Theorem for all monomials and compare with Taylor's expansion.

$$P(x) = \sum_{0 \leq l \leq d} p_l x^l$$

$$P(x + h) = \sum_{0 \leq l \leq d} p_l (x + h)^l = \sum_{0 \leq l \leq d} \sum_{0 \leq k \leq l} \binom{l}{k} p_l x^{l-k} h^k$$

$$= \sum_{0 \leq k \leq d} \left[\sum_{k \leq l \leq d} \binom{l}{k} p_l x^{l-k} \right] h^k$$

$$= P(x) + P'(x)h + \sum_{2 \leq k \leq d} \frac{P^{(k)}(x)}{k!} h^k$$

Problem 73. *Let $n \geq 4$ be even and let $P(x) \in \mathbf{Z}[x]$ be an integer polynomial such that*

(IV.3.18) $$P(x) \equiv Q_n(x) \pmod{\mathbf{Z}_n}$$

Prove that the derivative $P'(x)$ is divisible by 2.

Solution. Because of $n \geq 3$ and hence $\phi(n)$ is even. By lemma 11, the mapping $a \mapsto i(a) = n - a$ is an involution $G_n^* \mapsto G_n^*$ which has no fixed points. The Euler-type polynomial as given by definition (IV.3.1) may now be partitioned as follows

$$Q_n(x) = \{\prod (x - a_i) \,:\, 1 \leq i \leq \phi(n) \text{ and } \gcd(a_i, n) = 1\}$$

$$= \{\prod (x^2 - a_i^2) \,:\, 1 \leq i \leq \tfrac{\phi(n)}{2} \text{ and } \gcd(a_i, n) = 1\}$$

$$- n\{\prod (x - a_i) \,:\, 1 \leq i \leq \tfrac{\phi(n)}{2} \text{ and } \gcd(a_i, n) = 1\}$$

The first term is a polynomial of x^2 and hence has a derivative divisible by 2. The second term, and its derivative, too, are a multiple of n which is assumed to be even. Hence $Q_n'(x)$ is divisible by 2.

Any other polynomial $P(x) \equiv Q_n(x) \pmod{\mathbf{Z}_n}$ is a sum $P(x) = Q_n(x) + nR(x)$ with integer term $R(x) \in \mathbf{Z}[x]$. Hence $P'(x)$ is divisible by 2. $\square$

End of the proof of Theorem 12. Let $m \geq 1$, let p be a prime. We proceed by induction on $t \geq 1$. The induction start with $r = 1$ is a tautology.
Here is the induction step $r \mapsto r + 1$: From definition (IV.3.1) we use literally

$$Q_{mp^r}(x) = \{\prod (x - a_i) \,:\, 1 \leq i \leq \phi(mp^r)\} \in \mathbf{Z}[x]$$

as an integer polynomial. Lemma 67 provides the list

$$\{a_i + jmp^r : 1 \le i \le \phi(mp^r) \text{ and } 0 \le j \le p-1\}$$

of integers relatively prime to mp^{r+1} which represent the Euler group $G^*_{mp^{r+1}}$. Definition (IV.3.1) tells

$$Q_{mp^{r+1}}(x) = \{\prod(x - a_i - jmp^r : 1 \le i \le \phi(mp^r) \text{ and } 0 \le j \le p-1\} \in \mathbf{Z}[x]$$

$$(IV.3.19) \qquad Q_{mp^{r+1}}(x) \equiv \prod_{0 \le j \le p-1} Q_{mp^r}(x - jmp^r) \quad (\text{mod } \mathbf{Z}_{mp^{r+1}}[x])$$

Lemma 41 gives the expansion

$$\prod_{0 \le j \le p-1} Q_{mp^r}(x - jmp^r) = \prod_{0 \le j \le p-1} [Q_{mp^r}(x) - Q'_{mp^r}(x)jmp^r + \mathcal{D}(jmp^r)^2]$$

$$= (Q_{mp^r}(x))^p - \left[\sum_{0 \le j \le p-1} jmp^r\right] Q'_{mp^r}(x)(Q_{mp^r}(x))^{p-1} + \mathcal{D}(mp^r)^2$$

$$= (Q_{mp^r}(x))^p - \frac{(p-1)p}{2}mp^r \cdot Q'_{mp^r}(x)(Q_{mp^r}(x))^{p-1} + \mathcal{D}(mp^r)^2$$

An equivalence in $\mathbf{Z}_{mp^{r+1}}$ has to be obtained. Here we use $2r \ge r + 1$. In case that p is an odd prime $\frac{(p-1)p}{2}$ is divisible by p. Thus an additional factor of p is gained.

But in the special case $p = 2$ a factor of 2 is missing because of $\frac{(p-1)p}{2} = 1$. Luckily we may use problem IV.3.18. Since $mp^r \ge mp > 3$ is even, it has been shown by problem IV.3.18 that the derivative $Q'_{mp^r}(x)$ is divisible by 2. Thus the factor of 2 is regained. Finally

$$(IV.3.20) \qquad \prod_{0 \le j \le p-1} Q_{mp^r}(x - jmp^r) \equiv (Q_{mp^r}(x))^p \quad (\text{mod } \mathbf{Z}_{mp^{r+1}})$$

holds for any prime p including the case $p = 2$. Together with equation (IV.3.19) we obtain

$$(IV.3.21) \qquad Q_{mp^{r+1}}(x) \equiv (Q_{mp^r}(x))^p \quad (\text{mod } \mathbf{Z}_{mp^{r+1}})$$

Assumed to be known from the induction assumption is only a weaker equivalence

$$(IV.3.3) \qquad Q_{mp^r}(x) \equiv (P(x))^{p^{r-1}} \quad (\text{mod } \mathbf{Z}_{mp^r})$$

According to assumption (IV.3.9) we have put $P(x) \in \mathbf{Z}[x]$ in place of $Q_{mp}(x)$. In the more stringent equivalence

$$(IV.3.3) \qquad Q_{mp^r}(x) \equiv (P(x))^{p^{r-1}} + R(x)mp^r \quad (\mathrm{mod}\ \mathbf{Z}_{mp^{r+1}})$$

has still to be included an unknown term $R(x)$. Thus equation (IV.3.21) gives only

$$Q_{mp^{r+1}}(x) \equiv (Q_{mp^r}(x))^p \equiv [(P(x))^{p^{r-1}} + R(x)mp^r]^p \quad (\mathrm{mod}\ \mathbf{Z}_{mp^{r+1}})$$

Luckily enough one may use the binomial formula to save the result. Since the binomial coefficients $\binom{p}{k}$ for all $1 \leq k \leq p-1$ are divisible by the prime p, only the term with $k = 0$ matters.

$$((P(x))^{p^{r-1}} + R(x)mp^r)^p = \sum_{0 \leq k \leq p} \binom{p}{k} (P(x))^{(p-k)p^{r-1}} (R(x)mp^r)^k$$
$$\equiv (P(x))^{p^{1+r-1}} \quad (\mathrm{mod}\ \mathbf{Z}_{mp^{r+1}})$$

and equation (IV.3.3) implies

$$Q_{mp^{r+1}}(x) \equiv P(x)^{p^r} \quad (\mathrm{mod}\ \mathbf{Z}_{mp^{r+1}})$$

which is the instance of claim (IV.3.3) with $r + 1$. $\qquad\qquad\qquad\qquad\square$

Remark. I want to thank my former student Mr. Bolling to motivate me doing this research. My first proof was even much more involved, and nevertheless gave a weaker result.

IV.3.4   Examples

Problem 74. *Provide a list of the Euler-type polynomials $Q_m(x) \in \mathbf{Z}_m[x]$ for $m \leq 30$.*

m	$Q_m(x)$	$\phi(m)$
2	$x - 1$	1
3	$x^2 - 1$	2
4	$x^2 - 1$	2
5	$x^4 - 1$	4
6	$x^2 - 1$	2
7	$x^6 - 1$	6
8	$(x^2 - 1)^2$	4
9	$(x^2 - 1)^3$	6
10	$x^4 - 1$	4
11	$x^{10} - 1$	10
12	$(x^2 - 1)^2$	4
13	$x^{12} - 1$	12
14	$(x^2 - 1)(x^4 + 8x^2 + 1)$	6
15	$(x^2 - 1)^2(x^2 + 11)^2$	8

m	$Q_m(x)$	$\phi(m)$
16	$(x^2 - 1)^4$	8
17	$x^{16} - 1$	16
18	$(x^2 - 1)^3$	6
19	$x^{18} - 1$	18
20	$(x^4 - 1)^2$	8
21	$(x^2 - 1)^2(x^4 + 8x^2 + 1)^2$	12
22	$x^{10} + 11x^8 + 11x^2 - 1$	10
23	$x^{22} - 1$	22
24	$(x^2 - 1)^4$	8
25	$(x^4 - 1)^5$	20
26	$(x^4 - 1)(x^8 + 14x^4 + 1)$	12
27	$(x^2 - 1)^9$	18
28	$(x^2 - 1)^2(x^4 + 8x^2 + 1)^2$	12
29	$x^{28} - 1$	28
30	$(x^2 - 1)^2(x^2 + 11)^2$	8

IV.3.5   Connection to Fermat Primes

Problem 75. *Show that all binomial coefficients $\binom{2^r}{k}$ for $1 \leq k < 2^r$ are even.*

Proof. By induction on $r \geq 1$ one shows

$$(x + 1)^{2^r} \equiv x^{2^r} + 1 \quad (\mathrm{mod}\ \mathbf{Z}_2[x])$$

$\square$

Problem 76. *Show that if all binomial coefficients $\binom{n}{k}$ for $1 \leq k < n$ are even, then $n = 2^r$.*

Solution. The recursion

$$\binom{n - 1}{k - 1} + \binom{n - 1}{k} = \binom{n}{k}$$

shows that all binomial coefficients $\binom{n-1}{k}$ with any $0 \leq k \leq n - 1$ are odd. Clearly n is even and with $a = n/2$ holds

$$2\binom{n - 1}{a} = \binom{n - 1}{a - 1} + \binom{n - 1}{a} = \binom{n}{a}$$

Hence $2 \| \binom{n}{a}$ has a sing le prime factor of 2 . We write the addition $n = a + a$ in binary representation and conclude via Lucas's proposition **??** that there is exactly one carry-over. This is only possible if a and hence n are powers of 2. $\qquad\square$

Problem 77. *Let p be a Fermat prime. Find a very simple expression for $Q_{2p}(x) \in$* $\mathbf{Z}_{2p}[x]$.

Proof. For a Fermat prime holds $p - 1 = 2^r$,—indeed even with $r = 2^n$. Hence by problem 75

$$(x - 1)^{p-1} = x^{p-1} - 1 + \sum_{1 \le k < p-1} \binom{2^r}{k}(-x)^k \equiv x^{p-1} - 1 \quad (\mathrm{mod}\ \mathbf{Z}_2[x])$$

We get $Q_{2p}(x) \equiv x^{p-1} - 1$ both modulo $\mathbf{Z}_2[x]$ and $\mathbf{Z}_p[x]$. Hence

$$Q_{2p} \equiv x^{p-1} - 1 \quad (\mathrm{mod}\ \mathbf{Z}_{2p}[x])$$

holds if p is a Fermat prime. $\qquad\square$

Problem 78. *Show that*

(IV.3.22) $$Q_{2p} \equiv x^{p-1} - 1 \quad (\mathrm{mod}\ \mathbf{Z}_{2p}[x])$$

holds if and only if p is a Fermat prime.

Proof. Assume that equation (IV.3.22) holds. Hence

$$(x - 1)^{p-1} \equiv x^{p-1} - 1 \quad (\mathrm{mod}\ \mathbf{Z}_2[x])$$

which implies that all binomial coefficients $\binom{p-1}{k}$ for $1 \le k < p - 1$ are even. By problem 76 we conclude that $p - 1 = 2^r$. Hence p is a Fermat prime. The converse was shown in problem 77. $\qquad\square$

IV.4   Chebyshev Polynomials

Lemma 42. *For all natural numbers $n \ge 1$ holds*

(IV.4.1) $$z^n(z - 1)\prod_{k=1}^{n}\left(z + \frac{1}{z} - 2\cos\frac{2\pi k}{2n + 1}\right) = z^{2n+1} - 1$$

(IV.4.2) $$z^{n-1}(z - 1)(z + 1)\prod_{k=1}^{n-1}\left(z + \frac{1}{z} - 2\cos\frac{2\pi k}{2n}\right) = z^{2n} - 1$$

Proof. Both sides of identity (IV.4.1) are monic polynomials in $\mathbb{C}[z]$ of degree $2n+1$. Both sides have the zeros

$$z_k = \exp \frac{2\pi i \cdot k}{2n+1} \quad \text{with } 0 \le k \le 2n$$

Hence the polynomials on the left-hand and right-hand side are equal.

Similarly, both sides of identity (IV.4.2) are monic polynomials in $\mathbb{C}[z]$ of degree $2n$. Both sides have the zeros

$$\exp\left(\pm\frac{2\pi i \cdot k}{2n}\right) \quad \text{with } 1 \le k \le n-1 \text{ and } z = \pm 1$$

Hence the polynomials on the left-hand and right-hand side are equal. $\qquad\square$

Lemma 43. *For all natural numbers $n \ge 0$ and $z = \exp i\theta$ holds*

$$(\text{IV.4.3}) \qquad \frac{z^{2n+1} - 1}{z^n(z-1)} = \frac{\sin \frac{(2n+1)\theta}{2}}{\sin \frac{\theta}{2}}$$

$$(\text{IV.4.4}) \qquad \frac{z^{2n} - 1}{z^{n-1}(z-1)(z+1)} = \frac{\sin n\theta}{\sin \theta}$$

Proof. We may assume $n \ge 1$ and put $z = \exp i\theta$ and $\sqrt{z} = \exp \frac{i\theta}{2}$ into the left-hand side to obtain

$$\frac{z^{2n+1} - 1}{z^n(z-1)} = \frac{z^{\frac{2n+1}{2}} - z^{-\frac{2n+1}{2}}}{z^{\frac{1}{2}} - z^{-\frac{1}{2}}} = \frac{2i \sin \frac{(2n+1)\theta}{2}}{2i \sin \frac{\theta}{2}}$$

and similarly

$$\frac{z^{2n} - 1}{z^{n-1}(z-1)(z+1)} = \frac{z^n - z^{-n}}{z - z^{-1}} = \frac{2i \sin n\theta}{2i \sin \theta}$$

$$\qquad\square$$

Lemma 44. *For all natural numbers $n \ge 1$ holds the identity*

$$(\text{IV.4.5}) \qquad 2^n \prod_{k=1}^{n} \left(\cos\theta - \cos\frac{2\pi k}{2n+1}\right) = \frac{\sin \frac{(2n+1)\theta}{2}}{\sin \frac{\theta}{2}}$$

$$(\text{IV.4.6}) \qquad 2^{n-1} \prod_{k=1}^{n-1} \left(\cos\theta - \cos\frac{2\pi k}{2n}\right) = \frac{\sin n\theta}{\sin \theta}$$

Definition 26 (Chebyshev polynomials)**.** Let the index $n \geq 0$ be a natural number. There exist the *Chebyshev polynomials $T_n(x)$* and the *Chebyshev polynomials of second kind $U_n(x)$* satisfying the identities

$$(\text{IV}.4.7) \qquad\qquad T_n(\cos\theta) = \cos(n\theta)$$

$$(\text{IV}.4.8) \qquad\qquad U_n(\cos\theta) = \frac{\sin(n+1)\theta}{\sin\theta}$$

Proposition 47. *The Chebyshev polynomials $T_n(x)$ and the Chebyshev polynomials of second kind $U_n(x)$ are integer polynomials of degree n. They are even or odd, depending on the index n being even or odd. They satisfy the recursion relations*

$$(\text{IV}.4.9) \qquad\qquad T_1(x) = x \,, \; U_0(x) = 1$$

$$(\text{IV}.4.10) \qquad\qquad T_n(x) = xT_{n-1}(x) - (1 - x^2)U_{n-2}(x)$$

$$(\text{IV}.4.11) \qquad\qquad U_n(x) = T_n(x) + xU_{n-1}(x)$$

Moreover their nonzero coefficients have alternating signs.

Proof. Of course the recursion formulas can be checked by brute force using the addition theorems for the trigonometric functions. More inspiring is to use the generating function

$$(1 - \exp(i\,\theta)u) \sum_{n \geq 1} \left(T_n(\cos\theta) + i\sin\theta U_{n-1}(\cos\theta) \right) u^n$$

$$= (1 - \exp(i\,\theta)u) \sum_{n \geq 1} \exp(i\,n\theta)u^n = \exp(i\,\theta)u$$

and compare for all powers u^n the real- and imaginary parts of the coefficients. The other assertions follow direct by recursion.

The alternating signs follow by Descartes rule of signs. One needs to check that $T_n(x)$ has the $\lfloor \frac{n+1}{2} \rfloor$ positive zeros at $\cos\frac{(2k-1)\pi}{2n}$ for $1 \leq k \leq \frac{n+1}{2}$. But the polynomial $T_n(x)$ has maximally $\lceil \frac{n+1}{2} \rceil$ nonzero coefficients. By Descartes rule of signs, the number of positive zeros is at most the number of sign changes in the list of its coefficients. We see that this maximal number of zeros is achieved. Hence the maximal number of nonzero coefficients are the maximal number of sign changes need to occur.

Similarly, $U_n(x)$ has the $\lfloor \frac{n+1}{2} \rfloor$ positive zeros at $\cos\frac{k\pi}{n+1}$ for $1 \leq k \leq \frac{n+1}{2}$. Again Descartes' rule of signs enforces the sign changes. $\qquad\square$

Problem 79. *Prove the recursion formula $T_n(x) = 2xT_{n-1}(x) - T_{n-2}(x)$.*

Solution. One has to use the trigonometric identity

$$\cos n\theta + \cos(n-2)\theta = 2\cos\theta\cos(n-1)\theta$$

□

Problem 80. *Show that* $2T_n\left(\frac{x}{2}\right)$ *is an integer monic polynomial.*

Problem 81. *Prove the recursion formula* $U_n(x) = 2xU_{n-1}(x) - U_{n-2}(x)$.

Solution. One has to use the trigonometric identity

$$\sin(n+1)\theta + \sin(n-1)\theta = 2\cos\theta\sin n\theta$$

□

Problem 82. *Show that* $U_n\left(\frac{x}{2}\right)$ *is an integer monic polynomial.*

Problem 83. *Show for all* $n \geq 1$ *that* $U_{2n}\left(\sqrt{\frac{x}{4}}\right)$ *is an integer monic polynomial.*

Problem 84. *Show for all* $n \geq 1$ *that* $\frac{1}{\sqrt{x}}U_{2n-1}\left(\sqrt{\frac{x}{4}}\right)$ *is an integer monic polynomial.*

Problem 85. *Prove the identity*

$$2yU_{n-1}(2y^2 - 1) = U_{2n-1}(y)$$

Both sides are polynomials of degree $2n+1$ *with the highest coefficient* 2^{2n+1}.

Proof. Put $y = \cos\theta$ and $2y^2 - 1 = \cos 2\theta$. One gets from the definition

$$2yU_{n-1}(2y^2 - 1) = \frac{2\cos\theta\,\sin(n\cdot 2\theta)}{\sin 2\theta} = \frac{\sin 2n\theta}{\sin\theta} = U_{2n-1}(y)$$

□

Problem 86. *Prove the identity* $U_{2n+1}(x) = 2U_n(x)T_{n+1}(x)$.

Solution. Let $x = \cos\theta$.

$$\sin\theta\,U_{2n+1}(x) = \sin(2(n+1)\theta) = 2\sin((n+1)\theta)\,\cos((n+1)\theta) = 2\sin\theta\,U_n(x)T_{n+1}(x)$$

□

Lemma 45 (Rothe's formulas). *For all natural numbers $n \geq 1$ hold the identities*

$$(IV.4.12) \qquad \prod_{k=1}^{n} \left(u - 2\cos\frac{2\pi k}{2n+1} \right) = U_{2n}\left(\frac{\sqrt{u+2}}{2} \right)$$

$$(IV.4.13) \qquad \prod_{k=1}^{n-1} \left(u - 2\cos\frac{2\pi k}{2n} \right) = U_{n-1}\left(\frac{u}{2} \right) = \frac{1}{\sqrt{u+2}} U_{2n-1}\left(\frac{\sqrt{u+2}}{2} \right)$$

Note that the right-hand side is indeed a polynomial since $U_{2n}(x)$ is an even polynomial.

Proof. Put $u = 2\cos\theta$ into equation (IV.4.5). Use

$$U_{2n}\left(\cos\frac{\theta}{2} \right) = \frac{\sin\frac{(2n+1)\theta}{2}}{\sin\frac{\theta}{2}}$$

from the definition of the Tschebychef polynomial of second kind. Finally

$$\cos^2\frac{\theta}{2} = \frac{1+\cos\theta}{2}$$

$$\cos\frac{\theta}{2} = \pm\frac{\sqrt{u+2}}{2}$$

$$U_{2n}\left(\cos\frac{\theta}{2} \right) = U_{2n}\left(\frac{\sqrt{u+2}}{2} \right)$$

from standard trigonometry and a lucky cancelling of an ambigous sign.

The case with even index may be left to the reader. The last claim follows from problem 85 with $2y = \sqrt{u+2}$ and $u = 4y^2 - 2$. $\qquad\square$

Definition 27. Assume $m \geq 3$, put $z = x \pm iy$ with, $|x| \leq 1$ and $y = \sqrt{1-x^2}$ and $u = 2x$. We define the polynomial $\Psi_m(u)$ such that

$$\Psi_m(u) = \pm\sqrt{\Phi_m(x+iy)\Phi_m(x-iy)}$$

The sign is determined by the requirements that $\Psi_m(u)$ is a polynomial and the highest coefficient is positive. I leave $\Psi_1(u)$ and $\Psi_2(u)$ undefined for now.

Lemma 46. *For $m \geq 3$ holds*

$$(IV.4.14) \qquad z^{\frac{\phi(m)}{2}} \Psi_m(z + z^{-1}) = \Phi_m(z)$$

For all $m \geq 3$, the function $\Psi_m(u)$ is a monic polynomial of degree $\frac{\phi(m)}{2}$.

180

Proof. We put $z = x \pm iy$ with, $|x| \leq 1$ and $y = \sqrt{1 - x^2}$ and moreover

$$u = 2x = z + z^{-1}$$

From the definition

$$z^{\frac{\phi(m)}{2}} \Psi_m(z + z^{-1}) = \pm z^{\frac{\phi(m)}{2}} \sqrt{\Phi_m(z)\Phi_m(\overline{z})} = \pm\sqrt{\Phi_m(z)z^{\phi(m)}\Phi_m(z^{-1})}$$
$$= \pm\sqrt{\Phi_m^2} = \pm\Phi_m(z)$$

are polynomials with highest coefficient positive and even equal to one. Hence the positive sign holds in equation (IV.4.14). $\square$

Problem 87. *Prove that the polynomials $\Psi_m(u)$ for $m \geq 3$ are monic and irreducible in $\mathbf{Z}[u]$, and have degree $\frac{\phi(m)}{2}$.*

Solution. Assume a decomposition $\Psi_m(u) = A(u)B(u)$ into nonconstant monic polynomials from $\mathbf{Z}[u]$ and assume that $A(u)$ is irreducible. Let $a = \deg A \leq b = \deg B$. From Lemma IV.4.14 we know

$$\Phi_m(z) = z^{\frac{\phi(m)}{2}} \Psi_m(z + z^{-1}) = z^a A(z + z^{-1}) \cdot z^b B(z + z^{-1})$$

Now the factor $z^a A(z + z^{-1})$ is a monic integer polynomial of degree $2a$ Since $z^a A(z + z^{-1})$ is a divisor of $\Phi_m(x)$ which is irreducible and not of maximal degree, A is a constant and $a = 0$. Hence the polynomial $\Psi_m(u)$ is irreducible. $\square$

Proposition 48. *Assume $m \geq 3$. Indeed holds with the above suitable analytic choice of the sign*

$$(IV.4.15) \qquad \Psi_m(u) = \prod_{1 \leq k < \frac{m}{2} \text{ and } \gcd(k,m)=1} \left(u - 2\cos\frac{2\pi k}{m}\right)$$

and this is a monic polynomial with integer coefficients of degree $\frac{\phi(m)}{2}$ which is irreducible in the ring $\mathbf{Z}[u]$.

Proof. It is sufficient to check the identity obtained by substituting $u \mapsto z + z^{-1}$.

$$z^{\frac{\phi(m)}{2}} \prod_{1 \leq k < m/2 \text{ and } \gcd(k,m)=1} \left(z + z^{-1} - 2\cos\frac{2\pi k}{m} \right)$$

$$= \prod_{1 \leq k < m/2 \text{ and } \gcd(k,m)=1} \left(z^2 + 1 - 2z\cos\frac{2\pi k}{m} \right)$$

$$= \prod_{1 \leq k < m/2 \text{ and } \gcd(k,m)=1} \left(z - \exp\frac{2i\pi \cdot k}{m} \right) \left(z - \exp\frac{-2i\pi \cdot k}{m} \right)$$

$$= \prod_{1 \leq k < m \text{ and } \gcd(k,m)=1} \left(z - \exp\frac{2i\pi \cdot k}{m} \right) = \Phi_m(z)$$

The last line uses item(b) from definition 22 of the cyclotomic polynomials. Next we use lemma 46

$$\Phi_m(z) = z^{\frac{\phi(m)}{2}} \Psi_m(z + z^{-1})$$

$\square$

Problem 88. *A numerical check, say for $3 \leq m \leq 10$, is a nice mathematica exercise.*

Corollary 22. *Let $m = 2n + 1$ be odd. The polynomial $U_{m-1}\left(\frac{\sqrt{u+2}}{2}\right)$ has the irreducible factors $\Psi_d(u)$ where $d \geq 3$ runs through the divisors of m.*

(IV.4.16) $$U_{2n}\left(\frac{\sqrt{u+2}}{2}\right) = \prod_{3 \leq d \mid 2n+1} \Psi_d(u)$$

Proof. It is sufficient to check the identity obtained from (IV.4.16) by substituting $u \mapsto z + z^{-1}$. From lemma 46 the addition property from proposition 15 of the Euler totient function and the properties of the cyclotomic polynomial

$$z^{\frac{2n+1}{2}} \prod_{3 \leq d \mid 2n+1} \Psi_d(z + z^{-1}) = z^{\frac{1}{2}} \prod_{3 \leq d \mid 2n+1} z^{\frac{\phi(d)}{2}} \Psi_d(z + z^{-1})$$

$$= z^{\frac{1}{2}} \prod_{3 \leq d \mid 2n+1} \Phi_d(z) = z^{\frac{1}{2}} \frac{z^{2n+1} - 1}{z - 1}$$

$$\prod_{3 \leq d \mid 2n+1} \Psi_d(z + z^{-1}) = \frac{z^{2n+1} - 1}{z^n(z - 1)}$$

182

With $z = \exp i\theta$ and $u = 2\cos\theta = z + z^{-1}$, we next use identity (IV.4.3) and the identities from the proof of lemma 45 to get

$$\frac{z^{2n+1} - 1}{z^n(z-1)} = \frac{\sin\frac{(2n+1)\theta}{2}}{\sin\frac{\theta}{2}} = U_{2n}\left(\cos\frac{\theta}{2}\right) = U_{2n}\left(\frac{\sqrt{u+2}}{2}\right)$$

$$\prod_{3 \le d \mid 2n+1} \Psi_d(z + z^{-1}) = U_{2n}\left(\sqrt{\frac{2 + z + z^{-1}}{4}}\right)$$

$\square$

Corollary 23. *Let* $m = 2n$ *be even. The polynomial* $\frac{1}{\sqrt{u+2}}U_{m-1}\left(\frac{\sqrt{u+2}}{2}\right)$ *has the irreducible factors* $\Psi_d(u)$ *where* $d \ge 3$ *runs through the divisors of* m.

$$\text{(IV.4.17)} \qquad \frac{1}{\sqrt{u+2}}U_{m-1}\left(\frac{\sqrt{u+2}}{2}\right) = U_{\frac{m}{2}-1}\left(\frac{u}{2}\right) = \prod_{3 \le d \mid m} \Psi_d(u)$$

Proof. It is sufficient to check the identity obtained from (IV.4.17) by substituting $u \mapsto z + z^{-1}$. From lemma 46 the addition property of the Euler totient function and the properties of the cyclotomic polynomial

$$z^{\frac{2n}{2}} \prod_{3 \le d \mid 2n} \Psi_d(z + z^{-1}) = z \prod_{3 \le d \mid 2n} z^{\frac{\phi(d)}{2}} \Psi_d(z + z^{-1}) = z \prod_{3 \le d \mid 2n} \Phi_d(z)$$

$$= z\,\frac{z^{2n} - 1}{(z-1)(z+1)}$$

$$\prod_{3 \le d \mid 2n} \Psi_d(z + z^{-1}) = \frac{z^{2n} - 1}{z^{n-1}(z-1)(z+1)}$$

With $z = \exp i\theta$ and $u = 2\cos\theta = z + z^{-1}$, we next use identity (IV.4.4)

$$\frac{z^{2n} - 1}{z^{n-1}(z-1)(z+1)} = \frac{\sin n\theta}{\sin\theta} = U_{n-1}(\cos\theta) == U_{n-1}\left(\frac{u}{2}\right)$$

Finally the identity from problem 85 allows for the additional formula.
Putting $y = \frac{\sqrt{2+u}}{2}$ yields

$$2y\,U_{n-1}(2y^2 - 1) = U_{2n-1}(y)$$

$$\sqrt{u+2}\,U_{n-1}\left(\frac{u}{2}\right) = U_{2n-1}\left(\frac{\sqrt{u+2}}{2}\right)$$

All together we have shown

$$\prod_{3\leq d|2n} \Psi_d(z + z^{-1}) = U_{n-1}\left(\frac{u}{2}\right) = \frac{1}{\sqrt{u+2}} U_{2n-1}\left(\frac{\sqrt{2+u}}{2}\right)$$

$\square$

Remark. With the,—a bid strange,—definition $\Psi_2(u) = \sqrt{2+u}$, it is possible to avoid to distinguish the cases of even and odd m. Indeed holds for all $m \geq 2$

$$U_{m-1}\left(\frac{\sqrt{u+2}}{2}\right) = \prod_{2\leq d|m} \Psi_d(u)$$

Problem 89. *Assume that $P(x)$ is an irreducible polynomial from the ring $\mathbf{Z}[x]$ and its highest coefficient is positive. Prove that either*

(i) *the polynomial $P(x^2)$ is irreducible,—even in $\mathbf{Q}[x]$ except for constant factors;*

(ii) *or $P(x^2) = A(x)A(-x)$ where $A(x)$ is an integer irreducible polynomial. Moreover, the highest and lowest coefficients of $P(x)$ is a perfect square. The polynomial A does not have any pair of zeros $\pm\alpha$.*

Proof. Assume that the polynomial $P(x^2)$ is reducible in $\mathbf{Q}[x]$ with nonconstant factors. By Gauss' Lemma it is even reducible in $\mathbf{Z}[x]$. Assume that $P(x^2) = A(x)B(x)$ where $A(x)$ is an integer irreducible polynomial, and neither A nor B are of degree zero. We choose the highest coefficient of A to be positive. The greatest common factor $C(x^2) = \gcd(A(x), A(-x))$ is a symmetric polynomial and a factor of $P(x^2)$. In other words $C(x)$ is a divisor of $P(x)$. Since $P(x)$ is assumed to be an irreducible polynomial from the ring $\mathbf{Z}[x]$, we conclude that $C = \pm 1$. We may assume $C = 1$ by the choice of the sign of A.

Hence $A(-x)$ is divisor of $B(x)$ We did not need yet the assumption that A is irreducible. Indeed , we show similarly that $B(-x)$ is a divisor of $A(x)$. Hence $B(x) = \pm A(-x)$. Since the highest coefficient of $P(x)$ is assumed to be positive, we get $B(x) = A(-x)$. Moreover $A(x)$ is irreducible. $\square$

Problem 90. *Are the polynomials $\Psi_m(y^2 - 2)$ irreducible? Check that*

$$\Psi_m(y^2 - 2) = \Psi_{2m}(y) \qquad \text{if } m \text{ even}$$
$$\Psi_m(y^2 - 2) = \Psi_{2m}(y) * \Psi_{2m}(-y) \quad \text{if } m \text{ odd}$$

Proof. Define

$$y_k = 2\cos\frac{\pi \cdot k}{m} \qquad \text{with } 1 \le k < m \text{ and } \gcd(k,m) = 1.$$

All the $\phi(m)$ zeros of $\Psi_m(y^2 - 2)$ are $\pm y_k$.
One the other hand, all the $\frac{\phi(m)}{2}$ zeros of $\Psi_m(u)$ are

$$u_k = 2\cos\frac{2\pi \cdot k}{m} \qquad \text{with } 1 \le k < \frac{m}{2} \text{ and } \gcd(k,m) = 1.$$

- Assume $m \ge 4$ is even. All the $\phi(m) = \frac{\phi(2m)}{2}$ zeros of $\Psi_{2m}(y)$ are

$$y_k = -y_{m-k} = 2\cos\frac{\pi \cdot k}{m} \qquad \text{with } 1 \le k < m \text{ and } \gcd(k,2m) = 1$$

  The $\pm$-pairs are zeros of the same irreducible polynomial $\Psi_{2m}(y)$.
  Hence $\Psi_m(y^2 - 2) = \Psi_{2m}(y)$ is irreducible, too.

- Assume $m \ge 3$ is odd. All the $\frac{\phi(m)}{2} = \frac{\phi(2m)}{2}$ zeros of $\Psi_{2m}(y)$ are

$$y_k = 2\cos\frac{\pi \cdot k}{m} \qquad \text{with } 1 \le k < m \text{ and } \gcd(k,2m) = 1$$

  Now k is restricted to odd values, and $m-k$ is even. Their negatives $-y_k = y_{m-k}$ are excluded since $\gcd(m - k, 2m) = 2 \ne 1$. They are not zeros of $\Psi_{2m}(y)$ but only of $\Psi_{2m}(-y)$.

$\square$

This problem begs for a table:

m	$\Phi_m(y^2 - 2)$	$\Phi_{2m}(y)$
3	$(y-1)(y+1)$	$y - 1$
4	$y^2 - 2$	$y^2 - 2$
5	$(y^2 - y - 1)(y^2 + y - 1)$	$y^2 - y - 1$
6	$y^2 - 3$	$y^2 - 3$
7	$(y^3 - y^2 - 2y + 1)(y^3 + y^2 - 2y - 1)$	$y^3 - y^2 - 2y + 1$
8	$y^4 - 4y^2 + 2$	$y^4 - 4y^2 + 2$
9	$(y^3 - 3y - 1)(y^3 - 3y + 1)$	$y^3 - 3y - 1$

Problem 91. *Decompose the polynomials $2T_n(\frac{u}{2})$ into irreducible factors in the ring $\mathbf{Z}[u]$. Let $u = z + z^{-1}$ and $z = \exp i\theta$. Let $n = 2^s q$ with odd q. Decompose $2T_n(\frac{u}{2})$ at first as a function of z into cyclotomic polynomials.*

Solution. Let $u = z + z^{-1}$ and $z = \exp i\theta$. Let $n = 2^s q$ with odd q. Decompose at first as a function of z into cyclotomic polynomials. Then use formula (IV.4.14).

$$2T_n\left(\frac{u}{2}\right) = 2T_n(\cos\theta) = 2\cos(n\theta) = z^n + z^{-n} = z^{-n}\frac{z^{4n} - 1}{z^{2n} - 1}$$

$$= z^{-n}\left(\prod_{3 \le d \mid 4n} \Phi_d(z)\right)\left(\prod_{3 \le e \mid 2n} \Phi_e(z)\right)^{-1} = z^{-n}\prod_{1 \le m \mid q} \Phi_{2^{s+2}m}(z)$$

$$= z^{-n}\prod_{1 \le m \mid q} z^{\frac{\phi(2^{s+2}m)}{2}}\Psi_{2^{s+2}m}(z + z^{-1}) = z^g \prod_{1 \le m \mid q} \Psi_{2^{s+2}m}(z + z^{-1})$$

$$= \prod_{1 \le m \mid q} \Psi_{2^{s+2}m}(u)$$

I have used

$$g = -n + \sum_{1 \le m \mid q} \frac{\phi(2^{s+2}m)}{2} == -n + 2^s \sum_{1 \le m \mid q} \phi(m) = -n + 2^s q = 0$$

$\square$

Problem 92. *Decompose the polynomials $U_k(y)$ into irreducible factors in the ring* $\mathbf{Z}[y]$.

Solution. Take the case that $k = \frac{m}{2} - 1$ and $m = 2n$ is even. Hence $m = 2k + 2$. Put $u = 2y$. From the second part of equation (IV.4.17)

$$U_k(y) = \prod_{3 \le d \mid 2k+2} \Psi_d(2y)$$

which is a decomposition into irreducible factors. $\square$

Part V

Construction with Hilbert Tools

V.1   Constructions and Positivity

V.1.1   Minimal Fields

Definition 28 (Euclidian field). A *Euclidean field* is defined to be an ordered field with the property that

$$a \in K \text{ and } a > 0 \Rightarrow \sqrt{a} \in K$$

Definition 29 (Pythagorean field). A *Pythagorean field* is defined to be an ordered field with the property that

$$a, b \in K \Rightarrow \sqrt{a^2 + b^2} \in K$$

Definition 30 (Hilbert field). The *Hilbert field* Ω is the <u>smallest</u> real field with the properties

(1) $1 \in \Omega$.

(2) If $a, b \in \Omega$, then $a + b, a - b, ab \in \Omega$.

(2a) If $a, b \in \Omega$ and $b \neq 0$, then $\frac{a}{b} \in \Omega$.

(3) If $a \in \Omega$, then $\sqrt{1 + a^2} \in \Omega$.

Remark. The *Hilbert field* Ω is the smallest Pythagorean field.

Definition 31 (Built-up Hilbert field). Let B be any (possibly empty) set of totally positive algebraic numbers. The *Hilbert field* $\Omega(B)$ is defined to be the smallest real Pythagorean field such that $B \subseteq \Omega(B)$ We get especially $\Omega = \Omega(\emptyset)$.

Definition 32 (constructible field). The *constructible field* K is the <u>smallest</u> real field with the properties

(1) $1 \in K$.

(2) If $a, b \in K$, then $a + b, a - b, ab \in K$.

(2a) If $a, b \in K$ and $b \neq 0$, then $\frac{a}{b} \in K$.

(4) If $a \in K$ and $a \geq 0$, then $\sqrt{a} \in K$.

It is also called the *surd field*.

Remark. There are many very different Pythagorean fields, and many different Euclidean fields, among them the real number field $\mathbb{R}$. There exist non-Archimedean Pythagorean fields, as well as non-Archimedean Euclidean fields, too, but these are not the minimal fields. Only the Hilbert's field Ω and the surd field K are minimal and hence they are uniquely defined structures.

Since the Hilbert field Ω is smallest Pythagorean field, we have the following more direct characterization of Ω:

Proposition 49. *Any number $x \in \Omega$ in the Hilbert field can be obtained in finitely many steps. We start with a rational number—or even just 0 and 1.*

Each step constructs a new number using only the field operations and the operation provided by the assumption (3): $a > 0 \mapsto \sqrt{1 + a^2}$, applied to the numbers already obtained.

Note that the total number of construction steps depends on the number $x \in \Omega$ and hence has no global bound. The corresponding statement holds for the constructible field K.

Proposition 50. *Any number $k \in K$ in the constructible field can be obtained in finitely many steps as follows: We start with a rational number—or even just 0 and 1. Each step constructs a new number using only the field operations and the operation provided by the assumption (4): $a > 0 \mapsto \sqrt{a}$, applied to the numbers already obtained.*

Corollary 24. *Since there are only countably many such processes, the constructible field is countable. The Hilbert field is a subfield of the constructible field and countable, too.*

V.1.2   Towers in the Constructible Field

Lemma 47. *Let the field F have characteristic not equal two and let L/F be a field extension. Equivalent are*

(i) *L/F is a field extension of dimension two.*

(ii) *There exists $D \in F$ such that $L = F(\sqrt{D})$.*

(i) $\Rightarrow$ (ii). Suppose that $[L : F] = 2$ and $\alpha \in L \setminus F$. The extension is generated by α since no proper intermediate field $F \subset M \subset L$ can exist. Hence $L = F(\alpha)$. The minimal polynomial $p \in F[x]$ of α has degree 2. Let

$$p(x) = x^2 + bx + c$$

The quadratic equation $p(\alpha) = 0$ has the solution

$$\alpha_{1,2} = \frac{-b \pm \sqrt{b^2 - 4c}}{2}$$

The division by $2 \neq 0$ is possible in a field of characteristic not two. Hence $p(x) = (x - \alpha_1)(x - \alpha_2)$. It is impossible that $\alpha_1 = \alpha_2$: that would contradict the fact that p is irreducible. We can produce the same field extension by adjoining the square root of the discriminant $D = b^2 - 4c \in F$. In other words, $L = F(\alpha) = F(\sqrt{D})$. The converse is immediately true. $\qquad\square$

Since we may introduce coordinates in any Euclidean plane, the steps of any geometric construction may be translated to field extensions which take place within the constructible field.

Proposition 51. *For any number $k \in K$ in the constructible field, there exists a tower of finitely many two dimensional extensions*

$$\mathbf{Q} = \mathbf{F}_0 \subset \mathbf{F}_1 \subset \mathbf{F}_2 \subset \cdots \subset \mathbf{F}_n \ni k$$

Moreover, $\dim[\mathbf{Q}(k) : \mathbf{Q}]$ is a power of 2.

Proof of the last part. Since $\mathbf{Q} \subseteq \mathbf{Q}(k) \subseteq \mathbf{F}_n$, the tower theorem 23 implies

$$\dim[\mathbf{F}_n : \mathbf{Q}(k)] \cdot \dim[\mathbf{Q}(k) : \mathbf{Q}] = \dim[\mathbf{F}_n : \mathbf{Q}]$$

We know that the right-hand side is a power of 2, hence the factors on the left-hand side are powers of 2. $\qquad\square$

Remark. Especially the dimension $[\mathbf{Q}(k) : \mathbf{Q}]$ is a power of two for all constructible numbers k. But this is only a necessary condition for k to be constructible. Their exists a four dimensional extension of the rational numbers which is not contained in the constructible field. An example is given in problem 188 below with an elaborated solution.

Proposition 52. *Suppose that any number $k \in \mathbb{R}$ is obtained in a tower of two-dimensional extensions*

(V.1.1) $$\mathbf{Q} = \mathbf{F}_0 \subset \mathbf{F}_1 \subset \mathbf{F}_2 \subset \cdots \subset \mathbf{F}_n \ni k$$

Then the number k is in the constructible field.

Theorem 13 (Towers in the constructible field). *The constructible field consist of the real numbers $k \in \mathbb{R}$ for which a* real *tower of two-dimensional extensions as in equation (V.1.1) exists.*

Proof of proposition 51. Fix any number $k \in K$ in the constructible field. The numbers obtained up to the i-th step of the construction process generate a field $\mathbf{F}_i$. Clearly $\mathbf{F}_{i-1} \subseteq \mathbf{F}_i$ for all construction steps. Either the i-th step involve only field operation in which case $\mathbf{F}_{i-1} = \mathbf{F}_i$, or the i-th construction step is an operation of the type $a > 0 \mapsto \sqrt{a}$. In the latter case

$$a \in \mathbf{F}_{i-1} \,, a > 0 \,, \quad \mathbf{F}_i = \mathbf{F}_{i-1}(\sqrt{a}) \quad \text{and} \quad [\mathbf{F}_i : \mathbf{F}_{i-1}] = 2$$

Suppose construction of the number $k \in K$ involves N steps of the second kind. We have obtained a tower of N extensions

$$\mathbf{Q} = \mathbf{F}_0 \subset \mathbf{F}_1 \subset \mathbf{F}_2 \subset \cdots \subset \mathbf{F}_N \ni k$$

In this tower, all extensions are two-dimensional, generated by adjoining a single root of a quadratic equation. Hence $[\mathbf{F}_N : \mathbf{Q}] = 2^N$. $\qquad\square$

Remark. Here an example. Let $k = \sqrt{\sqrt{2} - 1}.$ one gets

$$\mathbf{Q} = \mathbf{F}_0 \subset \mathbf{F}_1 = \mathbf{Q}(\sqrt{2}) \subset \mathbf{F}_2 = \mathbf{Q}(\sqrt{2})(k) \ni k$$

Indeed the number k is a root of $(1 + x^2)^2 - 2 = x^4 + 2x^2 - 1$ and this polynomial is irreducible, hence the minimal polynomial. The construction is done in two steps, $N = 2$ and $\mathbf{F}_N = \mathbf{Q}(k)$.

Remark. Every number of the constructible field lies in a finite dimensional extension of the rational numbers. Nevertheless, the <u>entire</u> constructible field is an infinite dimensional extension of the rational numbers.

Proof of Proposition 52. Suppose that any number $k \in \mathbb{R}$ is obtained in a tower of two-dimensional extensions (V.1.1). Inductively we check that $\mathbf{F}_i \subset K + iK$ for all $0 \leq i \leq N$.

By the first part of Lemma 47 there exists $D_i \in \mathbf{F}_i$ such that $\mathbf{F}_{i+1} = \mathbf{F}_i(\sqrt{D_i})$. From the induction assumption we know that $D_i \in K + iK$. The formula for the complex square root

$$(\text{VIII.2.3}) \qquad \sqrt{x + iy} = \sqrt{\frac{\sqrt{x^2 + y^2} + x}{2}} + i\operatorname{sign}(y)\sqrt{\frac{\sqrt{x^2 + y^2} - x}{2}}$$

implies $\sqrt{D_i} \in K + iK$. Hence $\mathbf{F}_{i+1} \subset K + iK$.

We may intersect all fields $\mathbf{F}_i$ of the tower with the field $\mathbf{Q}(k)$ and discard the steps with equality $\mathbf{F}_{i-1} \cap \mathbf{Q}(k) = \mathbf{F}_i \cap \mathbf{Q}(k)$. Thus we obtain the tower of *real* two-dimensional extensions, and still holds $k \in \mathbf{F}_N \cap \mathbf{Q}(k)$.

The surprise is that by only assuming that k is constructible and real, we have obtained a tower of fields all within the constructible field, hence all fields in the tower are subsets of the reals. $\square$

Remark. If the field F is a subfield of the constructible field and $a \in F$ and $a < 0$, the field $F(\sqrt{a})$ is not a subfield of the constructible field.

Remark. We continue the example above. Let $k = \sqrt{\sqrt{2} - 1}$. One get the construction tower

$$\mathbf{Q} = \mathbf{F}_0 \subset \mathbf{F}_1 = \mathbf{Q}(\sqrt{2}) \subset \mathbf{F}_2 = \mathbf{Q}(\sqrt{2})(k) = \mathbf{Q}(k) \ni k$$

Indeed the number k is a root of $(1 + x^2)^2 - 2 = x^4 + 2x^2 - 1$ and this polynomial is irreducible, Hence it is the minimal polynomial, $N = 2$ and $\mathbf{F}_N = \mathbf{Q}(k)$.

Indeed the field F is a subfield of the constructible field. Too, we get $L = F(\sqrt{D})$ with $D = a > 0$. L is subfield of the constructible field, but it is not the splitting field of an integer polynomial since the conjugate $\sqrt{-\sqrt{2} - 1} \in L \setminus \mathbb{R}$ is not in the constructible field.

Problem 93. *Prove the following equivalence, continuing lemma 47. Let the field F have characteristic not equal two and let L/F be a field extension. Equivalent are*

(a) *F is a subfield of the constructible field, L/F is a field extension of dimension two and $L \subseteq \mathbb{R}$*

(b) *F is a subfield of the constructible field and there exists $D \in F$ such that $D > 0$ and $L = F(\sqrt{D})$.*

(c) *L/F is a field extension of dimension two and L is a subfield of the constructible field.*

(a) $\Rightarrow$ (b). The constructible field has characteristic zero. As in the first part, we conclude $L = F(\sqrt{D})$ with $D = b^2 - 4c \in F$. Finally $L \subseteq \mathbb{R}$ implies $\sqrt{D} \in \mathbb{R}$ and hence $D \geq 0$. But $[L : F] = 2$ implies that even $D > 0$. $\square$

(b) $\Rightarrow$ (c). The field extension $F(\sqrt{D})/F$ corresponds to a step $D > 0 \mapsto \sqrt{D}$ in Proposition 50. Hence $F(\sqrt{D})$ is contained in the constructible field. $\square$

(c) $\Rightarrow$ (a). Since L is assumed to be a subfield of the constructible field holds $L \subseteq \mathbb{R}$. $\square$

V.1.3   Totally Real Extensions

Definition 33 (Totally real and totally positive numbers). An algebraic number (over the rational field) is called *totally real* if all its conjugates are real. An algebraic number (over the rational field) is called *totally positive* if all its conjugates are positive and real.

Proposition 53. *All numbers in the Hilbert field are totally real algebraic numbers. In other words, they are roots of a irreducible polynomial of (possibly high) degree N with integer coefficients, <u>all</u> N roots of which are real.*

Lemma 48. *If the number x is totally real, then $\sqrt{1 + x^2}$ is totally real, too.*

Proof. Assume the number x is totally real. Hence there exists an integer (even irreducible) polynomial $P \in \mathbf{Z}[x]$ of degree N all roots $x = x_1, x_2 \ldots x_N$ of which are real. We define $R(y) = P(y^2 - 1)$, which is again an integer polynomial and has the roots

$$\pm\sqrt{1 + x_1^2},\ \pm\sqrt{1 + x_2^2} \cdots \pm \sqrt{1 + x_N^2}$$

These are all real numbers and hence $\sqrt{1 + x_1^2}$ is totally real, as to be shown. $\square$

Remark. It does not matter whether the polynomial $R(y)$ is irreducible or not. The factoring $R(y) = Q(y)Q(-y)$ with

$$Q(y) = \left(y - \sqrt{1 + x_1^2} \right) \left(y - \sqrt{1 + x_2^2} \right) \cdots \left(y - \sqrt{1 + x_N^2} \right)$$

need not be possible over the rational numbers. Of course, this factoring works for coefficients in the Hilbert field.

Proof of Proposition 53. Any number $x \in \Omega$ in the Hilbert field can be obtained via finitely many algebraic field extensions

$$\mathbf{Q} = \mathbf{F}_0 \subset \mathbf{F}_1 \subset \mathbf{F}_2 \subset \cdots \subset \mathbf{F}_N \ni x$$

In this tower, all extensions are two-dimensional and generated by adjoining a single root:

$$\sqrt{1 + x_i^2} \in \mathbf{F}_{i+1} \setminus \mathbf{F}_i \quad \text{with } x_i \in \mathbf{F}_i$$

for $i = 0 \ldots N - 1$. By the lemma, we see successively that all intermediate fields $\mathbf{F}_i$ are totally real. Hence x is totally real. $\square$

Corollary 25. *All number in the Hilbert field are totally real.*

Proof. The proof is left to the reader. ☐

Problem 94. *Show that the polynomial $x^4 + 2x^2 - 1$ is irreducible over the rational numbers. Find the algebraic conjugates of $\sqrt{\sqrt{2} - 1}$, and show that this algebraic integer is <u>not totally real</u>.*

Answer. The only rational zeros of the polynomial $x^4 + 2x^2 - 1$ could be ± 1, but they are obviously not zeros. Any factorization of the polynomial into two quadratics could only be of the form

$$(x^2 + ax + 1)(x^2 - ax - 1) = x^4 - a^2 x^2 - 2ax - 1 \neq x^4 + 2x^2 - 1$$

The factorization turns out to be impossible. Hence the polynomial is irreducible. Its zeros are $\pm\sqrt{\sqrt{2} - 1}$ and $\pm i \sqrt{\sqrt{2} + 1}$, which are algebraically conjugate to each other, but not all real.

Proposition 54 (A constructible number not in the Hilbert field). *The number $z = \sqrt{\sqrt{2} - 1}$ is not totally real. It is a constructible number not in the Hilbert field.*

Problem 95. *Can one factor the polynomial $x^4 + 2x^2 - 1$ over the Gaussian integers.*

Answer. The only possibility of factoring would be

$$x^4 + 2x^2 - 1 = (x^2 + ax + i)(x^2 - ax + i) = (x^2 + i)^2 - a^2 x^2 = x^4 + (2i - a^2)x^2 - 1$$

This leads to $a^2 = 2(1 + i)$ which has no integer solution. Hence the polynomial is irreducible over the Gaussians, too.

Problem 96. *These are the steps for the construction of the figure on page 196: We draw a unit square $\square ABCD$. Let E be the point on the ray $\overrightarrow{AB}$ such that the diagonal AC is congruent to the segment AE. Let M be the midpoint of the side AB. Draw the circle around M through the point E. This circle intersects the side BC at point Y.*

Calculate exact simplified root expressions for the lengths of (i) *segment ME;* (ii) *segment AF;* (iii) *segment AY. Decide whether these segments are constructible with Hilbert tools.*

Answer. Since the diagonal has the length $|AC| = \sqrt{2}$, we get

$$|ME| = |AE| - |AM| = \sqrt{2} - \frac{1}{2}$$

196

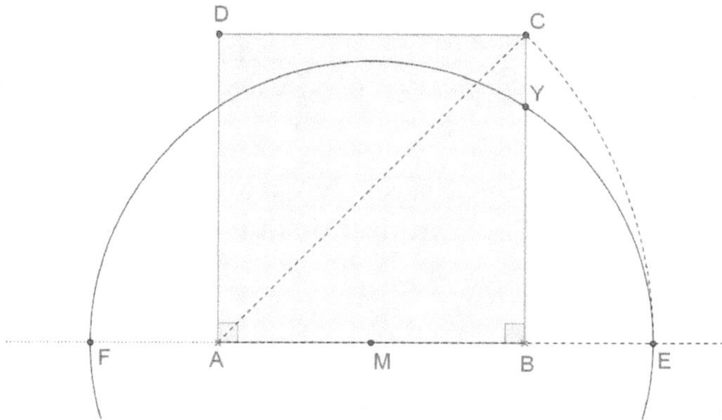

Figure 3: Find the length of segment BY, and AY exactly.

and Pythagoras' Theorem for triangle $\triangle MYB$ yields

$$BY^2 = MY^2 - MB^2 = \left(\sqrt{2} - \frac{1}{2}\right)^2 - \frac{1}{4} = 2 - \sqrt{2}$$

$$|BY| = \sqrt{2 - \sqrt{2}}$$

once more Pythagoras' Theorem, now for triangle $\triangle ABY$, yields

$$AY^2 = AB^2 + BY^2 = 1^2 + 2 - \sqrt{2} = 3 - \sqrt{2}$$

$$|AY| = \sqrt{3 - \sqrt{2}}$$

All three segment lengths are totally real and hence constructible with Hilbert tools. One see this since both expressions $\pm\sqrt{2} - \frac{1}{2}$ are real. Similarly, all four expressions $\pm\sqrt{2 \pm \sqrt{2}}$ are real, as well as all four expressions $\pm\sqrt{3 \pm \sqrt{2}}$ are real.

Problem 97. *Draw a unit square $\square ABCD$. Let E be the point on the ray $\overrightarrow{AB}$ such that the diagonal AC is congruent to the segment AE. Let N be the midpoint of the segment AE. Draw the circle with __diameter__ AE. This circle intersects the side BC at point F.*

Calculate exact simplified root expressions for the lengths of (i) *segment BF;* (ii) *segment AF. Decide whether these segments are constructible with Hilbert tools.*

Answer. Since the diagonal has the length $|AC| = \sqrt{2}$, we get $|BE| = \sqrt{2} - 1$. From

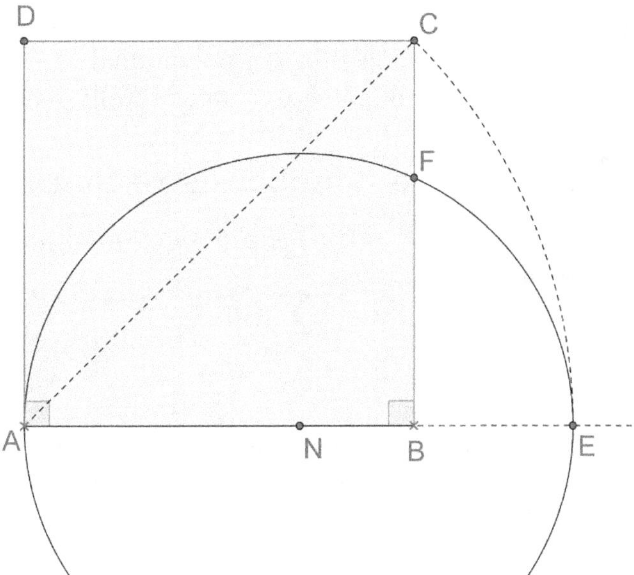

Figure 4: Find the length of segment BF exactly.

the altitude theorem for the right triangle $\triangle AFE$, one obtains

$$BF^2 = |AB| \cdot |BE| = 1 \cdot (\sqrt{2} - 1)$$
$$|BF| = \sqrt{\sqrt{2} - 1}$$

From Pythagoras' Theorem, now for triangle $\triangle ABY$, one obtains

$$AF^2 = AB^2 + BF^2 = 1^2 + \sqrt{2} - 1 = \sqrt{2}$$
$$|AF| = \sqrt[4]{2}$$

Neither one of these two segment lengths is totally real. Hence they are not constructible with Hilbert tools. But nevertheless, they are constructible with straight-edge and compass.

Take the example (i). One sees that not all four expressions $\pm\sqrt{\pm\sqrt{2} - 1}$ are real. To be completely accurate, one has still to check that these four numbers are *algebraic conjugate* to each other,—by checking that the polynomial $(x^2 + 1)^2 - 2$ is irreducible. This step completes the proof that the number $\sqrt{\sqrt{2} - 1}$ is not totally real.

Take the example (ii). One sees that not all four expressions $\pm\sqrt{\pm\sqrt{2}}$ are real. To be completely accurate, one has still to check that these four numbers are *algebraic conjugate* to each other,—by checking that the polynomial $x^4 - 2$ is irreducible. This step completes the proof that the number $\sqrt[4]{2}$ is not totally real.

V.1.4   Totally Positive Numbers Are Sums of Squares

Problem 98. *Take as base field $F = \mathbf{Q}$. Let the numbers a_i for $1 \le i \le m$ be totally real. Prove that the number*

$$b = \sum_{1 \le i \le m} a_i^2$$

is totally positive.

Solution. Let $[K/F]$ be a finite dimensional field extension such that K contains the splitting fields for all the numbers a_i for $1 \le i \le m$. Let $p \in K[x]$ be the polynomial which has all the numbers

$$(\text{V.1.2}) \qquad\qquad \sum_{1 \le i \le m} a_{j(i)}^2$$

where $a_{j(i)} \in K$ are the algebraic conjugates of a_i. Here j indeed marks a finite sequence $i \mapsto j(i)$. For each i, the index $j(i)$ may take any value from the finite set of respective conjugates. Fix any i. The coefficients of p are symmetric functions of the conjugates $a_{j(i)}$. Hence $p \in F[x]$, as one may confirm by induction on i with $1 \le i \le m$.

- By assumption b is a root of the polynomial p. Since $p \in F[x]$ this fact implies that all conjugates of b are roots of polynomial p. Hence the conjugates of b are among the numbers (V.1.2).

- Since the $a_i \in K$ are assumed to be totally real, all conjugates $a_{j(i)} \in K$ are real and nonzero. Hence all numbers (V.1.2) are strictly positive.

Together, these two facts imply that b is totally positive. $\qquad\qquad\square$

Problem 99. *Prove or disprove without using Artin's Theorem. If the number a is totally positive, then $\sqrt{a}$ is totally real.*

Proof. Let $a_j > 0$ with $1 \le j \le \deg P$ denote the conjugates of $a = a_1$ and $P(x) \in \mathbf{Z}[x]$ be the minimal polynomial

$$P(x) = \prod_{1 \le j \le \deg P} (x - a_j)$$

The polynomial $P(x^2)$ has the zeros $\pm\sqrt{a_j}$. Hence the minimal polynomial of $\sqrt{a}$ is a divisor of $P(x^2)$, which again has only real zeros. Hence $\sqrt{a}$ is totally real. $\quad\square$

Remark. By problem 89 there occur two possible cases:

- Either $P(x^2)$ is irreducible and hence all $\pm\sqrt{a_j}$ are conjugated;

- or $P(x^2) = A(x)A(-x)$ with $A(x) \in \mathbf{Z}[x]$ and no pair $\pm\sqrt{a_j}$ is conjugated.

The converse of this problem is much more interesting.

Theorem 14 (Emil Artin). *Let b be any totally positive algebraic number. Let $K/\mathbf{Q}$ be a Galois extension containing the one-element extension $\mathbf{Q}(b)$. Assume that $K \subseteq \mathbb{R}$. Then there exist finitely many totally real numbers $a_i \in K$ such that*

$$b = \sum a_i^2$$

Corollary 26 (Emil Artin). *Assume b is a totally positive algebraic number. Then there exist finitely many totally real algebraic numbers a_i such that*

$$b = \sum a_i^2$$

Proof. There exists a splitting (Galois) extension $K/\mathbf{Q}$ containing the one-element extension $\mathbf{Q}(b)$. $\quad\square$

Problem 100. *Explain why it is equivalent to write the totally positive b in the form*

$$b = r_0 + \sum r_i \cdot a_i^2$$

where the r_i are rational numbers and the a_i are totally real algebraic numbers.

Here are some easy examples: the number $10 - 2\sqrt{5}$ is totally positive. Indeed holds

$$10 - 2\sqrt{5} = 2^2 + (1 - \sqrt{5})^2$$

The number $34 - 2\sqrt{17}$ is totally positive. Indeed holds

$$34 - 2\sqrt{17} = 4^2 + (1 - \sqrt{17})^2$$

The number $85 - 19\sqrt{17}$ is totally positive. Indeed holds

$$85 - 19\sqrt{17} = \frac{1088 + \left(85 - 19\sqrt{17}\right)^2}{3230}$$

200

Problem 101. *Show that the discriminant*

$$\Delta(n,1,0) = \frac{F_n - \sqrt{F_n}}{2}$$

calculated in Problem (34) *is totally positive.*

Solution.

$$\left(\frac{1 - \sqrt{F_n}}{2}\right)^2 + \left(2^{2^{n-1}-1}\right)^2 = \frac{1 + F_n - 2\sqrt{F_n} + F_n - 1}{4} = \frac{F_n - \sqrt{F_n}}{2} = \Delta(n,1,0)$$

$\square$

Assembling the proof of Theorem 14. This proof follows mainly the ideas from Hartshorne's book, see [10] p.147. But several modifications were needed. By a *field-cone* I mean a subset of any field such that

(i) $1 \in S$ and $0 \notin S$;

(ii) If $a, b \in S$ then $a + b \in S$;

(iii) If $a, b \in S$ then $a \cdot b \in S$;

(iv) If $a \in S$ then $a^{-1} \in S$.

We define

(V.1.3) $\qquad S_1(K) = \{\sum a_i^2 : a_i \in K \text{ are any finitely many elements } a_i \neq 0\}$

Problem 102. *Check that $S_1(K)$ is a field-cone.*

Solution. Take any $a = \sum a_i^2$. By enlarging the fraction, the inverse may be written as a sum of squares

$$a^{-1} = \frac{a}{a^2} = \sum \left(\frac{a_i}{a}\right)^2$$

The other claims are even easier to check. $\square$

Problem 103. *Check that $b \in S_1(K)$ implies $b_j \in S_1(K)$ for all its algebraic conjugates. In other words, all elements of $S_1(K)$ are totally positive.*

Proof. Since $K/\mathbf{Q}$ is assumed to be a Galois extension, $b \in S_1(K) \subset K$ implies $b_j \in K$ for all its algebraic conjugates. Now we may prove that $b_j \in S_1(K)$ as done in the solution of problem 98. $\square$

Problem 104. *Using Zorn's lemma, prove there exists a __maximal__ field-cone such that $M \supseteq S_1(K)$ and $M \subset K$. Show that M induces a linear order of the field K.*

Solution. Let $\mathcal{F}$ be the family of all field-cones contained in the field K and containing $S_1(K)$. Any increasing chain $\mathcal{C} \subseteq \mathcal{F}$ has an upper bound. Indeed the union set $\cup \mathcal{C}$ is a field-cone and hence an upper bound.

Using (our belief in) Zorn's Lemma, there exists a maximal element $M \in \mathcal{F}$. Moreover $1 \in M$ and hence all positive rationals are in M.

To check that M induces a linear order of the field K, we need to confirm that for any $a \in K$ holds either $a = 0$ or $a \in M$ or $-a \in M$. To this end, assume $a \neq 0$ and $-a \notin M$. We define

$$M_a = \{x + ay : \ x \in M \text{ or } x = 0 \text{ and } y \in M \text{ or } y = 0, \text{ but } x, y \text{ are not both zero}\}$$
$$= (M \cup \{0\}) + a(M \cup \{0\}) \setminus \{0\}$$

(Almost) as in a problem above we check that M_a is a field cone. Hence the maximality implies $M_a = M$, and hence $a \in M_a = M$. $\qquad\square$

Lemma 49. *Assume that $b \notin S_1(K)$. Define*

$$\text{(V.1.4)} \qquad\qquad S_b = (S_1(K) \cup \{0\}) - b(S_1(K) \cup \{0\}) \setminus \{0\}$$

This set S_b is a field-cone containing $S_1(K)$.

Proof. Indeed $0 \in S_b$ would imply $b = s/t$ with $s, t \in S_1(K)$. Hence $b \in S_1(K)$ contrary to the assumption. Take any two elements $s - bt \in S_b$ and $s' - bt' \in S_b$. Their product is

$$(s - bt)(s' - bt') = (ss' + b^2 tt') - b(ts' + at') \in S_1 - bS_1$$

The inverse is
$$(s - bt)^{-1} = \frac{s}{(s - bt)^2} - b\frac{t}{(s - bt)^2} \in S_1 - bS_1$$

The remaining cases and claims are even easier to check. . $\qquad\square$

Problem 105. *Prove that the ordered field K occurring in Artin's theorem 14, with the order induced by the maximal cone M is Archimedean. In other words, for any $c \in M$ there exists a natural number n such that $n - c \in M$.* [12]

---

[12] It is very hard to see how this claim helps to prove Artin's theorem, as Hartshorne in his book [10] on p.147 suggests.

Proof. Assume that $c \notin \mathbf{Q}$, otherwise the claim holds anyway. Let c_j with $1 \le j \le m$ be the algebraic conjugates of c. Let $P \in \mathbf{Q}[x]$ be their minimal polynomial. Define another polynomial $Q \in \mathbf{Q}[x]$ by requiring

$$Q(x^2) = P(x)P(-x)$$

This polynomial has the positive zeros $b_j = c_j^2$. [13] We may arrange them decreasingly in their M- order. Moreover since $c_j \ne 0$, the total order induced by M implies $c_j^2 >_M 0$. Since the sum $\sum c_j^2$ is rational and positive, there exists a natural number n such that

$$0 < \sum_{1 \le j \le m} b_j < n \quad \text{and hence } b_m < \frac{n}{m}$$

We say that the Archimedean property holds for $b > 0$ if there exists a natural number such that $b < n$. Assume towards a contradiction that the Archimedean property holds for $b_m \ldots b_{p+1}$ but fails to hold for $b_p \ldots b_1$ where $1 \le p < m$. For any large enough natural number n would hold

$$2n < \sum_{1 \le j \le p} b_j$$

$$\sum_{1 \le j \le m} b_j < n$$

$$\text{Hence } \sum_{p < j \le m} b_j = \sum_{1 \le j \le m} b_j - \sum_{1 \le j \le p} b_j < -n$$

This is a contradiction unless the last sum is void. Hence all b_j and all c_j have the Archimedean property. $\qquad\square$

Problem 106. *Show that between any two numbers in K lies a rational number.*

Proof. Choose a positive integer q such that $\frac{1}{d-c} < q$. There exists a smallest integer p such that $qc < p$. Hence $p - 1 \le qc < p$ and

$$c < \frac{p}{q} \le c + \frac{1}{q} < d$$

as to be shown. $\qquad\square$

---

[13]In the case that conjugates c_j exist which are negative of each other, the polynomial Q is reducible.

Problem 107. *The maximal field-cone M is in fact not as mysterious as its construction based on the axiom of choice may suggest. With this goal in mind, check the following claims.*

(i) *Any field-cone M in the field K is maximal if and only if it induces a linear order of K.*

(ii) *Since $\mathbf{Q}^{+} \subset M$, the standard order of the rationals agrees with the order induced by the field-cone M.*

(iii) *For any $x \in K \supseteq \mathbf{Q}$, we define its Dedekind subclass to be*

$$S(x) = \{z \in \mathbf{Q} : x - z \in M\}$$

The order induced by the field-cone may be expressed by its Dedekind subclasses. Check that for any $x, y \in K$ holds:

$$x <_M y \Leftrightarrow y - x \in M \Leftrightarrow S(x) \subsetneq S(y)$$

(iv) *On the other hand, it is well know that the order given by the Dedekind subclasses is the standard order of the real numbers:*

$$x <_{\mathbb{R}} y \Leftrightarrow S(x) \subsetneq S(y)$$

Hence the order induced by M agrees with the standard order of the real numbers. Especially holds

$$M = \{x \in K : x >_{\mathbb{R}} 0\}$$

and hence the maximal field-cone is uniquely determined by the choice of the field K.

Hints to the solution. **(i)** Assume the field-cone $M \subset K$ induces a linear order of K. Take any possibly larger field-cone $M_1 \supseteq M$, and assume towards a contradiction that there exists an $a \in M_1 \setminus M$. By the linear order holds either $a = 0$ or $-a \in M$. Indeed holds $-a \in M$ since the first case is excluded by the assumption $a \in M_1$. Hence $-a \in M \subseteq M_1$ and $0 = a + (-a) \in M_1$ contradicting the assumption that M_1 is a field-cone.

Conversely, any maximal field-cone $M \subset K$ induces a linear order of K, as already has been checked above..

204

(iii) For any $x \in K \supseteq \mathbf{Q}$, we define its Dedekind subclass to be

$$S(x) = \{z \in \mathbf{Q} \ : \ x - z \in M\}$$

It is straightforward to check that for any $x, y \in K$ holds:

$$x <_M y \Rightarrow y - x \in M \Rightarrow S(x) \subset S(y)$$

Moreover, since $x <_M y$ implies $x \neq y$, problem 106 yields a rational number $z \in \mathbf{Q}$ for which hold $x <_M z <_M y$ and hence $z \in S(y) \setminus S(x)$ and $S(x) \subsetneq S(y)$. Similarly, $y <_M x$ implies $S(y) \subsetneq S(x)$. Since the order by induced by M is linear, the converse implications hold, too.

$\square$

We now take refuge to an indirect argument.

End of the proof of Theorem 14. Take any totally positive element $b \in K$, and assume towards a contradiction that $b \notin S_1(K)$. By Lemma 49 the set

$$S_b = (S_1(K) \cup \{0\}) - b(S_1(K) \cup \{0\}) \setminus \{0\}$$

is a field-cone containing $S_1(K)$. Let M be a maximal field-cone containing S_b. Hence $-b \in S_b \subseteq M$.

On the other hand, item (iv) of problem 107 yields

$$M = \{x \in K \ : \ x >_{\mathbb{R}} 0\}$$

and hence $b \in M$. Now both $b \in M$ and $-b \in M$ imply $0 \in M$, which is a contradiction.

We may conclude from this indirect proof that there cannot exist any totally positive element $b \in K$ such that $b \notin S_1(K)$.

$\square$

$\square$

Problem 108. *Convince yourself that for all elements $b \in \Omega$ of the Hilbert field, all its algebraic conjugates lie in Ω. Hence the following is true*

- *All elements $b \in \Omega$ are totally real.*

- *The Hilbert field contains the splitting extension of any element $b \in \Omega$.*

- *The extension $\Omega/\mathbf{Q}$ is a Galois extension containing the one-element extension $\mathbf{Q}(b)$ for any $b \in \Omega$.*

Corollary 27 (For Emil Artin). *Any totally positive element of the Hilbert field Ω may be written as a finite sum of squares of elements from Ω.*

Similarly, any totally positive element of the constructible field may be written as a finite sum of squares of elements from the constructible field.

A short proof of Corollary 27. We may choose $K = \Omega$ in Artin's Theorem 27 to be the totally real field K occur there. Hence there exist finitely many totally real numbers $a_i \in \Omega$ such that

$$b = \sum a_i^2$$

$\square$

Problem 109. *Is the field K occurring in Artin's theorem !14 always larger or equal to the Hilbert field Ω, or always smaller or equal to Ω?*

Proposition 55 (A bid more than Artin's Theorem). *Let $H/\mathbf{Q}$ be a splitting (Galois) extension and assume that H is a subfield of the constructible field. [14] Then $H \subset \Omega$ is even a subfield of the Hilbert field.*

Proof. Fix any number $k \in H$. Because of the universal splitting property, the splitting field H contains all conjugates k_j of $k = k_1$. Too, the field H is a totally real subfield of the constructible field. Thus the assumptions imply

$$\mathbf{Q}(k) \subseteq H \subseteq K \subset \mathbb{R}$$

We go through the construction process from Corollary 51, for all conjugates k_j. Begin with k_1. The numbers obtained up to the i-th step of the construction process generate a field $\mathbf{F}_i$. Clearly $\mathbf{F}_{i-1} \subseteq \mathbf{F}_i$ for all construction steps. Either the i-th step involve only field operation in which case $\mathbf{F}_{i-1} = \mathbf{F}_i$, or the i-th step is an operation of the type $a > 0 \mapsto \sqrt{a}$. In the latter case

$$a \in \mathbf{F}_{i-1}\,, a > 0\,, \quad \mathbf{F}_i = \mathbf{F}_{i-1}(\sqrt{a}) \quad \text{and} \quad [\mathbf{F}_i : \mathbf{F}_{i-1}] = 2$$

There exists a tower of finitely many two dimensional extensions

$$\mathbf{Q} = \mathbf{F}_0 \subseteq \mathbf{F}_1 \subseteq \mathbf{F}_2 \subseteq \cdots \subseteq \mathbf{F}_n \ni k_1$$

We may intersect all fields of this tower with the splitting field H and discard the steps with equality $\mathbf{F}_{i-1} \cap H = \mathbf{F}_i \cap H$. Next take k_2. We extend the field $\mathbf{F}_n$ by

---

[14]By Artin's Theorem any totally positive element of H may be written as a finite sum of squares of elements from H.

adjoining from the tower for k_2 the first $\sqrt{a} \notin \mathbf{F}_n$, next the second one, and so on. One obtains a (possibly) longer tower the last field from which contains both k_1 and k_2, One repeats the process for all conjugates k_j. Finally we have constructed a tower of subfields of H of finitely many two dimensional extensions

$$\mathbf{Q} = \mathbf{F}_0 \subset \mathbf{F}_1 \subset \mathbf{F}_2 \subset \cdots \subset \mathbf{F}_n = H \ni \text{ all conjugates of } k$$

I claim that for $1 \leq i \leq n$ holds $\mathbf{F}_i = \mathbf{F}_{i-1}(\sqrt{D_{i-1}})$ with $D_{i-1} \in \mathbf{F}_{i-1}$ and that D_{i-1} is totally positive.

Reason. From $\mathbf{F}_i = \mathbf{F}_{i-1}(\sqrt{a})$ we get $\sqrt{a} \in \mathbf{F}_i$ and $a > 0$ and $a \in \mathbf{F}_{i-1} \subset \mathbf{F}_{n-1}$.

Let a_j with $1 \leq j \leq \deg P$ denote the conjugates of $a = a_1$ and $P(x) \in \mathbf{Z}[x]$ the minimal polynomial

$$P(x) = \prod_{1 \leq j \leq \deg P} (x - a_j)$$

Hence $P(x^2)$ has the $2 \deg P$ roots, which are just $\pm\sqrt{a_j}$ with $1 \leq j \leq \deg P$. Since $\sqrt{a_1} \in \mathbf{F}_n = H$, the universal splitting property implies

$$\pm\sqrt{a_j} \in H \text{ for all } 1 \leq j \leq \deg P.$$

Hence all $a_j = (\sqrt{a_j})^2 \in H$ are squares and hence a is totally positive. $\qquad\square$

Inductively, we check that $\mathbf{F}_i \subset \Omega$ holds for all $0 \leq i \leq n$.
Induction start: Is it true that $F_0 = \mathbf{Q} \subset \Omega$ and $F_1 = \mathbf{Q}(\sqrt{D_0}) \subset \Omega$ since D_0 is a positive rational.
Induction step: Assume that $\mathbf{F}_{i-1} \subset \Omega$ for some $i \geq 1$. Is it true that $\mathbf{F}_i \subset \Omega$?

By construction $\mathbf{F}_i = \mathbf{F}_{i-1}(\sqrt{D_{i-1}})$ where D_{i-1} is totally positive and $D_{i-1} \in \mathbf{F}_{i-1} \subset \Omega$ by the induction assumption. By Corollary 27 "For Emil Artin" the discriminant D_{i-1} is a sum of squares from Hilbert field Ω. Hence the definition of Ω yields $\sqrt{D_{i-1}} \in \Omega$. Hence $\mathbf{F}_i = \mathbf{F}_{i-1}(\sqrt{D_{i-1}}) \subset \Omega$. $\qquad\square$

Problem 110. *Prove or disprove. If the number k is totally positive and constructible, then k and $\sqrt{k}$ are in the Hilbert field.*

My solution. Choose a definite root expression for k. There exists a polynomial $R(x) \in \mathbf{Z}[x]$ with the root k and all roots of $R(x)$ are obtained by sign changes in front of the square roots occurring inside the root expression for k. The minimal polynomial $P(x)$ of k is a divisor of $R(x)$. Hence all conjugates of k are obtainable by sign changes in front of the square roots in the root expression for k. [15]

---

[15] But there may be very well exist sign changes which produce an expression not conjugated.

Let H be the splitting field of P. By assumption all conjugates k_j are positive, hence real. Because of the above remark, all k_j are constructible. Hence H is a subfield of the constructible field. Now Proposition 55 implies that $k \in H \subset \Omega$. By Corollary 27 "For Emil Artin" the discriminant k is a sum of squares from Hilbert field Ω. Hence the definition of Ω yields $\sqrt{k} \in \Omega$. $\qquad\square$

Problem 111. *Prove or disprove. If a constructible number k is totally real, the splitting field H containing k and its conjugates is even contained in the Hilbert field.*

Proof. Let $H/\mathbf{Q}$ be a splitting (Galois) extension for k. By assumption holds $H \subset K \subset \mathbb{R}$.. By "A bid more than Artin's Theorem" 55 holds $H \subset \Omega$. $\qquad\square$

Here are more tricky examples. The number $5 - \sqrt{5} - \sqrt{10 - 2\sqrt{5}}$ is totally positive. Indeed holds

$$5 - \sqrt{5} - \sqrt{10 - 2\sqrt{5}} = \frac{3}{8} + 2\left(\frac{2 - \sqrt{5}}{4}\right)^2 + \left(1 - \sqrt{\frac{5 - \sqrt{5}}{2}}\right)^2$$

Problem 112. *The discriminant*

$$disrothe = 17 + 3s_1\sqrt{17} - s_1 \cdot s_2\sqrt{2(85 + 19s_1\sqrt{17})}$$

$$= 17 + 3s_1\sqrt{17} - s_2\sqrt{34 - 2s_1\sqrt{17}} - 2s_1s_2\sqrt{34 + 2s_1\sqrt{17}}$$

$$= 17 + 3s_1\sqrt{17} - \frac{s_2}{2}\sqrt{34 - 2s_1\sqrt{17}} + \frac{s_1s_2}{2}\sqrt{17(34 - 2s_1\sqrt{17})} - 4s_1s_2\sqrt{34 + 2s_1\sqrt{17}}$$

with the signs s_1, $s_2 = \pm 1$ is totally positive. Write this discriminant as a sum of squares.

Solution. Indeed holds

$$disrothe = \frac{3}{38}\left(\frac{19}{3} - s_1s_2\sqrt{2(85 + 19s_1\sqrt{17})}\right)^2 + \frac{47}{114}$$

$$= 114\left(\frac{19 - 3s_1s_2\sqrt{2(85 + 19s_1\sqrt{17})}}{114}\right)^2 + \frac{47}{114}$$

$\qquad\square$

The following mathematica file gives some hint how I solved these two examples.

```
In[1]:= 5 - Sqrt[5] - Sqrt[10 - 2 Sqrt[5]];

In[2]:= ExpandAll[(1-Sqrt[a])^2 + (-1+a)] /. {a -> (5-Sqrt[5])/2}
Out[2]= 5 - Sqrt[5] - Sqrt[2 (5 - Sqrt[5])]

In[3]:= (a - 1) /. {a -> (5 - Sqrt[5])/2}
Out[3]= -1 + 1/2 (5 - Sqrt[5])

In[4]:= (3-Sqrt[5])/2 ;;Expand[(1-Sqrt[5]/2)^2/2 + 3/8]
Out[4]= 1/2 (3 - Sqrt[5]) ;; 3/2 - Sqrt[5]/2

In[5]:= ExpandAll[
 3/8 + 2 (1/2 -Sqrt[5]/4)^2+ (1-Sqrt[1/2 (5-Sqrt[5])])^2]
Out[5]= 5 - Sqrt[5] - Sqrt[2 (5 - Sqrt[5])]

In[6]:= 34 - 6 Sqrt[17] - 2 Sqrt[2 (85 - 19 Sqrt[17])];

In[7]:= Expand[(1/x - x Sqrt[2 (85 - 19 Sqrt[17])])^2
 - 1/x^2 -2 x^2 (85-19 Sqrt[17]) + 34 - 6 Sqrt[17]]

Out[7]= 34 - 6 Sqrt[17] - 2 Sqrt[2 (85 - 19 Sqrt[17])]

In[8]:= Expand[-1/x^2 - 2 x^2 (85 - 19 Sqrt[17])
 +34-6 Sqrt[17]] /.  {x^2 -> 3/19, (1/x^2) -> 19/3}
Out[8]= 47/57

In[9]:= Expand[(Sqrt[19/3] -
    Sqrt[3/19] Sqrt[2 (85 - 19 Sqrt[17])])^2 + 47/57]

Out[9]= 34 - 6 Sqrt[17] - 2 Sqrt[2 (85 - 19 Sqrt[17])]

In[259]:=
rothe = (3/38) (19/3
  - s1*s2 Sqrt[2 (85 - s1*19 Sqrt[17])])^2 + 47/114;

In[263]:= Expand[rothe] /. {s1^2 -> 1, s2^2 -> 1}

Out[263]= 17 - 3 Sqrt[17] s1^3
```

```
-   Sqrt[2] s1 Sqrt[85 - 19 Sqrt[17] s1] s2
```

Lemma 50. *Make the assumptions from Artin's theorem, and assume that $f \in S_1(K)$,*
$2a \in S_1(K)$ and $a^2 - f \in S_1(K)$ and $\sqrt{f} \in K$. Then holds even $a \pm \sqrt{f} \in S_1(K)$.

Proof.
$$a + \sqrt{f} = (2a) \cdot \left(\frac{a + \sqrt{f}}{2a}\right)^2 + \frac{(2a) \cdot (a^2 - f)}{(2a)^2}$$

The right-hand side can easily be checked to be a sum of squares from the field K.
The same identity holds with $\sqrt{f}$ replaced by $-\sqrt{f}$. $\qquad\square$

Lemma 51 (Following an idea by Hilbert). *Assume that $f \in S_1(\Omega), 2a \in S_1(\Omega)$ and*
$a^2 - f \in S_1(\Omega)$ have been produced as sums of squares of totally real numbers. Then
$a \pm \sqrt{f} \in S_1(\Omega)$ is a sum of squares of totally real numbers.

Problem 113. *Let n be a natural number and $\frac{p}{q}$ be any rational number such that*
$q > 0$ and
$$\frac{p}{q} > \sqrt{n}$$

Show that $\frac{p}{q} - \sqrt{n}$ is totally positive. Hence Artin's theorem holds for the field $\mathbf{Q}(\sqrt{n})$.

Solution.
$$\frac{p}{q} - \sqrt{n} = 2pq \left[\left(\frac{p - q\sqrt{n}}{2pq}\right)^2 + \frac{(p^2 - nq^2)}{(2pq)^2}\right]$$

is a sum of squares of numbers from $\mathbf{Q}(\sqrt{n})$ since $p^2 - nq^2 \geq 0$ is a natural number. $\quad\square$

Problem 114. *Find a simpler solution of problem 112 by means of the above lemma 51.*
Again write the discriminant actually as a sum of squares.

Solution. Put the discriminant with signs $s_1 = s_2 = 1$

$$disrothe = 17 + 3s_1\sqrt{17} - s_1 \cdot s_2\sqrt{2(85 + 19s_1\sqrt{17})} =: a - \sqrt{f}$$

and check that $4a$ and $a^2 - f$ are sums of squares.

$$4a = 5 + 3(2 + \sqrt{17})^2$$

$$a^2 - f = (17 + 3\sqrt{17})^2 - 2(85 + 19\sqrt{17}) = \frac{16}{9}(4 + (9 + 2\sqrt{17})^2)$$

With a bid of care, we get $a - \sqrt{f}$ as a sum of squares:

$$a - \sqrt{f} = 2 \cdot (4a) \cdot \left(\frac{a - \sqrt{f}}{4a}\right)^2 + \frac{2 \cdot (4a) \cdot (a^2 - f)}{(4a)^2}$$

$$= 2 \cdot (5 + 3(2 + \sqrt{17})^2) \cdot \left(\frac{17 + 3\sqrt{17} - \sqrt{2(85 + 19\sqrt{17})}}{(5 + 3(2 + \sqrt{17})^2)}\right)^2$$

$$+ \frac{32 \cdot (5 + 3(2 + \sqrt{17})^2) \cdot (4 + (9 + 2\sqrt{17})^2)}{9(5 + 3(2 + \sqrt{17})^2)^2}$$

which now be distributed keeping itself as a sum of squares. $\square$

Remark. One may finally check the identity

$$disrothe = 114 \left(\frac{19 - 3s_1 s_2 \sqrt{2(85 + 19s_1\sqrt{17})}}{114}\right)^2 + \frac{47}{114}$$

$$= 2\left(5 + 3\left(2 + s_1\sqrt{17}\right)^2\right)\left(\frac{17 + 3s_1\sqrt{17} - s_1 s_2\sqrt{2\left(85 + 19s_1\sqrt{17}\right)}}{5 + 3\left(2 + s_1\sqrt{17}\right)^2}\right)^2$$

$$+ \frac{32\left(5 + 3\left(2 + s_1\sqrt{17}\right)^2\right)\left(4 + \left(9 + 2s_1\sqrt{17}\right)^2\right)}{9\left(5 + 3\left(2 + s_1\sqrt{17}\right)^2\right)^2}$$

It is an open secret that I have used mathematica for more effective simplification.

Problem 115. *Using the formulas*

$$S(n, 2, 0) = \frac{-1 + \sqrt{F_n} + \sigma(n, 1, 0)\sqrt{2F_n - 2\sqrt{F_n}}}{4}$$

$$S(n, 2, 2) = \frac{-1 + \sqrt{F_n} - \sigma(n, 1, 0)\sqrt{2F_n - 2\sqrt{F_n}}}{4}$$

$$S(n, 2, 1) = \frac{-1 - \sqrt{F_n} + \sigma(n, 1, 1)\sqrt{2F_n + 2\sqrt{F_n}}}{4}$$

$$S(n, 2, 2) = \frac{-1 - \sqrt{F_n} - \sigma(n, 1, 1)\sqrt{2F_n + 2\sqrt{F_n}}}{4}$$

obtained in Problem 34 to show that the quantities $S(n, 2, l) + 2^{2^n - 2}$ and $,2^{2^n - 2} - S(n, 2, l)$ with $l = 0, 1, 2, 3$ are totally positive, and write them as a sum of real squares.

Solution. I just take the case $l = 0$, the other cases are algebraically conjugated. Define a and f by

$$4S(n, 2, 0) + 2^{2^n} = F_n - 2 + \sqrt{F_n} + \sigma(n, 1, 0)\sqrt{2F_n - 2\sqrt{F_n}} =: a + \sigma(n, 1, 0)\sqrt{f}$$

Check that $a^2 - f$ and a are sums of squares

$$a^2 - f = (F_n - 2 + \sqrt{F_n})^2 - 2F_n + 2\sqrt{F_n} = 2^{2^n}\left(-4 + F_n + 2\sqrt{F_n}\right)$$

$$= 2^{2^n - 2}\left(-32 + 3 \cdot F_n + (4 + \sqrt{F_n})^2\right)$$

$$a = 2^{2^n} - 1 + \sqrt{F_n} = \frac{3 * (F_n - 4) + (2 + \sqrt{F_n})^2}{4}$$

Now the lemma 51 implies that $4S(n, 2, \{0, 2\}) + 2^{2^n} = a \pm \sqrt{f}$ are sums of squares.
  Similarly, we get the lower bound: This, time we define a and f by

$$2^{2^n} - 4S(n, 2, 0) = F_n - \sqrt{F_n} - \sigma(n, 1, 0)\sqrt{2F_n - 2\sqrt{F_n}} =: a - \sigma(n, 1, 0)\sqrt{f}$$

Check that $a^2 - f$ and a are sums of squares

$$a^2 - f = (F_n - \sqrt{F_n})^2 - 2F_n + 2\sqrt{F_n} = 2^{2^n - 2}\left(-16 + 3 \cdot F_n + (4 - \sqrt{F_n})^2\right)$$

$$a = F_n - \sqrt{F_n} = \frac{3 * F_n - 4 + (2 - \sqrt{F_n})^2}{4}$$

Now the lemma 51 implies that $2^{2^n} - 4S(n, 2, \{0, 2\}) = a \pm \sqrt{f}$ are sums of squares. $\square$

Problem 116 (The next hard problem). *Find for any arbitrary Fermat prime the representations of $1 \pm \cos\frac{2\pi l}{F_n}$ and $\sin^2\frac{2\pi l}{F_n}$ as a sum of squares from the Hilbert field.*

After I failed to solve problem 30 for an arbitrary Fermat prime, I now go ahead and tackle,—using the power of mathematica,—the above problem at least for the 17-gon. At first sight, one may guess that already repeated use of lemma 51 would lead to the solution. That is not the case, the appearance of sums of two or more boxed roots of the same type becomes a real nuisance. Already the solution of the simpler problem 112 depended on the reduction of my formula from Proposition (17), or alternatively Gauss' formula

(III.2.2)

$$16 \cos\frac{2\pi l}{17} = -1 + s_1\sqrt{17} + s_2\sqrt{34 - 2s_1\sqrt{17}}$$

$$+ 2s_2s_3\sqrt{17 + 3s_1\sqrt{17} - s_2\sqrt{34 - 2s_1\sqrt{17}} - 2s_1s_2\sqrt{34 + 2s_1\sqrt{17}}}$$

212

to the simpler formula

$$(\text{II}.0.5) \qquad 16\cos\frac{2\pi l}{17} = -1 + s_1\sqrt{17} + s_2\sqrt{34 - 2s_1\sqrt{17}}$$

$$+ \; 2s_2 s_3 \sqrt{17 + 3s_1\sqrt{17} - s_1 s_2 \sqrt{2(85 + 19s_1\sqrt{17})}}$$

Such a reduction I have only achieved with the help of mathematica, by means of the following trick: At first calculate the minimal polynomial for the respective sum of roots, and then solve for the roots of the minimal polynomial. Mathematica turns out to be clever enough to return to me a simpler formula for the respective sum of roots. This trick is in the present case only needed for equations of degree 4, where it works pretty well.

Proposition 56 (The 17-gon with Hilbert tools). *The notation $\pm l = 1\ldots 8$ and $s_1, s_2, s_3 = \pm 1$, is used as in Proposition 17. We obtain eight values as sums of squares*

$$16 + 16\cos\frac{2\pi l}{17} = 2a\left(\frac{2b\left(16 + (b + s_2\sqrt{g})^2\right)}{(ab)^2} + \left(\frac{a + s_2 s_3\sqrt{f}}{2a}\right)^2\right)$$

with

$$32a = 115 + 5\left(4 + s_1\sqrt{17}\right)^2 + \left(8 + 2s_2\sqrt{34 - 2s_1\sqrt{17}}\right)^2,$$

$$2b = 8 + (1 + s_1\sqrt{17})^2$$

$$f = 4 \cdot disrothe, \quad g = \frac{103 + (13 + 2\sqrt{17})^2}{2},$$

$$disrothe = 114\left(\frac{19 - 3s_1 s_2\sqrt{2(85 + 19s_1\sqrt{17})}}{114}\right)^2 + \frac{47}{114}$$

and similarly

$$16 - 16 \cos \frac{2\pi l}{17} = 2a'A \quad \text{with}$$

$$A := \frac{2b'\left(64 + 9\left(b' - s_2\sqrt{g'}\right)^2 + 16\left(9 - 2s_1\sqrt{17}\right)^2\right)}{(3a'b')^2} + \left(\frac{a' - s_2 s_3\sqrt{f}}{2a'}\right)^2$$

$$\text{with} \quad 32a' = 245 + 3\left(4 - s_1\sqrt{17}\right)^2 + \left(8 - 2s_2\sqrt{34 - 2s_1\sqrt{17}}\right)^2,$$

$$12\,b' = 55 + \left(9 - 2s_1\sqrt{17}\right)^2$$

$$f = 4 \cdot disrothe, \ g' = \frac{47 + \left(19 - 4s_1\sqrt{17}\right)^2}{4}$$

and hence get

$$256 \sin^2 \frac{2\pi l}{17} = \left(16 + 16 \cos \frac{2\pi l}{17}\right)\left(16 - 16 \cos \frac{2\pi l}{17}\right)$$

as a sum of squares, too.

Corollary 28. *The regular 17-gon is constructible with Hilbert tools. The actual steps for such a construction are straightforward to spell out using the formulas of Proposition 56.*

Artin's theorem is not restricted to the world of constructible numbers. Here an example with a polynomial of order 3, derived from proposition 80 below with $p = 4$, $q = 4$, $r = 4$. The polynomial equation $-8 + 36x - 12x^2 + x^3 = 0$ has the three roots

$$x_1 = 4 + 4\cos\frac{2\pi}{9}, \ x_2 = 4 - 2\sqrt{3}\sin\frac{2\pi}{9} - 2\cos\frac{2\pi}{9}, \ x_3 = 4 + 2\sqrt{3}\sin\frac{2\pi}{9} - 2\cos\frac{2\pi}{9}$$

which are positive and real. Indeed holds

$$x_i = \frac{1}{2}(x_{i+1} - 4)^2 \quad \text{for } i = 1, 2, 3 \mod 3$$

$$4 + 4\cos\frac{2\pi}{9} = \frac{1}{2}\left(2\sqrt{3}\sin\frac{2\pi}{9} + 2\cos\frac{2\pi}{9}\right)^2$$

$$4 - 2\sqrt{3}\sin\frac{2\pi}{9} - 2\cos\frac{2\pi}{9} = \frac{1}{2}\left(2\sqrt{3}\sin\frac{2\pi}{9} - 2\cos\frac{2\pi}{9}\right)^2$$

$$4 + 2\sqrt{3}\sin\frac{2\pi}{9} - 2\cos\frac{2\pi}{9} = 8\cos^2\frac{2\pi}{9}$$

Corollary 29. *There exist totally real algebraic numbers.which do not lie in the Hilbert field Ω.*

V.1.5   Coordinates of Regular Polygons as Sums of Squares

I want now to address the question to write the coordinates of regular polygons as sums of squares from a suitable field.

Lemma 52. *For any odd $m = 2n + 1 \geq 3$ or even $m = 2n \geq 4$, the splitting field of the integer monic polynomial $\Psi_m(u)$ is the one-element extension $\mathbf{Q}(\cos \frac{2\pi}{m})$. This vector space over $\mathbf{Q}$ is spanned, too, by the roots of $\Psi_m(u)$ which are*

$$2\cos\frac{2\pi k}{m} \quad for \ 1 \leq k < \tfrac{m}{2} \ and \ \gcd(k, m) = 1.$$

The dimension of this field extension is $[\mathbf{Q}(\cos\frac{2\pi}{m}) : \mathbf{Q}] = \frac{\phi(m)}{2}$.

Proof. Since the polynomial $\Psi_m(u)$ is irreducible, the one-element extension $\mathbf{Q}(\cos\frac{2\pi}{m})$ has the dimension $\deg \Psi_m = \frac{\phi(m)}{2}$. For all natural k, the numbers

$$\cos\frac{2\pi k}{m} = T_k\left(\cos\frac{2\pi}{m}\right)$$

lie in this field extension $\mathbf{Q}(\cos\frac{2\pi}{m})/\mathbf{Q}$. Especially, all the zeros of $\Psi_m(u)$ as obtained by equation (IV.4.15) lie in this field extension. Hence the one element extension is already a splitting extension of $\Psi_m(u)$.

Moreover, the zeros are linearly independent since they span the splitting extension, and their number is just the dimension of the extension. $\qquad\square$

Problem 117. *We may check linear independence directly, too. Assume with integer coefficients r_k holds*

$$\sum_{1 \leq k < \frac{m}{2} \ and \ \gcd(k,m)=1} r_k \cos\frac{2\pi k}{m} = 0$$

and prove that all $r_k = 0$.

Problem 118. *Find all values for $m \geq 3$ for which the above field extension has dimension $1, 2, 3$, respectively. Use Lemma 9 to confirm that your lists are complete.*

Solution.

$$[\mathbf{Q}(\cos\tfrac{2\pi}{m}) : \mathbf{Q}] = 1 \Leftrightarrow \phi(m) = 2 \Leftrightarrow m = 3, 4, 6$$

$$[\mathbf{Q}(\cos\tfrac{2\pi}{m}) : \mathbf{Q}] = 2 \Leftrightarrow \phi(m) = 4 \Leftrightarrow m = 5, 8, 10, 12$$

$$[\mathbf{Q}(\cos \tfrac{2\pi}{m}) : \mathbf{Q}] = 3 \Leftrightarrow \phi(m) = 6 \Leftrightarrow m = 7, 9, 14, 18$$

From Lemma 9 we know that $6 = \phi(m) \geq \sqrt{m}$. Hence it is enough to check the values $m \leq 36$. $\qquad\square$

Lemma 53. *Let $m \geq 3$. The numbers $\cos \tfrac{2\pi k}{m}$ for $1 \leq k < \tfrac{m}{2}$ and $\gcd(k, m) = 1$ are a complete set of algebraic conjugates.*

For any odd $m = 2n + 1 \geq 3$ or even $m = 2n \geq 4$, the splitting field of the integer monic polynomial $U_m \left(\tfrac{\sqrt{2+u}}{2} \right)$ is equal to the splitting field of its divisor $\Psi_m(u)$.

Corollary 30 ("Not enough to be famous"). *For any number $m \geq 3$ and all $1 \leq l < \tfrac{m}{2}$ there exist representations of $1 \pm \cos \tfrac{2\pi l}{m}$ and $\sin^2 \tfrac{2\pi l}{m}$ as a sum of squares from the splitting field of the integer monic polynomial $\Psi_m(u)$.*

Proof. Let K be the splitting field of $\Psi_m(u)$. Since $K/\mathbf{Q}$ is an algebraic Galois extension and $K \subseteq \mathbb{R}$, the universal splitting property implies that any $b \in K$ is totally real. By Artin's theorem 26, any totally positive $b \in K$ it has a representation

$$b = \sum a_i^2$$

with all finitely many totally real numbers $a_i \in K$. $\qquad\square$

Lemma 54. *If the splitting field K of Ψ_m is contained in the constructible field, then m is a product of different Fermat primes.*

Proof. There exist a tower of field extensions all of dimension two. The tower theorem and problem 28 imply that $\phi(m)$ is a power of two. By problem 28, this implies that m is a square-free product of Fermat primes. $\qquad\square$

Theorem 15 (Abstract constructibility with Hilbert tools). *Conversely, assume that m is a product of different Fermat primes. Then the splitting field H of Ψ_m is contained in the Hilbert field. There exists representations of $1 \pm \cos \tfrac{2\pi l}{m}$ and $\sin^2 \tfrac{2\pi l}{m}$ as a sum of squares from the Hilbert field.*

Hence we have shown that,—at least in principle,—the above given regular m-gons are constructible with Hilbert tools.

Proof. The Gauss-Wantzel Theorem 1 implies that the splitting field H of Ψ_m is contained in the constructible field.

The field H has as a basis the roots of $\Psi_m(u)$ which are

$$2 \cos \frac{2\pi k}{m} \quad \text{for } 1 \leq k < \tfrac{m}{2} \text{ and } \gcd(k, m) = 1.$$

as shown in lemma (52). The universal splitting property implies that H is a totally real field. Proposition 55 "A bid more than Artin's Theorem" yields even $H \subset \Omega$.

The numbers $1 \pm \cos \frac{2\pi l}{m}$ and $\sin^2 \frac{2\pi l}{m}$ for any $l \neq m/2$ are totally positive and contained in the Hilbert field. By Artin's Theorem there exists representations of $1 \pm \cos \frac{2\pi l}{m}$ and $\sin^2 \frac{2\pi l}{m}$ as a sum of squares from the Hilbert field. $\qquad\square$

Remark. Since the above proof of Artin's theorem gives no hint how to *actually construct* the representation of a totally positive algebraic number as a sum of squares,— and neither have I been able to find any algorithmic solution of this problem in general, Except for the pentagon and the 17-gon, the mere more than astronomic size of the problem makes it impossible to solve the problem by simply pointing out the solution. Therefore I have called theorem 15 "Abstract constructibility with Hilbert tools".

I allow myself to state the following.

Corollary 31 (The pessimistic conjecture). *For all Fermat primes F_n, the regular F_n-gon is constructible with Hilbert tools. But for the 257-gon or the $65\,537$-gon, the actual steps for such a construction can very likely not been spelled out.*

Part VI

Using Complex Numbers

VI.1   Arithmetic of the Complex Numbers

Definition 34 (Gaussian integer). A complex number with integer real- and imaginary part is called a *Gaussian integer*.

VI.1.1   Arctan Identities

What is remarkable about the following products?

(a) $(2+i) \cdot (3+i) = 5 + 5i$

(b) $(2+i)^2 \cdot (7-i) = 25 + 25i$

(c) $(3+i)^2 \cdot (7+i) = 50 + 50i$

(d) $(5+i)^4 \cdot (239-i) = 114244 + 114244i$

Each time, the real- and imaginary parts turn out to be equal. Do there exists more examples with this special property?

For any positive $x > 0$, the principal argument of the complex number $x + i$ is $\operatorname{Arg}(x+i) = \arctan \frac{1}{x}$. Too, $\operatorname{Arg}(1+i) = \arctan 1 = \frac{\pi}{4}$. Taking the arguments of the formulas (a) through (d), one gets:

$$\text{(a)} \qquad \arctan \frac{1}{2} + \arctan \frac{1}{3} = \frac{\pi}{4}$$

$$\text{(b)} \qquad 2 \arctan \frac{1}{2} - \arctan \frac{1}{7} = \frac{\pi}{4}$$

$$\text{(c)} \qquad 2 \arctan \frac{1}{3} + \arctan \frac{1}{7} = \frac{\pi}{4}$$

$$\text{(d)} \qquad 4 \arctan \frac{1}{5} + \arctan \frac{1}{239} = \frac{\pi}{4}$$

Of course, left- and right-hand side of these formulas could still differ by an integer multiple of 2π. But it is clear that this cannot happen in the given four examples. Do there exist more similar formulas?

Open Problem (May be difficult). *How many solutions has the equation*

$$\text{(VI.1.1)} \qquad\qquad c^2 + 1 = 2N^2$$

with natural numbers N and c?

Open Problem (May be difficult). *Are there more than four solutions to the equation*

$$(\text{VI.1.2}) \qquad \prod_{k=1}^{M}(c_k^2 + 1) = 2N^2$$

with natural numbers N and $M \geq 2$, and $c_k \geq 2$ for $k = 1, \ldots, M$?

VI.1.2 Gaussian Primes

It is straightforward to see that the Gaussian integers are a ring. The *units* in any ring are defined as its invertible elements. Clearly there are just four units $1, i, -1, -i$ in the ring of Gaussian integers. Furthermore, division with remainder is possible. Hence *the Euclidean algorithm works in the Gaussian integers.* Consequently, any Gaussian integer can be decomposed uniquely into a product of irreducible elements—unique up multiplication by the four units. Hence the Gaussian integers are a unique factorization domain, abbreviated UFD. The irreducible elements are called

Theorem 16 (Gaussian primes). *The Gaussian primes are:*

(a) *the number $1 + i$.*

(b) *the usual primes $p = 3, 7, 11, 19, \ldots$ which are $p \equiv 3 \mod 4$.*

(c) *all pairs $p + iq, p - iq$ where $p^2 + q^2 = r$ is a prime $r \equiv 1 \mod 4$.*

Indeed, for every usual prime $r = 5, 13, 17, 29, \ldots$ which is $r \equiv 1 \mod 4$ there exists a unique decomposition into a sum of two integer square.

The decomposition $r = p^2 + q^2$ becomes unique by the additional requirement that the (negative or positive) integer $p \equiv 1 \mod 4$ and q is even and positive.

Here is a table with the smallest examples. Too, I list the squares $a + ib = (p + iq)^2$.

r	p	q	a	b
5	1	2	-3	4
13	-3	2	5	-12
17	1	4	-15	8
29	5	2	21	20
37	1	6	-35	12
41	5	4	9	40
53	-7	2	45	-28
61	5	6	-11	60
73	-3	8	-53	-48
89	5	8	-39	80
97	9	4	65	72
101	1	10	-99	20
109	-3	10	-91	-60
113	-7	8	-15	-58
137	-11	4	105	-88

At that point, my heating system had been repaired, and I stopped.

Problem 119. *Prove that there are infinitely many primes $p \equiv 3 \mod 4$.*

Problem 120. *Prove that there are infinitely many primes $r \equiv 1 \mod 4$.*

Lemma 55. *Assume that $t = \frac{1}{\pi} \arctan \frac{p}{q}$ is rational with $q \geq 1$ and p, q integers. Then either $p = t = 0$ or $p = q$ and $t = \frac{1}{4}$ or $p = -q$ and $t = -\frac{1}{4}$.*

Proof. It is enough to consider the case with $p \geq 1$. It is now essential that we may take p and q to be relatively prime. Assume that $t = \frac{r}{s}$ with integers $r, s \geq 1$. Here is a sequence of calculations:

$$\frac{1}{\pi}\arctan\frac{p}{q} = \frac{r}{s}$$

$$\frac{p}{q} = \tan\frac{r\pi}{s}$$

$$\frac{p}{\sqrt{p^2+q^2}} = \sin\frac{r\pi}{s}$$

$$\frac{q}{\sqrt{p^2+q^2}} = \cos\frac{r\pi}{s}$$

$$\frac{q+ip}{\sqrt{p^2+q^2}} = \exp\frac{ir\pi}{s}$$

$$\left(\frac{q+ip}{\sqrt{p^2+q^2}}\right)^s = (-1)^r$$

$$(q+ip)^{2s} = (p^2+q^2)^s$$

$$(VI.1.3) \qquad (q+ip)^s = (q-ip)^s$$

We may now use the decomposition of $q+ip$ into Gaussian primes. Because p and q are relatively prime, there appear only the following prime factors:

- the unit,

- the factor $(1+i)$ with multiplicity one iff p and q are both odd, and zero otherwise,

- and any of the complex factors $u+iv$ where $u\cdot v \neq 0$ and u^2+v^2 is a real prime equivalent to 1 modulo 4. No conjugate complex prime factors are possible for p and q relatively prime.

Now the uniqueness of the factorization already exclude these latter prime factors $u+iv$ to occur, because of the last line (VI.1.3). We end up with the following two cases.

$p+q$ **is even** $p+iq = i^v(1+i)$ with $-1 \leq v \leq 2$. $p^2+q^2 = 2$ and hence $p = q = 1$. $t = \frac{1}{\pi}\arctan\frac{p}{q} = \frac{1}{4}$.

$p+q$ **is odd** $p^2+q^2 = 1$ and hence $pq = 0$ and $p+iq$ is a unit. This case leads to $p = 0$, and is obvious from the beginning.

$$\square$$

Here is just another reformulation of the same result.

Theorem 17 (The forgotten little theorem). *A Gaussian integer does only have an argument $\frac{i\pi k}{l}$ with integer k, l—an argument rational in degree measurement— if it lies on the x-axis or on the y-axis, or on one of the two lines of slope ± 1 through the origin.*

Solution. The Gaussian integer from the theorem satisfies

$$(VI.1.4) \qquad\qquad z = a + ib = \sqrt{a^2 + b^2}\, e^{\frac{i\pi k}{l}}$$

with real integers a, b, k, l. Let $a + ib = \gcd(a, b)(p + iq)$, and put $t = \frac{1}{\pi}\arctan\frac{p}{q}$. By the assumption of the little theorem, the number t is rational, and the fraction is in lowest terms since $\gcd(p, q) = 1$ Hence the claim follows from the lemma above. $\square$

Corollary 32. *Let $a + ib$ be a Gaussian integer with $a \neq 0, b \neq 0$.*

   Both real- and imaginary parts of $(a+ib)^l$ are nonzero if either l is odd or $|a| \neq |b|$. If $l > 0$ is even, the following happens:

l	$\Re(a + ib)^l$	$\Im(a + ib)^l$
$l \equiv 2 \mod 4$	$\Re(a + ib)^l = 0 \Leftrightarrow |a| = |b|$	$\Im(a + ib)^l \neq 0$
$l \equiv 0 \mod 4$	$\Re(a + ib)^l \neq 0$	$\Im(a + ib)^l = 0 \Leftrightarrow |a| = |b|$

Problem 121. *Prove that a natural number is a perfect square if and only if it has an odd number of divisors.*

VI.1.3   Geometry of Gaussian Primes

Theorem 18. *All primes congruent to 1 modulo 4 are the sum to two integer squares.*

Corollary 33. *A prime is not a sum of two squares if and only if it is congruent to 3 modulo 4.*

Proof. Assume that $p = a^2 + b^2$ and p is an odd prime. Then one number among a and b is even and the other one is odd. Assume that a is even and b is odd. Then $a^2 \equiv 0 \mod 4$ and $b^2 \equiv 1 \mod 8$ since $b = 2c + 1$ implies

$$b^2 = (2c + 1)^2 = 4c^2 + 4c + 1 = 8\,\frac{c(c + 1)}{2} + 1$$

where the fraction is an integer. Hence $a^2 + b^2 \equiv 1 \mod 4$. Hence a prime which is the sum of two squares is either equal to 2 or congruent to 1 modulo 4. Since $2 = 1^2 + 1^2$ and by theorem 18 the number 2 and the primes congruent to 1 modulo 4 are indeed the sum of two integer squares. The remaining primes are not a sum of two squares and are congruent to 3 modulo 4. $\square$

Proposition 57. *For each prime congruent to 1 modulo 4 there exists a number J such that*

$$1 + J^2 \equiv 0 \mod p$$

Proof based on quadratic reciprocity. The quadratic congruence $J^2 \equiv -1 \mod p$ is solvable if and only if

$$\left(\frac{-1}{p}\right) = (-1)^{\frac{p-1}{2}} = +1$$

Here one uses Euler's basic formula for the Legendre symbols with $a = -1$, and congruences turn out even to be equalities. The latter equality holds if and only if $p \equiv 1 \mod 4$. $\qquad\square$

Proof based on Wilson's Theorem. By Wilson's Theorem $1+(p-1)! \equiv 0 \mod p$ holds for all primes. In the factorial one puts pairs with sum p together:

$$1 + (p-1)! = 1 + \left(\frac{p-1}{2}\right)! \frac{p+1}{2} \cdots (p-2)(p-1)$$

$$\equiv 1 + \left(\frac{p-1}{2}\right)! \left(p - \frac{p-1}{2}\right) \cdots (p-2)(p-1)$$

$$\equiv 1 + (-1)^{\frac{p-1}{2}} \cdot \left[\left(\frac{p-1}{2}\right)!\right]^2 \mod p$$

Since $p \equiv 1 \mod 4$ is assumed and using Wilson's Theorem, one gets

$$0 \equiv 1 + (p-1)! \equiv 1 + J^2 \quad \text{with } J := \left(\frac{p-1}{2}\right)!$$

$$\square$$

Theorem 19 (Pick's Theorem). *The Euclidean plane is covered with a grid of congruent rhombuses of side length L and angle α. On this grid is given a simple (nonintersecting) closed polygon all vertices of which are grid points. The area of the interior domain of this polygon equals the number I of grid points in the interior of the polygon plus half the number B of grid points on the bounding polygon minus one.*

$$A = \left(I + \frac{B}{2} - 1\right) L^2 \sin \alpha$$

Proof of the sum of two squares theorem. The mapping

$$x + iy \in \mathbf{Z}_p + i\mathbf{Z}_p \mapsto \Phi(x+iy) = x + Jy \in \mathbf{Z}_p$$

is a ring homomorphism. By the basic theorem about homomorphisms there is an isomorphism

$$(VI.1.5) \qquad (\mathbf{Z}_p + i\mathbf{Z}_p)/\mathrm{Ker}\,\Phi \simeq \mathrm{Im}\,\Phi$$

The Gaussian integers and $\mathbf{Z}_p$ are both principal ideal domains. The reader may check that $\mathbf{Z}_p + i\mathbf{Z}_p$ is a principal ideal domain, too. The ideal $\mathrm{Ker}\,\Phi$ is generated by one element $a + ib$. This is indeed a solution of $a + Jb \cong 0 \mod p$ with $a^2 + b^2$ minimal. From the fact $a + Jb \equiv 0 \mod p$ one gets already

$$(VI.1.6) \qquad a^2 + b^2 \equiv (a + Jb)(a - Jb) \equiv 0 \mod p$$

Now we use the isomorphism VI.1.5 for counting. Φ restricted to the real $\mathbf{Z}_p$ with $y = 0$ is the identity mapping. Hence the image is $\mathrm{Im}\,\Phi = \mathbf{Z}_p$ and has p elements. Hence

$$|\mathrm{Ker}\,\Phi| = \frac{|\mathbf{Z}_p + i\mathbf{Z}_p|}{|\mathbf{Z}_p|} = \frac{p^2}{p} = p$$

Hence the generating element $a + ib \neq 0$ and hence $a^2 + b^2 = p \cdot f$ with $f \geq 1$. But we want to get the much stronger result that $f = 1$. From the generating property one gets

$$\mathrm{Ker}\,\Phi = \{(a + ib)(x + iy) \in \mathbf{Z}_p + i\mathbf{Z}_p : (x + iy) \in \mathbf{Z}_p + i\mathbf{Z}_p\}$$

From the properties of complex multiplication, we see that this is again a square grid $\mathcal{G}$. Now the basic square has vertices $0, a+ib, ia-b$ and the fourth vertex $a-b+i(a+b)$ and hence the area $L^2 = a^2 + b^2$. We use this grid for Pick's Theorem. Into the grid $\mathcal{G}$ is put the square Q with vertices $0, p, ip, p + ip$. How many grid points are in the interior of the square? These are the $a' + ib' \in \mathrm{Ker}\,\Phi$ with $0 < a' < p, 0 < b' < p$. On the boundary the four vertices of the square are grid points. I claim there are no further grid points on the boundary.

Reason. Assume $a' + ib' \in \mathrm{Ker}\,\Phi$ lies on the boundary of square Q. Using representatives for $\mathbf{Z}_p$, we may assume $0 \leq a', b' < p$. On the boundary holds either $a' = 0$ or $b' = 0$. Without loss of generality $b' = 0$. Hence $a' = a' + ib' \in \mathrm{Ker}\,\Phi$ and by the definition of the kernel $a' + Jb' = 0 \in \mathbf{Z}_p$. hence $a' = 0$ or $a' = p$. One gets a vertex of square Q once more. $\qquad\square$

Now Pick's formula gives with $I = p - 1$ and $B = 4$

$$p^2 = \mathrm{Area}\,Q = \left(I + \frac{B}{2} - 1\right) L^2 = |\mathrm{Ker}\,\Phi|(a^2 + b^2) = p(a^2 + b^2)$$

and hence $p = a^2 + b^2$, as claimed. $\qquad\square$

Remark. For a prime $p \equiv 1 \mod 4$ the equation $1 + J^2 \equiv 0 \mod p$ has exactly two solutions in $\mathbf{Z}_p$, these are J and $p - J$. For composite p there are cases with no solution and with more than two solutions.

Remark. For fixed J the equation $p = a^2 + b^2, a + Jb \equiv 0 \mod p$ has a unique solution with $a > 0, b > 0$. One sees this from the fact that $a + ib$ is the generator of the ideal $\mathrm{Ker}\Phi$. Hence any other generator $a' + ib'$ satisfies $a' + ib' = (a + ib)(x + iy)$ where the unit factor $x + iy$ is a Gaussian integer with $x^2 + y^2 = 1$. Hence $x + iy$ is one number among $1, i, -1, -i$ and the requirement $a > 0, b > 0, a' > 0, b' > 0$ implies $x + iy = 1$. To the second solution $p - J$ corresponds the pair $b + ia$.

VI.1.4   The Hexagonal Grid

The geometric idea above can be modified to the case of a grid of rhombuses with $60°$ angle which produce a grid covering the Euclidean plane, too. Usually, this grid is completed to the hexagonal grid, but we avoid to do that to keep the analog with the square grid more easy to see. As usual, we use the notation

$$\omega = \frac{-1 + i\sqrt{3}}{2}$$

for the generating third root of unity. Remember ω and ω^2 are the roots of the quadratic equation $x^2 + x + 1 = 0$, and they are conjugate complex. I begin with some facts about the ring $\mathbf{Z} + \omega\mathbf{Z}$.

Problem 122. *Prove that the ring $\mathbf{Z} + \omega\mathbf{Z}$ is a principal ideal domain.*

Proof. Take any ideal $\mathcal{I}$ and let $a + \omega b \in \mathcal{I}$ be a nonzero element with minimal absolute value. For any second element $x + \omega y \in \mathcal{I}$ we may do a complex division with remainder.

$$\frac{x + \omega y}{a + \omega b} = \frac{(x + \omega y)(a + \omega^2 b)}{a^2 - ab + b^2} = q_1 + \omega q_2 + r_1 + ir_2$$

We may chose the integers q_1, q_2 such that the absolute value of the remainder $|r_1 + ir_2|$ is as small as possible. In that case, the absolute value at most the distance of the center of an equilateral triangle of side 1 from is vertices. In other words

$$|r_1 + ir_2| \leq \frac{\sqrt{3}}{3} < 1$$

Multiplying the original division with the denominator yields

$$\mathcal{I} \ni x + \omega y - (q_1 + \omega q_2)(a + \omega b) = s_1 + is_2 := (r_1 + ir_2)(a + \omega b)$$

But the absolute value of the latter quality is bounded by

$$|s_1 + is_2| \leq \frac{\sqrt{3}}{3} |a + \omega b|$$

By the minimality of $|a+\omega b|$ this is only possible with $s_1+is_2 = 0$. Hence any arbitrary element $x+\omega y \in \mathcal{I}$ is a multiple $x+\omega y = (q_1+\omega q_2)(a+\omega b)$ of the generator. In other words any nonzero element of minimal absolute value is a generator of the ideal. $\square$

Remark. As a consequence, we see that the nonzero element of minimal absolute value is unique up to multiplication with the six units $\pm 1, \pm\omega, \pm\omega^2$.

We have already seen in Problem 122 that the ring $\mathbf{Z} + \omega\mathbf{Z}$ is a principle ideal domain. The main step in the proof is a division with remainder, which we have shown to be smaller than the denominator. Exploiting the same idea we get more interesting results for the ring $\mathbf{Z} + \omega\mathbf{Z}$.

The extended Euclidean algorithm works.

Euclid's lemma holds.

Each prime element is even irreducible.

Each prime ideal is maximal.

This is a unique factorization domain.

Lemma 56. *Suppose the real prime p decomposes in the ring $\mathcal{R} = \mathbf{Z} + \omega\mathbf{Z}$, and has in the ring the factor $a + \omega b$ which is not a unit, and $a \neq 0$ and $b \neq 0$ hold.*
Then $p = a^2 - ab + b^2$ and either $p = 3$ or $p \equiv 1 \mod 3$. Moreover $a + \omega b$ is irreducible.

Proof. Taking the conjugate complex, one gets that $a + \omega^2 b$ is a factor of p too. Moreover we conclude that $a + \omega b$ is irreducible. Indeed any nontrivial factoring would give a nontrivial factoring of the prime p, a contradiction.

Let g be the greatest common divisor of $a + \omega b$ and $a + \omega^2 b$ in the ring $\mathcal{R}$. One carefully starts the Euclidean algorithm in that ring and gets that g is a divisor of both $(\omega - 1)a$ and of $(1 - \omega)b$ (modulo units), and hence of $(1 - \omega)\gcd(a, b)$. Since

$a + \omega b$ is supposed to be a divisor of the real prime p, one gets $\gcd(a, b) = 1$. Hence either occurs case (i): $g = 1$ or case (ii): $g = 1 - \omega$.

In the first generic case (i) holds $g = 1$. Hence $a + \omega b$ and $a + \omega^2 b$ are relatively prime and the product $0 < a^2 - ab + b^2 = (a + \omega b)(a + \omega^2 b)$ is a divisor of p. We are back to real quantities and conclude that either $a^2 - ab + b^2 = 1$ or $a^2 - ab + b^2 = p$. In the first case one gets only the solutions $(a, b) = (1, 1)$ and $(a, b) = (-1, -1)$ and $a + \omega b$ is a unit. Nothing new. Under the assumption that $a + \omega b$ is not a unit, one gets $a^2 - ab + b^2 = p$.

We need still to clarify the second rare case (ii). Since the irreducible element $g = 1 - \omega$ divides $a + \omega b$ and this is irreducible, too, holds $a + \omega b = u(1 - \omega)$ with a unit u. Hence $a + \omega^2 b = \overline{u}(1 - \omega^2)$ and multiplication with the conjugate complex gives $a^2 - ab + b^2 = 3$. This has the solutions $(a, b) = (1, -1), (-1, 1), (2, 1), (-2, -1)$ and no other integer solutions. Hence $p = 3$.

Problem 123. *Why?*

Proof. From $a^2 + b^2 = 3 + ab \le 3 + (a^2 + b^2)/2$ is obtained $a^2 + b^2 \le 6$ and hence $|a|, |b| \le 2$. $a = 2$ implies $-2b + b^2 = -1$ and $(1 - b)^2 = 0$, hence $b = 1$. $\qquad \square$

Putting the cases (i) and (ii) together, Finally we get from $a^2 - ab + b^2 = p$ that either $p = 3$ or $p \equiv 1 \mod 3$. $\qquad \square$

Lemma 57. *The ring $\mathbf{Z} + \omega \mathbf{Z}$ contains the following irreducible elements*

- *the integer primes $p \equiv 2 \mod 3$;*

- *the elements $a + \omega b$ and $a + \omega^2 b$ with integers $a > 0$ and $b \ne 0$ for which $p = a^2 - ab + b^2$ is a prime. For $a + |b| > 3$ holds $p \ne 3$, and these are two distinct irreducible elements. But $1 - \omega$ and $1 - \omega^2$ are equal modulo a unit.*

Proof. The contrapositive of lemma 56 tells that all primes equivalent to 2 modulo 3 cannot be factored in the ring $\mathcal{R}$. In other words, they are irreducible elements of this ring.

$(1 - \omega)(1 - \omega^2) = 3$ and the factors have equal absolute value. Hence $|1 - \omega| = |1 - \omega^2| = \sqrt{3}$. What is their unit factor? Divide or simply check that $-\omega^2(1 - \omega) = 1 - \omega^2$.

Take any irreducible element $a + \omega b$ with integers $a > 0$ and $b \ne 0$. The product $0 < a^2 - ab + b^2 = (a + \omega b)(a + \omega^2 b)$ is real and can be factored into real primes. These primes may be factored further in the ring $\mathcal{R}$ into irreducible elements. Indeed, new factors may appear only for the prime 3 and the primes equivalent to 1 modulo 3, which occur in the real factoring,——the primes equivalent to 2 modulo 3 are already irreducible. But at least one among the first group needs to appear, otherwise the factoring would stay bluntly real. Therefore there exists a prime p, either $p = 3$ or $p \equiv 1 \mod 3$, for which $a + \omega b$ is a factor. Now lemma 56 proves that $p = a^2 - ab + b^2$ does hold. $\qquad \square$

The hard part is to show that an extra factoring does occur for *all primes $p \equiv 1$ mod 3*, and none among them remains irreducible in $\mathcal{R}$ for some strange overlooked reason.

Here comes in once more the geometric idea leading to the theorem 18 above. We modified to the case of a grid of rhombuses with 60° angle which produce a grid covering the Euclidean plane, too. I can hardly avoid to repeat the entire development in this more surprising variant.

Theorem 20. *All primes p congruent to 1 modulo 3 satisfy $p = a^2 - ab + b^2$ with integer a and b.*

Proposition 58. *Assume that $p = a^2 - ab + b^2$ and p is a prime. Then either $p = 3$ or $p \equiv 1 \mod 3$.*

Proof. One gets several cases:

either a or b are divisible by 3. They cannot be both divisible by three since otherwise $9|p$ and p is not a prime. Without loss of generality we may assume b is divisible by 3 and $a \equiv \pm 1 \mod 3$. Hence $p = a^2 - ab + b^2 \equiv a^2 \equiv 1 \mod 3$

neither a nor b are divisible by 3 and $a - b$ is divisible by 3. In that case one gets once more $p = a^2 - ab + b^2 \equiv a^2 \equiv 1$.

$a \equiv 2 \mod 3$ **and** $b \equiv 1 \mod 3$. In that case one gets $p = a^2 - ab + b^2 \equiv 4 - 2 + 1 \equiv 0 \mod 3$. Since p is assumed to be prime we conclude that $p = 3$.

$a \equiv 1 \mod 3$ **and** $b \equiv 2 \mod 3$. In that case one gets $p = a^2 - ab + b^2 \equiv 1 - 2 + 4 \equiv 0 \mod 3$. Since p is assumed to be prime we conclude once more that $p = 3$.

$\qquad \square$

Corollary 34. *A prime p does not satisfy the equation $p = a^2 - ab + b^2$ if and only if it is congruent to 2 modulo 3.*

Proposition 59. *For each prime congruent to 1 modulo 3 there exists a number Ω such that*

$$\Omega^2 + \Omega + 1 \equiv 0 \mod p$$

Proof based on quadratic reciprocity. We may multiply both sides of the congruence in question by 4. Since 4 is relatively prime to p, one gets the equivalent congruence

$$4\Omega^2 + 4\Omega + 4 \equiv 0 \quad \mod p$$
$$(2\Omega + 1)^2 \equiv -3 \quad \mod p$$

Necessary and sufficient for the quadratic congruence to be solvable is that the Legendre symbol is $+1$.

$$(VI.1.7) \qquad\qquad \exists z(z^2 \equiv -3 \quad \mod p) \Leftrightarrow \left(\frac{-3}{p}\right) = +1$$

One uses at first Euler's basic definition to simplify

$$\left(\frac{-3}{p}\right) = \left(\frac{-1}{p}\right)\left(\frac{3}{p}\right) = (-1)^{\frac{p-1}{2}}\left(\frac{3}{p}\right)$$

and then uses quadratic reciprocity

$$\left(\frac{-3}{p}\right) = (-1)^{\frac{p-1}{2}}\left(\frac{3}{p}\right) = (-1)^{\frac{p-1}{2}}(-1)^{\frac{(p-1)(3-1)}{4}}\left(\frac{p}{3}\right)$$
$$= \left(\frac{p \mod 3}{3}\right) = 1 \quad \text{provided that } p \equiv 1 \quad \mod 3$$

Hence the equivalence (VI.1.7) implies that the congruence $z^2 \equiv -3 \mod p$ is solvable. For an odd solution z, one puts $\Omega = \frac{z-1}{2}$. For an even solution z, one puts $\Omega = \frac{z+p-1}{2}$. In both cases one get an integer Ω satisfying $\Omega^2 + \Omega + 1 \equiv 0 \mod p$. $\quad\square$

Proof of the $a^2 - ab + b^2$ theorem. One goes on very similar to the two-squares theorem. But now the third root of unity ω takes the role of the imaginary unit i. The mapping

$$x + \omega y \in \mathbf{Z}_p + \omega\mathbf{Z}_p \mapsto \Phi(x + \omega y) = x + \Omega y \in \mathbf{Z}_p$$

is a ring homomorphism. By the basic theorem about homomorphisms there is an isomorphism between the quotient of the domain with the kernel and the image space.

$$(VI.1.8) \qquad\qquad (\mathbf{Z}_p + \omega\mathbf{Z}_p)/\mathrm{Ker}\Phi \simeq \mathrm{Im}\Phi$$

Problem 124. *Prove that the ring $\mathbf{Z}_p + \omega\mathbf{Z}_p$ with any prime p is a principal ideal domain.*

Solution. $\qquad\qquad\qquad\qquad\qquad\qquad\qquad\qquad\qquad\qquad\qquad\qquad\qquad\square$

The ideal $\mathrm{Ker}\Phi$ is generated by one element $a + \omega b$. This is, as seen above, a solution of $a + \Omega b \equiv 0 \mod p$ and $|a + \omega b|^2 \neq 0$ with minimal absolute value. From the fact $a + \Omega b \equiv 0 \mod p$ one gets already

$$(\mathrm{VI.1.9}) \qquad a^2 - ab + b^2 \equiv (a + \Omega b)(a + \Omega^2 b) \equiv 0 \mod p$$

Of course, we have claimed more. Now we use the isomorphism (VI.1.8) for counting. Φ restricted to the real $\mathbf{Z}_p$ with $y = 0$ is the identity mapping. Hence the image is $\mathrm{Im}\Phi = \mathbf{Z}_p$ and has p elements. Hence

$$|\mathrm{Ker}\Phi| = \frac{|\mathbf{Z}_p + \omega \mathbf{Z}_p|}{|\mathbf{Z}_p|} = \frac{p^2}{p} = p$$

The generating element $a + \omega b \neq 0$ and hence the minimal absolute value square

$$|a + \omega b|^2 = (a + \omega b)(a + \omega^2 b) = a^2 - ab + b^2$$

From the generating property one gets

$$\mathrm{Ker}\Phi = \{(a + \omega b)(x + \omega y) \in \mathbf{Z}_p + \omega \mathbf{Z}_p : (x + \omega y) \in \mathbf{Z}_p + \omega \mathbf{Z}_p\}$$

From the properties of complex multiplication, we see that this is again a rhombus grid $\mathcal{R}h$. (Strictly speaking lying on a torus.) Now the basic rhombus has vertices 0, $a + \omega b$, $\omega a + \omega^2 b = \omega a - b - \omega b$ and the fourth vertex $a + \omega a - b$. The area of this rhombus is

$$L^2 \sin 60° = |a + \omega b|^2 \frac{\sqrt{3}}{2} = \frac{(a^2 - ab + b^2)\sqrt{3}}{2}$$

We use this grid for Pick's Theorem. Into the grid $\mathcal{R}h$ is put the rhombus R with vertices $0, p, \omega p, (1 + \omega)p$. How many grid points are in the interior of this rhombus? These are the $a' + \omega b' \in \mathrm{Ker}\Phi$ with $0 < a' < p, 0 < b' < p$. On the boundary the four vertices of the rhombus are grid points. I claim there are no further grid points on the boundary.

Reason. Assume $a' + \omega b' \in \mathrm{Ker}\Phi$ lies on the boundary of the rhombus R. Using representatives for $\mathbf{Z}_p$, we may assume $0 \leq a', b' < p$. On the boundary holds either $a' = 0$ or $b' = 0$. Without loss of generality $b' = 0$. Hence $a' = a' + \omega b' \in \mathrm{Ker}\Phi$ and by the definition of the kernel $a' = a' + \Omega b' = 0 \in \mathbf{Z}_p$. hence $a' = 0$ or $a' = p$. One gets a vertex of the rhombus R once more. $\qquad\square$

Now Pick's formula gives with $I = p - 1$ and $B = 4$

$$p^2 \sin 60° = \mathrm{Area}R = \left(I + \frac{B}{2} - 1\right) L^2 \sin 60° = |\mathrm{Ker}\Phi|(a^2 - ab + b^2) \sin 60°$$

$$= p \cdot (a^2 - ab + b^2) \sin 60°$$

and hence $p = a^2 - ab + b^2$, as claimed. $\qquad\square$

Remark. For a prime $p \equiv 1 \mod 3$ the equation $\Omega^2 + \Omega + 1 \equiv 0 \mod p$ has exactly two solutions in $\mathbf{Z}_p$, these are Ω and $p - 1 - \Omega$, as the reader is helped to check here:

$$(p - 1 - \Omega)^2 + (p - 1 - \Omega + 1) \equiv (1 + \Omega)^2 - \Omega \equiv \Omega^2 + \Omega + 1 \equiv 0 \mod p$$

For composite p there are cases with no solution and with more than two solutions.

Remark. For fixed Ω the equation $p = a^2 - ab + b^2, a + \Omega b \equiv 0 \mod p$ has a unique solution with $a > 0, b > 0$. One sees this from the fact that $a + \omega b$ is the generator of the principal ideal $\mathrm{Ker}\Phi$. Hence any second generator $a' + \omega b'$ satisfies $a' + \omega b' = (a + \omega b)(x + \omega y)$ where the unit factor $x + \omega y$ has absolute value one. Hence $x + \omega y$ is one among the six numbers $\pm 1, \pm \omega, \pm \omega^2$. Now the requirements $a' + \omega b' = (a + \omega b)(x + \omega y)$ with $a > 0, b > 0, a' > 0, b' > 0$ imply a contradiction in all five cases except $x + \omega y = 1$. For example

$$a' + \omega b' = \omega(a + \omega b)$$
$$a' + \omega b' = \omega a - (1 + \omega)b$$
$$a' = -b \text{ and } b' = a - b$$

To the second solution $p - 1 - \Omega$ corresponds the pair $b + \omega a$ since

$$b + (p - 1 - \Omega)a \equiv b + \Omega^2 a \equiv \Omega^2(a + \Omega b) \equiv 0 \mod p$$

Proposition 60. *There exist infinitely many primes $p \equiv 1 \mod 3$.*

Proof. Let $p_1 = 7, p_2, \ldots, p_N$ be the smallest primes $p \equiv 1 \mod 3$. Define

$$P = \prod_{i=1}^{N} p_i \quad \text{and} \quad Q = 4P^2 - 2P + 1$$

From $Q \equiv 1 \mod 3$ and Q odd, we see that 2 or 3 are not divisors of Q, and the number of prime factors of Q which are equivalent to -1 modulo 3 is even. But we want to get more. Does the number Q have a prime factor equivalent to 1 modulo 3?

To get further information, one decomposes in the domain $\mathbf{Z} + \omega\mathbf{Z}$:

$$4P^2 - 2P + 1 = (2P + \omega)(2P + \omega^2)$$

We decompose the factor $2P + \omega$ further into irreducible elements of the domain $\mathbf{Z} + \omega\mathbf{Z}$. Among them cannot be any real primes r since $2P + \omega = r(A + B\omega)$ would imply $1 = rB$ and hence $r = B = 1$. Neither can $1 - \omega$ be an irreducible factor of $2P + \omega$. This would imply that $1 - \omega^2$ is a irreducible factor of $2P + \omega^2$ and hence $3 = (1 - \omega)(1 - \omega^2)$ is a factor of $4P^2 - 2P + 1$, what is false. Hence one gets a decomposition

$$2P + \omega = u \prod_{j=1}^{K}(A_j + \omega^{e_j} B_j) \quad \text{with unit } u \text{ and } e_j = 1, 2 \text{ and } |A_j| + |B_j| > 2$$

One multiplies with the conjugate complex equation and gets

$$2P + \omega^2 = \overline{u} \prod_{j=1}^{K}(A_j + \omega^{2e_j} B_j)$$

$$4P^2 - 2P + 1 = \prod_{j=1}^{K}(A_j^2 - A_j B_j + B_j^2) \quad \text{since } 1 + \omega^{e_j} + \omega^{2e_j} = 0$$

The last equation is the prime decomposition of $4P^2 - 2P + 1$ into real primes, and confirms that all these prime factors are congruent to 1 modulo 3. Since $\gcd(4P^2 - 2P + 1, P) = 1$ all these prime factors are different from those of P and hence larger than the prime factors of P. Hence there exists a prime which is congruent to 1 modulo 3 and larger than the smallest of these primes $p_1, \ldots, p_N$. $\square$

234

VI.1.5 Numerical Calculation of π and Machin-Like Formulas

Because the series is simpler, it is natural to exploit the arctan series for the calculation of π. It is called the *Gregory series*, too.

$$\arctan x = x - \frac{x^3}{3} + \frac{x^5}{5} - \frac{x^7}{7} + \frac{x^9}{9} + O\left(x^{11}\right)$$

Using Gregory's Series

For $x = 1$ the series is only conditionally convergent, and actually converges too slowly. For better convergence, one needs to use as small values of x as possible. One simple idea is to use the formula

$$\frac{\pi}{6} = \arctan \frac{1}{\sqrt{3}}$$

which follows from the fact that $\tan 30° = \frac{1}{\sqrt{3}}$. Easier to handle than the Gregory series is the related series

$$\frac{\arctan \sqrt{x}}{\sqrt{x}} = 1 - \frac{x}{3} + \frac{x^2}{5} - \frac{x^3}{7} + \frac{x^4}{9} + O\left(x^{10}\right)$$

With $x = 1/3$ this series is conveniently simplified to

$$\frac{\pi}{\sqrt{3}} = \sum_{n=1}^{\infty} \frac{16n}{3^{2n-1}(4n-3)(4n-1)}$$

Ten terms of this series already give 10 accurate digits of π.

A second, even more successful approach involving the Gregory series is using the addition theorem for tangent

$$\tan(\alpha + \beta) = \frac{\tan \alpha + \tan \beta}{1 - \tan \alpha \tan \beta}$$

From this is obtained

$$\arctan u + \arctan v \in \arctan \frac{u + v}{1 - uv} + \{0, \pm\pi\}$$

The extra term $\pm\pi$ may only occur if $|u| > 1$ or $|v| > 1$. If one plugs $u = \frac{1}{2}$ and $v = \frac{1}{3}$ into the addition theorem, one gets

$$\arctan \frac{1}{2} + \arctan \frac{1}{3} = \arctan 1 = \frac{\pi}{4}$$

Now one needs to calculate the arctan series with $x = \frac{1}{2}$ and $x = \frac{1}{3}$. For these instances, the series converges already faster, and one needs only a few terms to get ten, or even twenty digits of π. This method was already known to Machin around 1700 and has lateron been propagated by Euler.

Around 1706 Machin came up with an even better idea. He tried the fraction $\frac{1}{5}$. Since the arctan function is already nearly linearly for these small values, one needs to multiply by four. Now one does the following calculations

$$2 \arctan \frac{1}{5} = \arctan \frac{2/5}{24/25} = \arctan \frac{5}{12}$$

$$2 \arctan \frac{5}{12} = \arctan \frac{5/6}{119/144} \overset{!}{\to} = \arctan \frac{120}{119}$$

Because the latter fraction is so close to 1, we try to solve the equation

$$\arctan 1 + \arctan \frac{1}{x} = \arctan \frac{x+1}{x-1} = \arctan \frac{120}{119}$$

for x. After multiplying up and down with 2, one gets a solution $x + 1 = 240$. Hence we have obtained the identity

$$4 \arctan \frac{1}{5} - \arctan \frac{1}{239} = \arctan 1 = \frac{\pi}{4}$$

Around 1706 Machin has used this identity to calculate 100 decimal points for π.

A Few Machin-Like Expressions of Length Two

What is remarkable about the following products?

(a) $(2 + i) \cdot (3 + i) = 5 + 5i$

(b) $(2 + i)^2 \cdot (7 - i) = 25 + 25i$

(c) $(3 + i)^2 \cdot (7 + i) = 50 + 50i$

(d) $(5 + i)^4 \cdot (239 - i) = 114244 + 114244i$

Each time, the real- and imaginary parts turn out to be equal. Do there exists more examples with this special property?

For any positive $x > 0$, the principal argument of the complex number $x + i$ is $\text{Arg}\,(x+i) = \arctan \frac{1}{x}$. Too, $\text{Arg}\,(1+i) = \arctan 1 = \frac{\pi}{4}$. Taking the arguments of the formulas (a) through (d), one gets:

(a)
$$\arctan \frac{1}{2} + \arctan \frac{1}{3} = \frac{\pi}{4}$$

(b)
$$2 \arctan \frac{1}{2} - \arctan \frac{1}{7} = \frac{\pi}{4}$$

(c)
$$2 \arctan \frac{1}{3} + \arctan \frac{1}{7} = \frac{\pi}{4}$$

(d)
$$4 \arctan \frac{1}{5} + \arctan \frac{1}{239} = \frac{\pi}{4}$$

Of course, left- and right-hand side of these formulas could still differ by an integer multiple of 2π. But it is clear that this cannot happen in the given four examples. Do there exist more similar formulas?

How to Find Machin-Like Expressions

In this essay, I explain how one can construct Machin-like expressions

$$a \arctan \frac{1}{r} + b \arctan \frac{1}{s} + c \arctan \frac{1}{t}$$

with integers a, b, c, r, s, t such that this integer combination turns out to be equal to an integer multiple of $\frac{\pi}{4}$. Of course, we need $r, s, t \geq 2$ and want them to be as large as possible. Too, I shall be able to handle such expression with only two summands, or four or even more summands. We use the complex logarithm for which holds

$$\log(x + iy) = \frac{1}{2} \log(x^2 + y^2) + i \arg(x + iy) \mod 2\pi i$$

The argument is only determined modulo 2π. The value which is used for actual calculations has to be fixed by some convention. It is customary to use the convention $-\pi < \text{Arg}\, z \leq \pi$, and indicate this value by the capitals Log and Arg. But the prize is that a difficulty arises in the addition theorem.

$$\text{Log}\,(u \cdot v) = \text{Log}\, u + \text{Log}\, v \mod 2\pi i$$

This identity holds only modulo $2\pi i$.

The logarithm is useful since Machin's expression is the imaginary part of

$$(\text{VI.1.10}) \quad \Im \text{Log}\,(r+i)^a (s+i)^b (t+i)^c \equiv a \arctan \frac{1}{r} + b \arctan \frac{1}{s} + c \arctan \frac{1}{t} \mod 2\pi$$

To get a Machin-like formula, one has to satisfy the requirement

$$(\text{VI.1.11}) \qquad \Im \operatorname{Log}(r+i)^a(s+i)^b(t+i)^c \text{ is an integer multiple of } \frac{\pi}{4}.$$

Here occur complex numbers with integer real and imaginary parts. These numbers are known as the *Gaussian integers*. It turns out that there exists a unique prime decomposition in the ring of Gaussian integers. The primes, or irreducible elements of this ring are

- the integer primes which are congruent to 3 modulo 4,

- the complex number $1+i$, and

- the complex numbers $u \pm iv$ where $u^2 + v^2$ is an odd integer prime. The latter prime is always congruent to 1 modulo 4.

Additionally, there are the four *units* $1, i, -1, -i$.

In this essay are used the factorizations of $r+i$. To get them by hand for small values of r, it is best to prime factor at first $r^2 + 1$. Let me explain the procedure with the example $r = 8$. One gets $8^2 + 1 = 65 = 5 \cdot 13$. Next one needs to find sums of squares that add up to 5 and to 13. This is easy: $5 = 1^2 + 2^2$ and $13 = 2^2 + 3^2$. Hence one can write down the expressions

$$(8+i)(8-i) = (1+2i)(1-2i)(2+3i)(2-3i)$$

Both sides are equal to 65 and are the product of two conjugate complex expressions. One wants to separate the factors in such a way to obtain two conjugate complex identities. There are only a few choices. Which one is successful one cannot tell in advance. Indeed, either the expression $(1+2i)(2+3i)$ or the expression $(1+2i)(2-3i)$ has to be one among the eight expressions $\pm 8 \pm i$ or $\pm 1 \pm 8i$. Hence one calculates

$$(1+2i)(2+3i) = -4 + 7i$$
$$(1+2i)(2-3i) = 8 + i$$

In the second attempt we have already obtained a prime factorization for $8+i$. Such Gaussian prime factorizations are now used for the expression $(r+i)^a(s+i)^b(t+i)^c$. Next one takes the logarithm on both sides. In order for the imaginary part of the logarithm to be a rational multiple of π, we need to satisfy

$$(\text{VI.1.12}) \quad (r+i)^a(s+i)^b(t+i)^c \in \{\pm 1, \pm i, \pm 1 \pm i\} \cdot R \text{ where } R > 0 \text{ is rational}$$

Lemma 58. *The requirements* (VI.1.11) *and* (VI.1.28) *are equivalent.*

It is easy enough to convince oneself that (VI.1.28) $\Rightarrow$ (VI.1.11). The converse holds, too. I shall give a proof only later in the solution of problem VI.1.5 since we shall not use that harder part of the above lemma.

Let me assume that the numbers r, s, t have already been given. The next task is the appropriate choice of integers a, b, c. Each Gaussian prime factor of the left-hand side of requirement (VI.1.28), except of $1 + i$ and the units $1, i, -1, -i$ has to be cancelled by the choice of a, b, c. Thus are obtained several linear equations for the three unknowns $a, b.c$. To be sure to get a nontrivial solution $(a, b, c) \neq (0, 0, 0)$, for three unknowns, we want to have *at most two* equations. Hence we require that the numbers $r + i, s + i, t + i$ should contain at most too additional Gaussian prime factors, beyond $1 + i$.

Lemma 59. *Assume that the numbers* $r + i, s + i, t + i$ *contain at most too additional Gaussian prime factors, beyond* $1 + i$. *Then requirement* (VI.1.28) *has a nontrivial integer solution* $(a, b, c) \in \mathbf{Z}^3 \setminus \{(0, 0, 0)\}$.

Reason. By linear algebra, a homogeneous system with less equations than unknowns has a real nonzero solution. Since this solution can be found by Gaussian elimination, and all equations have integer coefficients. we get even an *integer* nontrivial solution. $\qquad\square$

Lemma 60. *Assume that the numbers* $r + i, s + i, t + i$ *contain at most too additional Gaussian prime factors, beyond* $1 + i$. *From a nontrivial integer solution* $(a, b, c) \in \mathbf{Z}^3 \setminus \{(0, 0, 0)\}$ *of requirement* (VI.1.11) *is obtained an identity*

$$(\text{VI.1.13}) \qquad k\frac{\pi}{4} = a \arctan \frac{1}{r} + b \arctan \frac{1}{s} + c \arctan \frac{1}{t}$$

with integer k.

Remark. The integer k may turn out to be zero, as in the example

$$0 = \arctan \frac{1}{3} - \arctan \frac{1}{5} - \arctan \frac{1}{8}$$

Too, it can be rather large as in the example

$$\frac{7\pi}{4} = 12 \arctan \frac{1}{2} - 4 \arctan \frac{1}{57} + \arctan \frac{1}{239}$$

Explanation for an example. Let me explain the example with $r = 5, s = 8, t = 57$. One gets the Gaussian prime factorizations

$$(5 + i) = -i(1 + i)(2 + 3i)$$
$$(8 + i) = -i(1 + 2i)(3 + 2i)$$
$$(57 + i) = -(1 + i)(2 + i)^3(2 + 3i)$$

I have used the convention of mathematica, that real and imaginary parts of the complex Gaussian primes are *positive*. This convention determines the unit factors. To do the above linear algebra, one needs to eliminate half of the factors $u + iv$ with $uv \neq 0$. In the above example occur four such factors: $(2+i), (1+2i), (3+2i), (2+3i)$. To get a nonzero solution, we may only require two equations.

We may indeed eliminate the factors $u + iv$ with $uv \neq 0$ and $v > u$, by means of the formula

$$(\text{VI.1.14}) \qquad\qquad u + iv = i(u^2 + v^2)(v + iu)^{-1}$$

With this trick we half the number of complex Gaussian primes besides $1 + i$ from 4 to 2. The extra positive factors and units do not disturb. In that way are obtained the formulas

$$(5 + i) = 13 \cdot (1 + i)(3 + 2i)^{-1}$$
$$(8 + i) = 5 \cdot (2 + i)^{-1}(3 + 2i)$$
$$(57 + i) = 13 \cdot (-i)(1 + i)(2 + i)^3(3 + 2i)^{-1}$$

We may now calculate the product $(5 + i)^a(8 + i)^b(57 + i)^c$ in question, and collect the terms for the different complex Gaussian primes.

$$(5 + i)^a(8 + i)^b(57 + i)^c = 13^{a+c}5^b \left(\frac{1 + i}{3 + 2i}\right)^a \left(\frac{3 + 2i}{2 + i}\right)^b \left(-\frac{i(1 + i)(2 + i)^3}{3 + 2i}\right)^c$$

$$(\text{VI.1.15})$$
$$(5 + i)^a(8 + i)^b(57 + i)^c = 5^b(-i)^c13^{a+c}(1 + i)^{a+c}(2 + i)^{3c-b}(3 + 2i)^{-a+b-c}$$

To get rid to the two remaining complex Gaussian primes $(2 + i)$ and $3 + 2i$, one has to solve the system

$$-b + 3c = 0$$
$$-a + b - c = 0$$

One needs a nonzero integer solution, which should be as small as possible. This solution is $a = 2, b = 3, c = 1$. We may plug this choice of coefficients into our formula (VI.1.15) (or even the line above that) and get

$$(5 + i)^2(8 + i)^3(57 + i) = 5^3 13^3(-i)^c(1 + i)^{a+c} = 5^3 13^3 \cdot 2(1 + i)$$

Remember that we have done all these calculations to get the *phases* of these complex numbers. Hence the positive factors do not matter. It makes no sense to multiply out the positive factors,—this is hard to tell mathematica but quite important to avoid hugh numbers. For the phase we use the formula

(VI.1.16) $\arg(x + iy) = \arctan \dfrac{y}{x}$ for $x > 0$

One gets

$$\text{Arg}\,(1 + i) = 2\text{Arg}\,(5 + i) + 3\text{Arg}\,(8 + i) + \text{Arg}\,(57 + i) \quad \bmod 2\pi$$

(VI.1.17) $\dfrac{\pi}{4} = 2\arctan\dfrac{1}{5} + 3\arctan\dfrac{1}{8} + \arctan\dfrac{1}{57} \quad \bmod 2\pi$

since the phase of $1 + i$ is $\frac{\pi}{4}$. From the example above is clear how to prove lemma 59 in the general case. The essential tools are the identities (VI.1.14) and (VI.1.16). After using linear algebra, one gets the formula (VI.1.17). $\square$

Remark. It takes some care to remove the nuisance with the additional multiples of 2π. My exact approach will be given at the end of this essay.

Implementation with mathematica

Here is my present procedure implementing these ideas.

```
makeMachinG5 = Function[mydom,
   dophases = Function[z,
     fiz = FactorInteger[z, GaussianIntegers -> True];
     withunitfac =
      If[Abs[First[fiz][[1]] ] == 1, fiz, Prepend[fiz, {1, 1}]];
     phs = {First[withunitfac][[2]]* Arg[First[withunitfac][[1]] ]};
      For[m = 2, m <= Length[withunitfac], m++,
      primpow = withunitfac[[m]];
      x = Re[primpow[[1]]]; y = Im[primpow[[1]]];
      argsideeffect = If[0 <= y <= x, AppendTo[arctanyx, ArcTan[y/x]],
```

```
      AppendTo[arctanyx, ArcTan[x/y]]] ;
    arg = If[0 <= y <= x, ArcTan[y/x], Pi/2 - ArcTan[x/y]] ;
    AppendTo[phs, primpow[[2]]*arg]]; phs;
   (*  The end of procedure dophases0 without winding correction. *)
   (*  The winding correction *)
   unit0 = First[withunitfac];
   winding = FullSimplify[Total[phs] - Arg[z]];
   oldunitph = Which[unit0 == {1, 1}, 0, unit0 == {I, 1}, Pi/2,
      unit0 == {-1, 1}, Pi,
      unit0 == {-I, 1}, -Pi/2, True, Indeterminate];
   newphs = Prepend[Rest[phs],
      oldunitph - winding]; newphs];

 unknowns = Array[a, Length[mydom]];
 rhs = unknowns.Arg[mydom + I];
 arctanyx = {};
 totalphase = unknowns.(Total[dophases[#]] & /@ (mydom + I));
 arctanyx = DeleteDuplicates[arctanyx];
 arctanyx1 = Complement[arctanyx, {Pi/4}];
 reduced =
  Reduce[Coefficient[totalphase, arctanyx1] == 0, Integers];
 solved = (Solve[reduced] /. {C[1] -> 1})[[1]];
 lhs = Simplify[totalphase /. solved];
 If[lhs == rhs /. solved,
   With[{lhs = Simplify[totalphase /. solved]},
    HoldForm[lhs] == rhs /. solved], Nothing]];
```

I begin by explaining the subprocedure dophases. Take any Gaussian integer, for example $z = 239 + i$. It has the Gaussian prime factorization

```
{{I, 1}, {1 + I, 1}, {2 + 3 I, 4}}
```

which has the phases and the sum

```
phs = dophases[239 + I]
{-((3 \[Pi])/2), \[Pi]/4, 4 (\[Pi]/2 - ArcTan[2/3])}
```

```
Total[phs]
 (3 \[Pi])/4 + 4 (\[Pi]/2 - ArcTan[2/3])
```

One needs to do these phase calculations for all three items of the list. Take a generic triplet. Find the phases of the product of powers. Sum up to the total phase, at least modulo 2π.

```
Clear[r, s, t]; triple = {r, s, t}; triple + I
Out[862]= {I + r, I + s, I + t}
```

```
Clear[a, b, c]; Times @@ ((triple + I) ^{a, b, c})
Out[863]= (I + r)^a (I + s)^b (I + t)^c
```

```
{a, b, c}.Arg[triple + I]
Out[1053]= a Arg[I + r] + b Arg[I + s] + c Arg[I + t]
```

which gives for $r > 0, s > 0, t > 0$ the right-hand side

```
rhs = {a, b, c}.(ArcTan[1/#] & /@ triple)
Out[1071]= a ArcTan[1/r] + b ArcTan[1/s] + c ArcTan[1/t]
```

I proceed to the right-hand side, where the Gaussian prime factorizations are implemented. Take the same example

```
Clear[r, s, t]; example = {r -> 8, s -> 57, t -> 239};
mydom = triple /. example; mydom + I
Out[1068]= {8 + I, 57 + I, 239 + I}
```

By means of the Gaussian prime decomposition for the three items of $mydom{+}I$, their phases become sums of integer multiples of the phases for the respective Gaussian prime factors. In the procedure are subtracted the respective winding corrections.

```
(Total[dophases[#]] & /@ (mydom + I))
Out[1060]= {-ArcTan[1/2] + ArcTan[2/3],
                -(\[Pi]/4) + 3 ArcTan[1/2] - ArcTan[2/3],
                -((5 \[Pi])/4) + 4 (\[Pi]/2 - ArcTan[2/3])}
```

thus one gets the integer combination, which is the total phase

```
arctanyx = {};
totalphase = {a, b, c}.(Total[dophases[#]] & /@ (mydom + I))
```

```
Out[1061]=
a (-ArcTan[1/2] + ArcTan[2/3])
  + b (-(\[Pi]/4) + 3 ArcTan[1/2] - ArcTan[2/3])
  +  c (-((5 \[Pi])/4) + 4 (\[Pi]/2 - ArcTan[2/3]))
```

Here one needs to collect the multiples of $\frac{\pi}{4}$ and the two occurring arctan$[u/v]$ terms.

```
arctanyx = DeleteDuplicates[arctanyx]
Out[1062]= {ArcTan[1/2], ArcTan[2/3], \[Pi]/4}
```

```
Collect[totalphase, arctanyx]
Out[1063]= - (b \[Pi])/4  + (3 c \[Pi])/4
               + (-a + 3 b) ArcTan[1/2] + (a - b - 4 c) ArcTan[2/3]
```

The integer valued unknowns a, b, c have to be chosen such that the latter expression becomes a rational multiple of π. There exists a nonzero solution since we have three unknowns which need to satisfy only two equations.

```
arctanyx1 = Complement[arctanyx, {Pi/4}];
    reduced =
    Reduce[Coefficient[totalphase, arctanyx1] == 0, Integers]
```

```
C[1] \[Element] Integers && a == 6 C[1] && b == 2 C[1] && c == C[1]
```

In most cases, one gets only one free parameter, which by mathematica is called $C[1]$. To get a solution with small integer values, I put $C[1]$ equal to 1. Such a solution is unique, up to a common sign.

```
    solved = (Solve[reduced] /. {C[1] -> 1})[[1]]
```

```
Out[1065]= {a -> 6, b -> 2, c -> 1}
```

To get a general expression for the π-term left is awkward enough, but not very inspiring anyway. But notice that there exist cases where the Machin-like expression turns out to be just zero. In other words, $k = 0$ in lemma 60 is possible. In the present case mathematica is mischievous enough to force me to simplify.

```
lhs = Simplify[totalphase /. solved]
Out[1066]= \[Pi]/4
```

On the right-hand side, too, one has to plug in the a, b, c from the solution of the homogeneous system. All calculations for the right-hand side are done without factoring. As simple as the idea is, there occur two complications:

1) Mathematica tends now to simplify too much and evaluate the right-hand side, too, to a multiple of π. This is not what we want to happen.

2) Mathematica uses always blindly the principle values. But a turn by any multiple of 2π may always sleek into the addition theorem for the logarithm.

```
lhs = Simplify[totalphase /. solved]
Out[1075]= \[Pi]/4


In[1078]:= rhs /. example /. solved
Out[1078]= ArcTan[1/239] + 2 ArcTan[1/57] + 6 ArcTan[1/8]


In[1073]:= With[{lhs = Simplify[totalphase /. solved]},
 HoldForm[lhs] == rhs /. solved]


Out[1073]= FractionBox["\[Pi]", "4"]
    == 6 ArcTan[1/8] + 2 ArcTan[1/57] + ArcTan[1/238]
```

After having solved problem 1, I have obtained a formula as expected. But there may sleek in a multiple of 2π to cause a difference between the left-hand and right-hand sides. The reason is problem 2. For confirmation, we may check that the difference of the left-hand and right-hand side is a multiple of 2π. I leave out this part.

Problem 125. *Actually requirement* (VI.1.28) *is equivalent to*

$$(\mathrm{VI.1.18}) \qquad (r + \mathrm{sign}\,(a)i)^{|a|}(s + \mathrm{sign}\,(b)i)^{|b|}(t + \mathrm{sign}\,(c)i)^{|c|} \in \{\pm 1, \pm i, \pm 1 \pm i\} \cdot R$$

where $R > 0$ *is rational. Explain the reason for the latter claim.*

F.C.M. Størmer constructed in 1896 the Machin-like formula

$$\frac{\pi}{4} = 44 \arctan \frac{1}{57} + 7 \arctan \frac{1}{239} - 12 \arctan \frac{1}{682} + 24 \arctan \frac{1}{12\,943}$$

Checking this formula with the above procedure, we obtain

```
makeMachinG5[{57, 239, 682, 12\,943}]


Out[1088]= \!\(\*
TagBox[
FractionBox["\[Pi]", "4"],
HoldForm]\) ==
  24 ArcTan[1/12943] - 12 ArcTan[1/682]
  + 7 ArcTan[1/239] +   44 ArcTan[1/57]
```

We primefactor

```
FactorInteger[1 + #^2] & /@ {57, 239, 682, 12943}
```

```
Out[1089]= {{{2, 1}, {5, 3}, {13, 1}},
               {{2, 1}, {13, 4}}, {{5, 3}, {61, 2}},
               {{2, 1}, {5, 4}, {13, 3}, {61, 1}}}
```

and see that the primes involved are $\{2, 5, 13, 61\}$. The homogeneous linear system to be solved is

```
Simplify[Coefficient[totalphase, arctanyx1] == 0]
```

```
Out[1094]= {3 a[1] + 3 a[3] - 4 a[4],
         -a[1] - 4 a[2] + 3 a[4], -2 a[3] - a[4]} == 0
```

This is how I get a small integer valued solution

```
In[1091]:= reduced =
 Reduce[Coefficient[totalphase, arctanyx1] == 0, Integers]
Out[1091]=
C[1] \[Element] Integers && a[1] == 44 C[1] && a[2] == 7 C[1] &&
 a[3] == -12 C[1] && a[4] == 24 C[1]
```

```
In[1092]:=    solved = (Solve[reduced] /. {C[1] -> 1})[[1]]
Out[1092]= {a[1] -> 44, a[2] -> 7, a[3] -> -12, a[4] -> 24}
```

In recent years there have been found other bigger Machin-like identities. See for example

```
https://en.wikipedia.org/wiki/Machin-like_formula
```

But as a computation effective method to get thousands and lately even millions digits of π, other methods turned out to be more effective. There are the famous formulas by Ramanujan.

The most astonishing discovery I could find is the Bailey-Borwein-Plouffe formula

```
https:// en.wikipedia.org/wiki/
   Bailey%E2%80%93Borwein%E2%80%93Plouffe_formula
```

$$\pi = \sum_{k=0}^{\infty} \left[\frac{1}{16^k} \left(\frac{4}{8k+1} - \frac{2}{8k+4} - \frac{1}{8k+5} - \frac{1}{8k+6} \right) \right]$$

246

The Main Result and its Proof

Here is the, a bid more general result, I am going to prove.

Theorem 21. *Take any natural numbers $r_{j+1} > r_j \geq 2$ for $1 \leq j < J$. Assume the decompositions of $r_j^2 + 1$ into real primes for $1 \leq j \leq J$ involve together at most $J - 1$ odd primes. Then a Machin-like identity*

$$(\text{VI.1.19}) \qquad \sum_{1 \leq j \leq J} a_j \arctan \frac{1}{r_j} = k \frac{\pi}{4}$$

holds for some integer a_j, k not all equal to zero.

Proof. In the following, all Gaussian primes $u + iv$ are assumed to be in *Maple style*. By this is meant that $u \equiv 1 \mod 4$ and v is even. The good thing about Maple's convention is that the product

$$U + iV = \prod_{3 \leq k} (u + iv)^{\mu_k}$$

of any such Maple-style Gaussian primes satisfies $U \equiv 1 \mod 4$ and V even, too. I now get the factorization of $r_j + i$, following these conventions.

$$(\text{VI.1.20}) \qquad r_j + i = i^{m_{j,1}}(1 + i)^{m_{j,2}} \prod_{3 \leq k \leq K} (u_k + iv_k)^{m_{j,k}} = i^{m_{j,1}}(1 + i)^{m_{j,2}} f_j$$

The factors $f_j \in \mathcal{G}auss0$ have to be cancelling in the end. Since $r_j + i$ have relatively prime real- and imaginary parts holds $m_{j,2} \in \{0, 1\}$, no prime-factor $u_k + iv_k$ in (VI.1.20) is a rational integer, and conjugate Gaussian primes do not occur for any fixed j.

Indeed for any real prime $p_k = u_k^2 + v_k^2 \equiv 1 \mod 4$ holds

p_k^m is the highest prime power dividing $r_j^2 + 1$ if and only if either $(u_k + iv_k)^m$
or $(u_k - iv_k)^m$ is the highest power of a Gaussian prime dividing $r_j + i$.

The Maple normalized Gaussian primes $u_k + iv_k$ need to run over the union set from those primes occurring in any one of the $r_j + i$ to be put into the Machin-like formula. Moreover, it is required that all conjugate complex pairs $u_k \pm iv_k$ appear together. Thus there are K^* complex Gaussian primes $u_k + iv_k$ with $v_k > 0$ involved. The same number K^* of integer odd primes occur in the factorizations of the quantities $r_j^2 + 1$

for $1 \leq j \leq J$. Obviously holds $K = 2 + 2K^*$. We define an *involution* $k \mapsto k^*$ on the domain $3 \leq k \leq K$ by putting $u_{k^*} + iv_{k^*} = u_k - iv_k$. Define

$$(\text{VI.1.21}) \qquad M_{j,k} = m_{j,k} - m_{j,k^*} \quad \text{with } v_k > 0.$$

We are now ready to set up the Gaussian integer relevant for the construction of the Machin-like formulas. This is

$$(\text{VI.1.22}) \qquad q + ip := \prod_{1 \leq j \leq J} (r_j + i)^{a_j}$$

with the integer valued exponents a_j to be chosen appropriately. We plug into this expression the decompositions into Gaussian primes and obtain

(VI.1.23)
$$\prod_{1 \leq j \leq J} (r_j + i)^{a_j} = i^n (1+i)^m \prod_{3 \leq k \leq K} (u_k + iv_k)^{\sum_{1 \leq j \leq J} a_j m_{j,k}} \quad \text{for } 1 \leq j \leq J \text{ and}$$

$$(\text{VI.1.24}) \qquad n = \sum_{1 \leq j} a_j m_{j,1}$$

$$(\text{VI.1.25}) \qquad m = \sum_{1 \leq j} a_j m_{j,2}$$

One wants an integer combination that makes the product of the terms with $3 \leq k \leq K$ on the right-hand side to become a positive rational number. The imaginary part of the logarithm is the argument function, denoted by Arg. Taking the argument on both sides of relation (VI.1.23) produces the following equivalence modulo 2π:

$$\sum_{1 \leq j} a_j \arctan \frac{1}{r_j} = \sum_{1 \leq j} a_j \mathrm{Arg}\,(r_j + i) \equiv \sum_{1 \leq j} a_j m_{j,1} \frac{\pi}{2} + \sum_{1 \leq j} a_j m_{j,2} \frac{\pi}{4}$$

$$+ \mathrm{Arg}\left(\prod_{3 \leq k \leq K} (u_k + iv_k)^{\sum_{1 \leq j \leq J} a_j m_{j,k}} \right) \quad \mod 2\pi$$

In order to produce a Machin-like identity the last summand has to vanish. This is of course equivalent to the product inside the argument function to be a positive rational number. Thus one has to solve the equations

$$\sum_{1 \leq j} \sum_{3 \leq k < K} a_j m_{j,k} \mathrm{Arg}\,(u_k + iv_k) \overset{!}{=} 0$$

Because of the appearance of conjugate complex terms, it is equivalent to the system

$$(\text{VI.1.26}) \qquad \sum_{1 \leq j} a_j M_{j,k} = 0 \quad \text{for all } k \geq 3 \text{ where G-primes satisfy } v_k > 0.$$

This is a homogeneous linear system of J equations with K^* unknowns. Since the matrix M has integer entries, there exists a basis of integer solutions. The number of free parameters occurring in the solution equal the dimension of the kernel and hence by the main theorem of linear algebra is equal to $J - \operatorname{rank} M$. Since $\operatorname{rank} M \leq \min(J, K^*)$, the assumption that $K^* \leq J - 1$ is just strong enough to get a one-dimensional space of nonzero solution vectors a_j.

If this system has a nontrivial solution, one has achieved to get

$$(\text{VI.1.27}) \qquad \sum_{1 \leq j} \sum_{3 \leq k} a_j m_{j,k} \operatorname{Arg}\left(u_k + i v_k\right) = 0$$

To restate the salient point, equivalent requirements for the vector a_j are

(a) the product $\prod_{3 \leq k \leq K}(u_k + i v_k)^{\sum_{1 \leq j \leq J} a_j m_{j,k}}$ is a positive rational number;

(b) the vector a_j satisfies equation (VI.1.34);

(c) the vector a_j is a solution of system (VI.1.26).

From a nonzero solution of anyone of these three requirements, one has obtained a Machin-like identity

$$\sum_{1 \leq j} a_j \arctan \frac{1}{r_j} = k\frac{\pi}{4} \quad \text{with the integer}$$

$$k \equiv 2n + m = \sum_{1 \leq j} a_j(2m_{j,1} + m_{j,2}) \quad \bmod 8$$

$\square$

An Open Problem, and Some Easy Consequences

Remark. [Open problem] Is it true or not that the matrix M above has always maximal rank? Under which additional conditions does it have maximal rank?

Corollary 35. *Suppose that for a Machin-like identity the sum $\sum a_j r_j$ is odd. Then k is odd, and hence the Machin-like sum is an odd multiple of $\frac{\pi}{4}$ and hence is nonzero and explicitly contains the number π.*

Corollary 36. *Suppose that for a Machin-like identity the sum $\sum a_j r_j$ is even. Then k is even, and hence the Machin-like sum is an integer multiple of $\frac{\pi}{2}$.*

Reason. k is odd iff $\sum_{1 \leq j} a_j m_{j,2}$ is odd. For all $1 \leq j \leq J$, $m_{j,2}$ is odd iff r_j is odd. Hence $\sum_{1 \leq j} a_j m_{j,2}$ is odd iff $\sum_{1 \leq j} a_j r_j$ is odd $\qquad\square$

The following slight generalization is a consequence of lemma 55 from above.

Proposition 61. *If a Machin-like expression adds up to rational multiple of π, it even adds up to a an integer multiple of $\frac{\pi}{4}$.*

A Strictly Constructive Result is Proved

It takes some care to remove the nuisance with the additional multiples of 2π. Of course, one may take resort to numerical computations. I do not consider such an approach as strictly constructive. For the strictly constructive approach, the reasoning may not suddenly jump to something obtained from outside, like numerical results are. Instead, one needs to use some nice properties of the Maple convention for the Gaussian primes, and, most important, the Riemann surface for the complex logarithm.

Let $\mathcal{G}auss00$ be set of all Gaussian integers $x + iy$ with $x \equiv 1 \mod 4$ and even y.

Lemma 61. *There exists a convention for the phase $\operatorname{marg}(x + iy)$ of the complex logarithm $\operatorname{mlog}(x + iy)$ such that for all Gaussian integers $x + iy$ in the set $\mathcal{G}auss00$, the addition theorem is simply*

$$\operatorname{mlog}(z \cdot w) = \operatorname{mlog} z + \operatorname{mlog} w$$
$$\operatorname{marg}(z \cdot w) = \operatorname{marg} z + \operatorname{marg} w$$

without an added multiple of 2π. For the Gaussian primes satisfying the Maple convention holds $\operatorname{mlog}(z) = \operatorname{Log} z$ and $\operatorname{marg}(z) = \operatorname{Arg} z$.

Indication of reason. Considered in the lemma are only the Gaussian integers $U + iV$ in the set $\mathcal{G}auss00$. By definition holds $U \equiv 1 \mod 4$ and V is even. These Gaussian integers have a factorization

$$U + iV = \prod_{3 \leq k} (u_k + iv_k)^{\mu_k}$$

with Gaussian primes $u_k + iv_k$ in Maple style and different from $1+i$. No unit $i, -i, -1$ factor may be permitted, and the factor $1+i$ is not permitted to occur. This situation is indicated by requiring $k \geq 3$. The factorization is strictly unique.

The functions mlog and marg are defined inductively on the number of prime factors of $U+iV$. For one and two prime factors this definition is obvious. To join an additional prime factor, we may choose several different placements for parentheses. By the associative law for multiplication these all give the same product. By the associative law for addition these all give the same phases and logarithms, which are hence well-defined. $\qquad\square$

With the convention of the lemma one gets for example $\mathrm{marg}\,(5) = 0$, but $\mathrm{marg}\,(9) = 2\pi$.

The functions mlog and marg cannot be extended to all Gaussian integers, but a slight extension is still possible. The functions mlog and marg can be extended to the larger domain. Let

$$\mathcal{G}auss00 \subsetneqq \mathcal{G}auss0 := \{(1+i)^{\mu_2} \prod_{3 \leq k}(u_k + iv_k)^{\mu_k}\}$$

with Gaussian primes $u_k + iv_k$ in Maple style and $\mu_k \geq 0$. Thus the Gaussian integers are partitioned into the four disjoint sets $\mathcal{G}auss0$, $i\mathcal{G}auss0$, $-\mathcal{G}auss0$, $-i\mathcal{G}auss0$. The smaller domain $\mathcal{G}auss00$ is symmetric to the real axis. Because of the inclusion of the powers of $(1+i)$, the domain is $\mathcal{G}auss0$ is not symmetric to the real axis. For example $(1+i) \in \mathcal{G}auss0$, but $(1-i) \notin \mathcal{G}auss0$.

Problem 126. *Multiply the sequence of natural numbers with a unit to get members of $\mathcal{G}auss0$, and enumerate this sequence.*

Answer.

$$1, 2i, -3, -4, 5, \quad -6i, -7, -8i, 9, 10i, \quad -11, 12, 13, -7i, -15, \quad 16, 17, 18i, -19, -20,$$
$$21, -22i, -23, 24i, 25, \quad 29, -30i, -31, 32i, 33, \quad 34i, -35, -36, 37$$

$\qquad\square$

Problem 127. *Enumerate the ten lowest natural numbers in $\mathcal{G}auss0$, and give their above defined phases.*

Answer. These and their phases divided by 2π are

$$1, 5, 9, 12, 13, \quad 16, 17, 21, 25, 29, \quad 33, 37,$$
$$0, 0, 1, 1, 0, \quad 1, 0, 1, 0, 0, \quad 1, 0,$$

$\qquad\square$

Problem 128. *Prove the little theorem. All Gaussian integers $X + iY$ which are not real and satisfy $X \equiv 1 \mod 4$ and Y even, have a phase which is not a multiple of $\frac{\pi}{4}$.*

Answer. Because of the assumptions $X \equiv 1 \mod 4$ and Y even, the function $\operatorname{marg}(X + iY)$ is well-defined.

Assume towards a contradiction that $X + iY$ is not real and has a phase $k\frac{\pi}{4}$ with integer k. The fourth power $U + iV = (X + iY)^4$ has $\operatorname{marg}(X + iY) = k\pi$. Hence $U + iV$ has a phase a multiple of π and is real. Moreover $U \equiv 1 \mod 4$ and $U = 1 + 4n$.

$$1 + 4n = X^4 - 6X^2Y^2 + Y^4$$
$$0 = X^3Y - XY^2$$

Since $X \neq 0$ one gets either $Y = 0$ or $X^2 = Y^2$. The latter contradicts the assumptions $X \equiv 1 \mod 4$ and Y even. Hence $Y = 0$ and $X + iY$ is real. $\qquad\square$

Problem 129. *Deduct from this statement that* (VI.1.11) $\Rightarrow$ (VI.1.28).

Proof. Assume that $\Im\operatorname{Log}(r + i)^a(s + i)^b(t + i)^c$ is an integer multiple of $\frac{\pi}{4}$. There exist units $i^\alpha, i^\beta, i^\gamma$ such that $i^\alpha(r + i), i^\beta, (s + i), i^\gamma(t + i)$ are members of $\mathcal{G}auss0$. Hence

$$X + iY := \left(i^\alpha(r + i)\right)^a \left(i^\beta(s + i)\right)^b \left(i^\gamma(t + i)\right)^c \in \mathcal{G}auss0$$

By the previous problem holds either $Y = 0$ or $\operatorname{Arg}(X + iY)$ is an integer multiple of $\frac{\pi}{4}$. In both cases holds $X + iY = (1 + i)^\epsilon \cdot R$ with $\epsilon \geq 0$ integer and $R > 0$ rational.

$$(r + i)^a(s + i)^b(t + i)^c = i^{-(a\alpha + b\beta + c\gamma)}(X + iY)$$

(VI.1.28) $\qquad (r + i)^a(s + i)^b(t + i)^c \in \{\pm 1, \pm i, \pm 1 + \pm i\} \cdot \mathbf{Q}$

$$\square$$

Let me go back to the factorization of the $r_j + i$ which are needed for the construction of the Machin-like formulas.

Lemma 62. *The first two factors of the decomposition*

$((\text{VI.1.20}))$ $\qquad\qquad r_j + i = i^{m_{j,1}}(1 + i)^{m_{j,2}} \prod_{3 \leq k \leq K} (u_k + iv_k)^{m_{j,k}}$

252

have the exponents

$$(\text{VI.1.29}) \qquad (m_{j,1}, m_{j,2}) = \begin{cases} (1,0) & \text{if } r_j \text{ even} \\ (0,1) & \text{if } r_j \equiv 1 \mod 8 \\ (-1,1) & \text{if } r_j \equiv 3 \mod 8 \\ (2,1) & \text{if } r_j \equiv 5 \mod 8 \\ (1,1) & \text{if } r_j \equiv -1 \mod 8 \end{cases}$$

Reason. For Gaussian integers $x + iy$ with either $x + y$ odd and $\gcd(x, y) = 1$ or xy odd and $\gcd(x, y) = 1$ one gets factorizations

$$x + iy = i^{m_1}(1 + i)^{m_2} \prod_{3 \le k} (u + iv)^{m_k}$$

with $u^2 + v^2 \equiv 1 \mod 4$ primes, and no conjugate complex ones appear.

For even x and odd y, we use $x + iy = i(U + iV) = -V + iU$. The reader may check the exponents for the unit factor

$$(\text{VI.1.30}) \qquad m_{j,1} = \begin{cases} 0 & \text{if } x \equiv 1 \mod 4 \text{ and } y \text{ even,} \\ 2 & \text{if } x \equiv 3 \mod 4 \text{ and } y \text{ even,} \\ 1 & \text{if } x \text{ even and } y \equiv 1 \mod 4, \\ -1 & \text{if } x \text{ even and } y \equiv 3 \mod 4, \end{cases}$$

For $x = r$ even and $y = 1$, one gets the third case. Hence

$$r + i = i \prod_{3 \le k} (u + iv)^{m_k}$$

For odd r and $y = 1$, there is an obvious factorization.

$$r + i = (1 + i) \left(\frac{1 + r}{2} + \frac{1 - r}{2} i \right)$$

Put $r = 8q + s$ with $s \in \{1, 3, 5, 7\}$ and go through these cases.

$$8q + s + i = (1 + i) \left(4q + \frac{1 + s}{2} + \left(-4q + \frac{1 - s}{2} \right) i \right)$$

$$8q + 1 + i = (1 + i)(4q + 1 - 4qi)$$
$$8q + 3 + i = (-i)(1 + i)(4q + 1 + (4q + 2)i)$$
$$8q + 5 + i = (-1)(1 + i)(-4q - 3 + (4q + 2)i)$$
$$8q - 1 + i = i(1 + i)(-4q + 1 - 4qi)$$

$\square$

Of course, all these details of the Maple conventions are not enough to resolve the ambiguity of the argument, and the imaginary part of the logarithm. To this end, we put these Gaussian integers as points onto the *unwound Riemann surface* for the logarithm. Moreover, we define free entities $\mathcal{I}$ and $\mathcal{I}p$. These entities are called *free* because of the assumptions

$$\mathcal{I}^n = 1 \text{ and } n \in \mathbf{Z} \text{ implies } n = 0$$
$$\mathcal{I}p^m = 1 \text{ and } m \in \mathbf{Z} \text{ implies } m = 0$$

In the above factorization for an Gaussian integer are done the replacements $i \mapsto \mathcal{I}$ and $1 + i \mapsto \mathcal{I}p$. These replacements give calculations for the Riemann surface. The mathematician say the quantities and calculations are "lifted" onto the Riemann surface. The lifted argument becomes a real number, and is no longer taken modulo 2π. For the set of Gaussian integers $\mathcal{G}auss00$, the lifted argument is already given by marg from the above lemma 61. The function marg and mlog can now be extended to the entire unwound Riemann surface. This fact is essential to get equation (VI.1.33).[16]

Lifting on the Riemann surface. Put

$$(\text{VI.1.31}) \qquad r_j + i = \mathcal{I}^{m_{j,1}} \mathcal{I}p^{m_{j,2}} f_j = \mathcal{I}^{m_{j,1}} \mathcal{I}p^{m_{j,2}} \prod_{3 \le k} (u_k + iv_k)^{m_{j,k}}$$

as a point onto the unwound Riemann surface for the logarithm. In the third factor right-hand side

$$\prod_{3 \le k} (u_k + iv_k)^{m_{j,k}} \in \mathcal{G}auss00$$

has been used the Maple convention for all the Gaussian primes $u+iv$ with $u^2+v^2 \ge 5$ occurring. The argument for the latter product is well defined *on the Riemann surface*, and its logarithm and phase are obtained by the above lemma 61. On the Riemann surface is obtained from (VI.1.31) the unwound argument

$$(\text{VI.1.32}) \qquad \mathrm{marg}\,(r_j + i) = m_{j,1}\frac{\pi}{2} + m_{j,2}\frac{\pi}{4} + \mathrm{marg}\, f_j$$

I define the winding number by setting

$$2\pi\mathrm{winding}\,(r_j) := \mathrm{marg}\,(r_j + i) - \mathrm{Arg}\,(r_j + i)$$

---

[16]I also simply say that we "calculate with unwound i".

254

and get

$$\arctan \frac{1}{r_j} = \operatorname{Arg}(r_j + i) = \operatorname{marg}(r_j + i) - 2\pi \operatorname{winding}(r_j)$$

$$= m_{j,1}\frac{\pi}{2} + m_{j,2}\frac{\pi}{4} + \operatorname{marg} f_j - 2\pi \operatorname{winding}(r_j)$$

$$(\text{VI.1.33}) \qquad \operatorname{marg} f_j = \sum_{3 \le k} m_{j,k} \operatorname{Arg}(u_k + iv_k)$$

$$\log(1 + r_j^2) = \sum_{3 \le k} m_{j,k} \log(u_k^2 + v_k^2)$$

We are now in a position to do linear algebra as usual. One wants an integer combination that makes the terms with $k \ge 3$ on the right-hand side vanish.

$$\sum_{1 \le j} a_j \arctan \frac{1}{r_j} = \sum_{1 \le j} a_j m_{j,1} \frac{\pi}{2} + \sum_{1 \le j} a_j m_{j,2} \frac{\pi}{4} - 2\pi \sum_{1 \le j} a_j \operatorname{winding} r_j$$

$$+ \sum_{1 \le j} \sum_{3 \le k} a_j m_{j,k} \operatorname{Arg}(u_k + iv_k)$$

Note that we have obtained an *equality* in place of a congruence modulo 2π. One needs to solve the same homogeneous linear system (VI.1.26) as above. For any (nontrivial) solution, one has achieved to get

$$(\text{VI.1.34}) \qquad \sum_{1 \le j} \sum_{3 \le k} a_j m_{j,k} \operatorname{Arg}(u_k + iv_k) = 0$$

and hence one has obtained a Machin-like identity

$$\sum_{1 \le j} a_j \arctan \frac{1}{r_j} = k\frac{\pi}{4} \quad \text{with the integer}$$

$$k = 2n + m - 8 \sum_{1 \le j} a_j \operatorname{winding} r_j = \sum_{1 \le j} a_j(2m_{j,1} + m_{j,2}) - 8 \sum_{1 \le j} a_j \operatorname{winding} r_j$$

Finally we have obtained equation (VI.1.19) with the exact the value for k. Too, the integer combination of equation (VI.1.33) on the Riemann surface and equation (VI.1.34) together imply

$$\sum_{j\geq 1} a_j \mathrm{marg}\, f_j = \sum_{j\geq 1} a_j \sum_{3\leq k} m_{j,k}\, \mathrm{Arg}\,(u_k + iv_k) = 0$$

$$\mathrm{marg}\left(\prod_{j\geq 1} f_j^{\,a_j}\right) = 0$$

$$\prod_{j\geq 1} f_j^{\,a_j} > 0 \ \text{ is a rational number.}$$

$\square$

As ugly as this proof is, there are some nice consequences.

Problem 130. *Prove that*

(VI.1.35)
$$\prod_{1\leq j}\left(\frac{1+r_j^2}{2:\ \ if\ r_j\ is\ odd}\right)^{a_j}$$

is the square of an odd integer.

Explanation. From the real part is obtained the additional equation

$$\sum_{1\leq j} a_j \log(1+r_j^2) = \sum_{1\leq j} a_j m_{j,2} \log 2 + \sum_{3\leq k}\sum_{1\leq j} a_j m_{j,k} \log(u_k^2 + v_k^2)$$

$$= \sum_{1\leq j\ \text{and}\ r_j\ \text{odd}} a_j \log 2 + S$$

$$S := 2 \sum_{3\leq k\ \text{and}\ v_k>0}\sum_{1\leq j} a_j m_{j,k*} \log(u_k^2 + v_k^2)$$

The last sum S is the logarithm of a rational square. The product from the problem is e^S, and it is an odd integer. Any quantity which is both a rational square and an odd integer is the square of an odd integer. $\square$

Part VII

Solving Polynomial Equations

VII.1   The Square Root of a Complex Number

Proposition 62. *A branch with $\Re\sqrt{z} \geq 0$ for the square root of any complex number $z = x + iy$ is*

$$(\text{VII.1.1}) \qquad \sqrt{x + iy} = \sqrt{\frac{\sqrt{x^2 + y^2} + x}{2}} + i\operatorname{sign}(y)\sqrt{\frac{\sqrt{x^2 + y^2} - x}{2}}$$

The square root $w = \sqrt{z}$ of any complex number $z = x + iy$ has two branches. The second branch is $w_2 = -w$.

Reason. Name the square root in question $\sqrt{x + iy} =: u + iv$. Squaring yields $x + iy = (u + iv)^2$. Separate the real- and imaginary part to get

$$x = u^2 - v^2 \, , \; y = 2uv$$

The absolute value squared is

$$x^2 + y^2 = |x + iy|^2 = |u + iv|^4 = (u^2 + v^2)^2$$

Add and substrate

$$u^2 + v^2 = \sqrt{x^2 + y^2}$$
$$u^2 - v^2 = x$$
$$u^2 = \frac{\sqrt{x^2 + y^2} + x}{2}$$
$$v^2 = \frac{\sqrt{x^2 + y^2} - x}{2}$$

Of the last two expression, one takes real square roots. $u \geq 0$ has been assumed. One needs still to determine the sign of v. But $y = 2uv$ and $u > 0$ imply $\operatorname{sign} y = (\operatorname{sign} u)(\operatorname{sign} v) = \operatorname{sign} v$. In the special case $u = 0$, both signs of v give a correct result for the square root. This special case corresponds to $y = 0$ and $x \leq 0$—the negative numbers $z \leq 0$. $\qquad\square$

Problem 131. *Calculate the square roots:*

(a) $\qquad\qquad\qquad \sqrt{-3 + 4i} \, , \; \sqrt{-12 + 16i} \, , \; \sqrt{-48 + 64i}$

(b) $\qquad\qquad\qquad \sqrt{-8 - 6i} \, , \; \sqrt{-32 - 24i} \, , \; \sqrt{-128 - 96i}$

(c) $\qquad\qquad\qquad \sqrt{-4 - 3i} \, , \; \sqrt{-16 - 12i} \, , \; \sqrt{-64 - 48}$

(d) $\qquad\qquad\qquad \sqrt{-6 + 8i} \, , \; \sqrt{-24 + 32i} \, , \; \sqrt{-96 + 128i}$

(e) $\qquad\qquad\qquad\qquad \sqrt{2i} \, , \; \sqrt{8i} \, , \; \sqrt{32i}$

A Gaussian integer that is the square of another Gaussian integer is called a perfect square. Which of the numbers under the square roots are perfect squares, which are not?

VII.2   About Cubic Equations

VII.2.1   The Solution of the Reduced Cubic Equation

Given is the equation

$$(\text{VII.2.1}) \qquad\qquad x^3 + b\,x + c = 0$$

Since the quadratic term is lacking, it is called a *reduced* or *depressed cubic.* We want to find all three solutions, including the complex ones. Putting $x := u - \frac{b}{3u}$ leads to the equation

$$(\text{VII.2.2}) \qquad\qquad u^3 - \left(\frac{b}{3u}\right)^3 + c = 0$$

Hence $z = u^3$ satisfies the quadratic resolvent equation

$$(\text{VII.2.3}) \qquad\qquad z^2 + cz - \frac{b^3}{27} = 0$$

The two solutions of equation (VII.2.3) are

$$(\text{VII.2.4}) \qquad\qquad z_{1,2} = -\frac{c}{2} \pm \sqrt{\frac{c^2}{4} + \frac{b^3}{27}}$$

We get a resolvent equation with real or complex solutions, depending on whether the discriminant

$$(\text{VII.2.5}) \qquad\qquad D = \frac{c^2}{4} + \frac{b^3}{27}$$

is positive or negative. In the first case, the original cubic has only one or (exceptionally) two real solutions. The second case with complex solutions of the resolvent equation (VII.2.3) is called the *casus irreducibilis.* In that case, the original cubic has three real solutions. Hence one is forced to go through some complex arithmetic —just in the case where all final results are real. The solution procedure turns out to be different for these two cases.

Case 1: $D \geq 0$ The cubic has one real solution— or in the special case $D = 0$—two real solutions. From Viëta's formula

$$(\text{VII.2.6}) \qquad\qquad \begin{aligned} z_1\, z_2 &= -\frac{b^3}{27} \\[2mm] -\frac{b}{3\sqrt[3]{z_1}} &= \sqrt[3]{z_2} \end{aligned}$$

Hence the real solution of the cubic (VII.2.1) is

(VII.2.7)
$$x_1 = u - \frac{b}{3u} = \sqrt[3]{z_1} + \sqrt[3]{z_2}$$
$$= \sqrt[3]{-\frac{c}{2} + \sqrt{\frac{c^2}{4} + \frac{b^3}{27}}} + \sqrt[3]{-\frac{c}{2} - \sqrt{\frac{c^2}{4} + \frac{b^3}{27}}}$$

which is just Cardano's formula. The two complex solutions are

(VII.2.8)
$$x_2 = \sqrt[3]{z_1}\,\omega + \sqrt[3]{z_2}\,\omega^2$$
$$x_3 = \sqrt[3]{z_1}\,\omega^2 + \sqrt[3]{z_2}\,\omega$$

where

(VII.2.9)
$$\omega = \frac{-1 + i\sqrt{3}}{2} = \cos\frac{2\pi}{3} + i\sin\frac{2\pi}{3}$$

is the primitive third root of unity. Simplifying (VII.2.7) and (VII.2.8) yields

(VII.2.10)
$$x_{2,3} = -\frac{\sqrt[3]{z_1} + \sqrt[3]{z_2}}{2} \pm \frac{\sqrt[3]{z_1} - \sqrt[3]{z_2}}{2}\sqrt{3}\,i$$

Case 2: $D < 0$ The cubic has three real solutions. In this case, the resolvent equation (VII.2.3) has complex solutions. As already Rafael Bombelli discovered (1572), this leads to a third root of a complex number in Cardano's formula (VII.2.7).

Actually, this case can only occur for $b < 0$. To cover the general case, extraction of the third root has to be done via polar coordinates. Viëta's formula (VII.2.6) and equation (VII.2.4) imply

(VII.2.11)
$$|z_1|^2 = z_1 z_2 = \frac{-b^3}{27}$$
$$z_{1,2} = -\frac{c}{2} \pm i\sqrt{\frac{-b^3}{27} - \frac{c^2}{4}}$$
$$= \sqrt{\frac{-b^3}{27}}(\cos\theta \pm i\sin\theta)$$

The argument θ can be calculated more simple from the real parts. One gets $\cos\theta$, and finally θ:

(VII.2.12)
$$-\frac{c}{2} = \sqrt{\frac{-b^3}{27}}\cos\theta$$
$$\theta = \text{arc}\cos\frac{-c}{2\sqrt{\frac{-b^3}{27}}}$$

Hence equations (VII.2.11) and (VII.2.9) imply

$$\sqrt[3]{z_1} = \sqrt{\frac{-b}{3}}\left(\cos\frac{\theta}{3} + i\sin\frac{\theta}{3}\right)$$

(VII.2.13)

$$\sqrt[3]{z_1}\,\omega = \sqrt{\frac{-b}{3}}\left(\cos\frac{\theta+2\pi}{3} + i\sin\frac{\theta+2\pi}{3}\right)$$

and equations (VII.2.6) and (VII.2.7) imply that

$$x_1 = 2\sqrt{\frac{-b}{3}}\,\cos\frac{\theta}{3}$$

(VII.2.14)

$$x_2 = 2\sqrt{\frac{-b}{3}}\,\cos\frac{\theta+2\pi}{3}$$

$$x_3 = 2\sqrt{\frac{-b}{3}}\,\cos\frac{\theta-2\pi}{3}$$

are the three <u>real</u> solutions of the original cubic.

Problem 132. *Calculate the roots in the limiting case $D \searrow 0+$.*

Answer.

$$x_1 = 2\sqrt[3]{-\frac{c}{2}}\ , \quad x_2 = x_3 = -\frac{x_1}{2}$$

Problem 133. *Calculate the roots in the limiting case $D \nearrow 0-$ in terms of c. Distinguish the cases $c < 0$ and $c > 0$. In both cases $b < 0$.*

Answer. In the approach $D \nearrow 0-$, one gets

$$\frac{c^2}{4} = -\frac{b^3}{27}\ , \quad \sqrt[3]{\frac{|c|}{2}} = \sqrt{\frac{-b}{3}}$$

$$\frac{|c|}{2} = \sqrt{\frac{-b^3}{27}}$$

$$-\frac{c}{2} = \sqrt{\frac{-b^3}{27}}\,\cos\theta$$

Distinguish the cases $c < 0$ and $c > 0$.

Case $c < 0$.   One gets $\cos\theta = 1$, and hence $\theta = 0$. The result (VII.2.14) implies

$$x_1 = 2\sqrt[3]{-\frac{c}{2}}\ , \quad x_2 = x_3 = -\frac{x_1}{2}$$

Case $c > 0$. One gets $\cos\theta = -1$, and hence $\theta = \pi$. The result (VII.2.14) implies

$$x_2 = -2\sqrt[3]{-\frac{c}{2}}\ ,\quad x_1 = x_3 = -\frac{x_1}{2}$$

Problem 134. *Use Cardano's formula and Bombelli's method to get one solutions of $x^3 - 30x - 36 = 0$.*

Answer. Cardano's formula for a solution of $x^3 + bx + c = 0$ yields in the present case

(VII.2.15)
$$\begin{aligned}
x_1 &= \sqrt[3]{-\frac{c}{2} + \sqrt{\frac{c^2}{4} + \frac{b^3}{27}}} + \sqrt[3]{-\frac{c}{2} - \sqrt{\frac{c^2}{4} + \frac{b^3}{27}}} \\
&= \sqrt[3]{18 + 26i} + \sqrt[3]{18 - 26i}
\end{aligned}$$

To calculate the third root, let $\sqrt[3]{18 + 26i} = u + iv$ and $18 + 26i = (u + iv)^3$. We compare real- and imaginary parts and get the system

$$\begin{aligned}
18 &= u^3 - 3uv^2 = u(u^2 - 3v^2) \\
26 &= 3u^2v - v^3 = (3u^2 - v^2)v
\end{aligned}$$

With the help of the factoring, one finds the integer solution $u + iv = 3 + i$. Hence Cardano's formula gives $x_1 = 6$.

Problem 135. *Find the two other solutions of the equation $x^3 - 30x - 36 = 0$.*

Answer. The simplest way is division of the polynomial. One gets $(x^3 - 30x - 36)/(x - 6) = x^2 + 6x + 6$. Hence the two other solutions are $x_{2,3} = -3 \pm \sqrt{3}$.

Remark. Too, we can find the two other roots by introducing the third root of unity $\omega = \frac{-1 + i\sqrt{3}}{2}$ into Cardano's formula. This has to be done in a consistent way and produces the solutions

$$\begin{aligned}
x_2 &= \omega\sqrt[3]{18 + 26i} + \omega^2\sqrt[3]{18 - 26i} = \frac{-1 + i\sqrt{3}}{2}(3 + i) + \frac{-1 - i\sqrt{3}}{2}(3 - i) - 3 - \sqrt{3} \\
x_3 &= \omega^2\sqrt[3]{18 + 26i} + \omega\sqrt[3]{18 - 26i} = \frac{-1 - i\sqrt{3}}{2}(3 + i) + \frac{-1 + i\sqrt{3}}{2}(3 - i) = -3 + \sqrt{3}
\end{aligned}$$

which is the same answer as above.

Remark. Formulas (VII.2.7) and (VII.2.8) can be used for the reduced cubic with complex coefficients, too. But in that more general setting, the choice of the square roots and third roots needs clarification. One may avoid part of the confusion by using a formula with a *denominator*:

$$x_1 = u - \frac{b}{3u} = \sqrt[3]{-\frac{c}{2} + \sqrt{\frac{c^2}{4} + \frac{b^3}{27}}} - \frac{b}{3\sqrt[3]{-\frac{c}{2} + \sqrt{\frac{c^2}{4} + \frac{b^3}{27}}}}$$

$$\text{(VII.2.16)} \qquad x_2 = \omega\sqrt[3]{-\frac{c}{2} + \sqrt{\frac{c^2}{4} + \frac{b^3}{27}}} - \frac{\omega^2 b}{3\sqrt[3]{-\frac{c}{2} + \sqrt{\frac{c^2}{4} + \frac{b^3}{27}}}}$$

$$x_3 = \omega^2\sqrt[3]{-\frac{c}{2} + \sqrt{\frac{c^2}{4} + \frac{b^3}{27}}} - \frac{\omega b}{3\sqrt[3]{-\frac{c}{2} + \sqrt{\frac{c^2}{4} + \frac{b^3}{27}}}}$$

Any choice of the third root and square root in the first term u, has to be repeated in all the terms above in *the same way*. Under this agreement, one obtains all three solutions of the reduced cubic (VII.2.1), each solution occurring twice. This is the style for Cardano's formula in use in several computer languages, as e.g. mathematica.

The denominator is really needed for such a simplification to work properly. If one would make independent choices for the third root and square root in the first term u and the second term $\frac{b}{3u}$, one would obtain up to 36 different values, of which only 6 ones actually are solutions of the cubic equation (VII.2.1). I would not go as far as stating Cardano's formula to be useless, but in any case clarification is needed, once one goes beyond the classical case with real coefficients and only one real solution.

VII.2.2   Casus Irreducibilis

I am interested in solving the equation

$$\text{(VII.2.1)} \qquad\qquad x^3 + b\,x + c = 0$$

in the *casus irreducibilis*, the case where there exist of three real solutions, and with a natural tendency for positive integers. From the previous section, we know Cardano's formula

$$x = \sqrt[3]{-\frac{c}{2} + \sqrt{\frac{c^2}{4} + \frac{b^3}{27}}} + \sqrt[3]{-\frac{c}{2} - \sqrt{\frac{c^2}{4} + \frac{b^3}{27}}}$$

To avoid the denominators, I put $b := -3p$ and $c := 2q$, Of special interest is the equation

$$(VII.2.17) \qquad P(x) := x^3 - 3px + 2q = 0$$

the case for positive integer p and $q \neq 0$. The discriminant $D = -p^3 + q^2$ is obtained from formula (VII.2.5) above. The same value is found under the small square roots in Cardano's formula. (Of course the latter method is only a good hint, there may be some factor lost.)

Remark. In general, the discriminant for a third order polynomial is defined as

$$(VII.2.18) \qquad \Delta(P) := \prod_{1 \leq i < k \leq 3} (x_i - x_k)^2$$

A rather lengthy calculation, best done with mathematica gives $\Delta(P)$ in terms of the elementary symmetric functions to be $s1^2 s2^2 - 4s2^3 - 4s1^3 s3 + 18s1s2s3 - 27s3^2$ One obtains with Viëta's formulas $s1 = 0, s2 = -3p, s3 = 2q$ and hence Δ in terms of the coefficients r, q, p to be

$$(VII.2.19) \qquad \Delta(P) = 4 \cdot 27(p^3 - q^2) = -108D$$

Lemma 63. *If $D < 0$ and $q > 0$, there exist three real solution for equation (VII.2.17). One is negative and the two others are positive. In the case $D < 0$ and $q < 0$, one gets two negative and one positive root.*

Proof. Every clever school boy should be able to prove this lemma. Only it is very difficult to prove existence of such a school boy. $\qquad \square$

Proof. I take the case with $D < 0$ and $q > 0$. From $\lim_{x \to -\infty} P(x) = -\infty$ and $P(0) = 2q > 0$, the intermediate value theorem gives existence of a negative root $x_1 < 0$. The extrema of $P(x)$ are the zeros of $P'(x) = 3x^2 - 3p$ hence they occur for $\pm\sqrt{p}$. The critical value at the positive root is $P(+\sqrt{p}) = p\sqrt{p} - 3p\sqrt{p} + 2q = 2(-p\sqrt{p} + q) \leq 0$. This is negative or zero because of the assumption $p^3 - q^2 = d^2 \geq 0$. The value zero occurs for $d = 0$ in which case one gets a double root of P. For $d \neq 0$, the intermediate value theorem now shows existence of a root $0 < x_2 < p\sqrt{p}$, because of the sign change in that interval. Since $\lim_{x \to +\infty} P(x) = +\infty$ one gets a third root $x_3 > p\sqrt{p}$. $\qquad \square$

By the way, the point $p\sqrt{p}$ is a local minimum and the point $-p\sqrt{p}$ is a local maximum for the function $x \mapsto P(x)$.

The Cardano formula gives the roots of a third order polynomial only in convoluted form, going over complex numbers and their third roots. May be that has inspired Törleß and Beineberg to their discussions in Robert Musil's *The confusions of the young Törleß*. A bid earlier, it has led Bombelli to the famous example already given in the last section.

Born: 1526, Borgo Panigale, Italy Died: 1573, Rome, Italy Education: University of Bologna

`https://mathshistory.st-andrews.ac.uk/Biographies/Bombelli/`

The following remark gives a vague idea of his abilities.

> At the bottom of the page 188 from Bombelli's algebra book he demonstrates cubing the complex number $3 + 4i$, by first squaring it, using distributivity, to get $-7 + 24i$, which is then multiplied by another factor of $3 + 4i$ to yield $(3 + 4i)^3 = -117 + 44i$.

VII.3   The Solution of the Depressed Quartic

It is actually easier to treat first the general depressed polynomial

$$h(x) = r + qx + px^2 + x^4$$

I make the assumptions that $r \neq 0$ and $q \neq 0$. I prefer to go on with Descartes' method and use a factoring

(VII.3.1) $$r + qx + px^2 + x^4 = (b + ax + x^2)(c - ax + x^2)$$

The new coefficients have to satisfy the equations

(VII.3.2) $$r = b \cdot c, \; q = -a(c - b), \; p = -a^2 + b + c$$

Elimination of b and c yields

$$R(a^2) = 0 \quad \text{with the Descartes-resolvent}$$
(VII.3.3) $$R(v) = -q^2 + (p^2 - 4r)v + 2pv^2 + v^3$$

One gets now a cubic equation for a^2, which is at least some progress.

For the determination of (a, b, c) one needs to solve the equation $R(a^2) = 0$ with Descartes' resolvent (IX.1.6) Let the three solutions be v_1, v_2, v_3, allowing for multiple solutions. We put $a := \sqrt{v_1}$, with a choice of the square root fixed from now on. How are the other resolvent roots v_2 and v_3 related to (a, b, c)? From a guess that $\sqrt{v_2} - \sqrt{v_3}$ may be $\sqrt{a^2 - 4b}$ and $\sqrt{v_2} + \sqrt{v_3}$ may be $\sqrt{a^2 - 4c}$, one is lead to check that indeed

$$R\left(\frac{(\sqrt{a^2 - 4b} \pm \sqrt{a^2 - 4c})^2}{4}\right) = 0$$

can be confirmed. By possibly switching v_2 and v_3 and appropriate choice of the signs of the square roots of v_2 and v_3, it is possible to satisfy

$$\sqrt{v_2} = \frac{\sqrt{a^2 - 4b} + \sqrt{a^2 - 4c}}{2}, \quad \sqrt{v_3} = \frac{\sqrt{a^2 - 4b} - \sqrt{a^2 - 4c}}{2},$$

and hence holds

$$\sqrt{a^2 - 4b} = \sqrt{v_2} + \sqrt{v_3} \quad \text{and} \quad \sqrt{a^2 - 4c} = \sqrt{v_2} - \sqrt{v_3}$$

One gets now for the four roots in equation (IX.1.7) the formulas

$$(\text{VII.3.4}) \qquad [x_i] = \tfrac{1}{2}(-\sqrt{v_1} - \sqrt{v_2} + \sqrt{v_3}),\ \tfrac{1}{2}(-\sqrt{v_1} + \sqrt{v_2} - \sqrt{v_3}),$$
$$\tfrac{1}{2}(\sqrt{v_1} - \sqrt{v_2} - \sqrt{v_3}),\ \tfrac{1}{2}(\sqrt{v_1} + \sqrt{v_2} + \sqrt{v_3})$$

There remains an important point to clarify. How can one see in the end that the choice of the roots of the v_i has been done appropriately?

To get the correct roots with Descartes' method one has to satisfy $q = a(c - b)$. For the three roots this implies

$$4q = 4a(c - b) = \sqrt{v_1}((\sqrt{v_2} - \sqrt{v_3})^2 - (\sqrt{v_2} + \sqrt{v_3})^2) = -4\sqrt{v_1}\sqrt{v_2}\sqrt{v_3}$$

Indeed two of the roots may be chosen arbitrarily in formula (VII.3.4). But the third root has to satisfy

$$(\text{VII.3.5}) \qquad \sqrt{v_1}\sqrt{v_2}\sqrt{v_3} = -q$$

Proposition 63. *The four roots of the depressed cubic $h(x) = r + qx + px^2 + x^4$ are obtained from the roots of the resolvent*

$$(\text{IX.1.6}) \qquad R(v) = -q^2 + (p^2 - 4r)v + 2pv^2 + v^3$$

by means of the formulas

$$(\text{VII.3.4}) \qquad [x_i] = \tfrac{1}{2}(-\sqrt{v_1} - \sqrt{v_2} + \sqrt{v_3}),\ \tfrac{1}{2}(-\sqrt{v_1} + \sqrt{v_2} - \sqrt{v_3}),$$
$$\tfrac{1}{2}(\sqrt{v_1} - \sqrt{v_2} - \sqrt{v_3}),\ \tfrac{1}{2}(\sqrt{v_1} + \sqrt{v_2} + \sqrt{v_3})$$

where the three square roots have to chosen such that

$$(\text{VII.3.5}) \qquad \sqrt{v_1}\sqrt{v_2}\sqrt{v_3} = -q$$

Problem 136. *Check that the proposition remains true in the case $q = 0$.*

Solution. In that case the solutions of the depressed quartic $r + px^2 + x^4$ are easily obtained to be

$$\pm\sqrt{\frac{-p \pm \sqrt{p^2 - 4r}}{2}}$$

The resolvent $R(v) = v((p^2 - 4r) + 2pv + v^2)$ has the zeros

$$v_1 = 0,\ v_{2,3} = -p \pm \sqrt{4r}$$

and one just checks the general formula (VII.3.4) to give

$$\pm\tfrac{1}{2}(\sqrt{v_2} - \sqrt{v_3}),\ \pm\tfrac{1}{2}(\sqrt{v_2} + \sqrt{v_3})$$

which still look somehow different. The squared values are

$$(\sqrt{v_2} \pm \sqrt{v_3})^2/4 = (v_2 + v_3)/4 \pm \sqrt{v_2}\sqrt{v_3}/2 = -p/2 \pm \sqrt{p^2 - 4r}/2$$

as it has to be. $\square$

One wants still to get information about the discriminants

$$(\text{VII.3.6}) \qquad\qquad \Delta(h) := \prod_{1 \le i < k \le 4} (x_i - x_k)^2$$

$$(\text{VII.3.7}) \qquad\qquad \Delta(R) := \prod_{1 \le i < k \le 3} (v_i - v_k)^2$$

A rather lengthy calculation, best done with mathematica gives $\Delta(R)$ in terms of the elementary symmetric functions. In the end, one obtains with Viëta's formulas $\Delta(R)$ in terms of the coefficients r, q, p to be

$$(\text{VII.3.8}) \qquad \Delta(R) = -4p^3q^2 - 27q^4 + 16p^4r + 144pq^2r - 128p^2r^2 + 256r^3$$

An even more lengthy calculation leads now to the result.

Proposition 64. *The discriminants of the depressed quartic and its resolvent are equal:*

$$\Delta(h) = \Delta(R)$$

More inspired is the following observation

270

Lemma 64 (Gerd Fisher). *The Descartes resolvent has the zeros*

$$[v_1, v_2, v_3] = [x_1x_2 + x_3x_4 - p, \ x_1x_3 + x_2x_4 - p, \ x_1x_4 + x_2x_3 - p]$$

Proof. One uses Viëta's $-2p = v_1 + v_2 + v_3$ and the result (VII.3.4) and calculates. The order comes out as stated. The sum gives a further check,— now via Viëta's formula for the coefficient p in the depressed quartic h. $\qquad\square$

Shorter proof of proposition 64.

$$v_1 - v_2 = x_1x_2 + x_3x_4 - x_1x_3 - x_2x_4 = (x_1 - x_4)(x_2 - x_3)$$
$$v_1 - v_3 = x_1x_2 + x_3x_4 - x_1x_4 - x_2x_3 = (x_1 - x_3)(x_2 - x_4)$$
$$v_2 - v_3 = x_1x_3 + x_2x_4 - x_1x_4 - x_2x_3 = (x_1 - x_4)(x_2 - x_3)$$

$$\square$$

VII.4   Towers of Fields From Square and Third Roots

Definition 35 (extension of a tower). The tower

$$\widetilde{\mathbf{F}_0} \subseteq \widetilde{\mathbf{F}_1} \subseteq \widetilde{\mathbf{F}_2} \subseteq \cdots \subseteq \widetilde{\mathbf{F}_l}$$

is called an *extension* of the tower

$$\mathbf{F}_0 \subset \mathbf{F}_1 \subset \mathbf{F}_2 \subset \cdots \subset \mathbf{F}_n$$

if and only if $\widetilde{\mathbf{F}_0} = \mathbf{F}_0$ and for all $0 \leq i \leq n$ there exists $0 \leq k \leq l$ such that $\mathbf{F}_i \subseteq \widetilde{\mathbf{F}_k}$.

Proposition 65 (Towers of fields in steps of at most four). *Assume that there exists a tower of extensions of dimension at most four*

(VII.4.1) $$\mathbf{Q} = \mathbf{F}_0 \subset \mathbf{F}_1 \subset \mathbf{F}_2 \subset \cdots \subset \mathbf{F}_n \subset \mathbb{R}$$

such that $x \in \mathbf{F}_n$. Then there exists a tower of real extensions of dimension at most three

(VII.4.2) $$\mathbf{Q} = \widetilde{\mathbf{F}_0} \subset \widetilde{\mathbf{F}_1} \subset \widetilde{\mathbf{F}_2} \subset \cdots \subset \widetilde{\mathbf{F}_l} \subset \mathbb{R}$$

such that $x \in \widetilde{\mathbf{F}_l}$ and each extension is of one of the following three types

(i) *The step from the i-th field to the $i+1$-th field is a one-element extension by a real square root:* $\widetilde{\mathbf{F}_{i+1}} = \widetilde{\mathbf{F}}_i(\sqrt{x_i})$ *with* $x_i \in \widetilde{\mathbf{F}}_i$*;*

(ii) *The step from the i-th field to the $i+1$-th field is a one-element extension by a real cube root:* $\widetilde{\mathbf{F}_{i+1}} = \widetilde{\mathbf{F}}_i(\sqrt[3]{x_i})$ *with* $x_i \in \widetilde{\mathbf{F}}_i$*;*

(iii) *The step from the i-th field to the $i+1$-th field is a trisection of an angle. In other words, there exists a real angle θ such that $\cos\theta \in \widetilde{\mathbf{F}}_i$, and the next field is the one element extension* $\widetilde{\mathbf{F}_{i+1}} = \widetilde{\mathbf{F}}_i(\cos\frac{\theta}{3})$*.*

In the case of item (iii) I say that we have an extension by *angle trisection*.

Proof. The given tower is extended in several steps. Note that an extension of an extension is again an extension. In finitely many steps is constructed the tower (VIII.3.1) as claimed. For the proof the tildes may be omitted.

For the case that $\dim[\mathbf{F}_{i+1} : \mathbf{F}_i] = 2$, the extension is a one-element extension $\mathbf{F}_i(\alpha)$ and the minimal polynomial $p \in \mathbf{F}_i[x]$ of α has degree 2. The zeros of the quadratic polynomial

$$p(x) = x^2 + bx + c$$

are given by the midnight formula

$$\alpha_{1,2} = \frac{-b \pm \sqrt{b^2 - 4c}}{2}$$

The extension is obtained by the square root of the discriminant $D = b^2 - 4c$. In the case of a real tower holds indeed $D > 0$, one is ready.

For the case that $\dim[\mathbf{F}_{i+1} : \mathbf{F}_i] = 3$, we have again a one-element extension $\mathbf{F}_i(\alpha)$ since there cannot be any intermediate fields. The minimal polynomial $p \in \mathbf{F}_i[x]$ of α has degree 3. We may assume the reduced form

(VII.2.1)
$$p(x) = x^3 + bx + c = 0$$

The discriminant

$$D = \frac{c^2}{4} + \frac{b^3}{27}$$

is either positive or negative. Since the minimal polynomial is irreducible, the case $D = 0$ with a multiple zero cannot occur.

Case 1: $D > 0$ The cubic has one real solution x_1. From the computer-ready formula VII.2.16 given above we get

$$x_1 = \sqrt[3]{-\frac{c}{2} + \sqrt{D}} - \frac{b}{3\sqrt[3]{-\frac{c}{2} + \sqrt{D}}}$$

One needs one additional two dimensional extension to adjoin $\sqrt{D}$, and a three dimensional extension to adjoin the real value of the third root.

Case 2: $D < 0$ The cubic has three real solutions. In this case, the resolvent equation (VII.2.3) has complex solutions. As explained above, in this case necessarily holds $b < 0$. One needs at first do a two dimensional extension adjoining $\sqrt{-b}$. The angle θ is defined such that

$$(\text{VII.4.3}) \qquad\qquad -\frac{c}{2} = \sqrt{\frac{-b^3}{27}} \cos\theta$$

and one gets in the real case three real solutions from Cardano's formula. Of course we need to adjoin only one of them, say

$$x_1 = 2\sqrt{\frac{-b}{3}} \cos\frac{\theta}{3}$$

and get the other two roots from a quadratic polynomial $p(x)/(x - x_1)$. Alternatively, one can do two quadratic extensions by $\sqrt{3}$ and $\sin\theta$, and use the addition theorem for cosine to get all three roots from formula (VII.2.14).

For the case that $\dim[\mathbf{F}_{i+1} : \mathbf{F}_i] = 4$, we may have an intermediate field, and are back to using two dimensional extensions twice following what is described above.

Under the assumption that no intermediate field exists, we get a four dimensional one element extension,—with a minimal polynomial of degree four. The minimal polynomial can be put in the depressed form

$$h(x) = r + qx + px^2 + x^4$$

Necessarily holds $r \neq 0$. In the case $q = 0$, a more simple approach via quadratic extensions is possible. I may now assume that $r \neq 0$ and $q \neq 0$, and proceed as explained above, in the section about equations of fourth order. One needs the roots of the resolvent

$$(\text{IX.1.6}) \qquad\qquad R(v) = -q^2 + (p^2 - 4r)v + 2pv^2 + v^3$$

the roots of which lie in extensions already constructed above. Sorry, I have tried to cheat a bid. We shall need *all three roots* of the resolvent. In the case of positive discriminant of the resolvent, we have used above only the real root, and did not use the further two conjugate complex ones. We now need to get all three roots from the computer-ready Cardano formulas (VII.2.16), which we need to cut down into formulas containing only real quantities. I leave it as an exercise for the reader to do this calculation, best with mathematica.

The next step is given by the formulas. They contain only square roots, but these may be complex ones. For them are used the formula for the complex square root

$$(\text{VIII.2.3}) \qquad \sqrt{x+iy} = \sqrt{\frac{\sqrt{x^2+y^2}+x}{2}} + i\,\text{sign}\,(y)\sqrt{\frac{\sqrt{x^2+y^2}-x}{2}}$$

In the end we get several two or three dimensional extensions of the above types (i), (ii) or (iii), all only among real fields. $\qquad\qquad\square$

Theorem 22 (Nicomedes through Hartshorne). *Equivalent are*

(a) *There exists a tower of extensions of dimension at most four*

$$(\text{VII.4.4}) \qquad\qquad \mathbf{Q} \subseteq \mathbf{F}_0 \subset \mathbf{F}_1 \subset \mathbf{F}_2 \subset \cdots \subset \mathbf{F}_n \subset \mathbb{R}$$

such that $x \in \mathbf{F}_n$.

(b) *There exists an extension tower of real extensions of dimension at most three*

$$(\text{VII.4.5}) \qquad\qquad \mathbf{Q} \subseteq F = \widetilde{\mathbf{F}_0} \subset \widetilde{\mathbf{F}_1} \subset \widetilde{\mathbf{F}_2} \subset \cdots \subset \widetilde{\mathbf{F}_l} \subset \mathbb{R}$$

such that $x \in \widetilde{\mathbf{F}_l}$ *and each extension is of one of the three types given in the proposition 65 above.*

(c) *In the Euclidean coordinate field* F^2*, the quantity* x *is constructible by means of compass and a two-marked straightedge,—with the restriction that the marks of the straightedge are only allowed to be put on straight lines, not on circles.*

Definition 36 (restricted marked straightedge, Solid construction). Let *restricted marked straightedge* denote a straightedge with two marks, which are only to be put at a given point or onto a given line, but not onto a given circle. A quantity constructed from the base field F as explained in the above theorem 22 is said to be obtained by *solid construction*.

Proof. Above has already been shown that assumption (a) implies assumption (b). The square root extensions are constructible with straightedge and compass. By the construction procedures given by Nicomedes, the third root extensions and the angle trisection are contructible by a two-marked straightedge and compass. The entire details are given by Hartshorne [10] and by Rothe [21]. In this way is shown that assumptions (b) imply assumptions (c).

Now assume conversely that the quantity x is constructible from the quantities in the Euclidean plane F^2, using only compass and the two-marked straightedge, to be put only among straight lines.

Construction steps by compass and unmarked straightedge produce quantities in a quadratic field extension. We now convert a construction step with two-marked straightedge into algebraic equations. Let the straightedge have marks of distance d, to be put onto the lines $y = mx$ and $y = b$. To put the marks on parallel lines cannot produce any definite new point. Hence $m \neq 0$. If one puts one mark P on the line $y = b$ and let the origin $(0,0)$ lie on the straightedge, the other mark Q of the straightedge turns out to lie on the *conchoid* with the equation

$$(x - \tfrac{b}{m})^2 + (y - b)^2 = d^2$$

If we achieve to put mark Q onto the line $y = mx$, its coordinate y satisfies the equation

$$(x^2 + y^2)(y - b)^2 = x^2(mx - b)^2 + y^2(y - b)^2 = d^2 y^2$$

This is a forth order polynomial of y with coefficients in the field $\mathbf{F}_i$, in the case with input quantities $b, d, m \in \mathbf{F}_i$. The constructed field extension $\mathbf{F}_i(y)$ is a one-element extension of degree at most 4. Hence any use of the accepted tools leads stepwise to quantities in field extensions specified as in item (a). $\qquad\square$

VII.5 Remarks and Computations of Gaussian Periods

VII.5.1 The Algebras $\mathcal{S}$ and $\mathcal{G}$

Let p be an odd prime. I use the vector space $\mathcal{S}$ the elements of which are the $p-1$-tuples $((r_1, 1), (r_2, 2), \ldots, (r_{p-1}, p-1))$ where the r_i are arbitrary rational numbers. Clearly $\mathcal{S}$ is a $p-1$-dimensional vector space over the rational numbers with the basis $(1, 1), (1, 2), \ldots, (1, p-1)$. Its operation of addition I call "adtion" defined to be

$$(\text{adtion} , ((r_1, 1), (r_2, 2), \ldots, (r_{p-1}, p-1)) \, ((s_1, 1), (s_2, 2), \ldots, (s_{p-1}, p-1)))$$
$$:= ((r_1 + s_1, 1), (r_2 + s_2, 2), \ldots, (r_{p-1} + s_{p-1}, p-1))$$

Its operation of scalar multiplication I call "scalarcation" which is defined to be

$$(\text{scalarcation } q \, ((r_1, 1), (r_2, 2), \ldots, (r_{p-1}, p-1)))$$
$$:= ((q\,r_1, 1), (q\,r_2, 2), \ldots, (q\,r_{p-1}, p-1))$$

Mathematically spoken the space $\mathcal{S}$ is just a $p-1$-dimensional extension of the rational field. I introduce the extra terms

$$(\text{VII.5.1}) \qquad ((q, 0)) := ((-q, 1), (-q, 2), \ldots, (-q, p-1))$$

I call these the *constant* terms. The procedure proj0 maps any redundently written element by setting

$$(\text{proj0} \, ((r_0, 0), (r_1, 1), (r_2, 2), \ldots, (r_{p-1}, p-1)))$$
$$:= ((r_1 - r_0, 1), (r_2 - r_0, 2), \ldots, (r_{p-1} - r_0, p-1))$$

Mathematically speaking this procedure proj0 is the identity mapping, [17] on the vector space $\mathcal{S}$. But that does not hold in the sense of a computational procedure. The procedure proj011 maps any (redundently written) element as follows

$$s := ((r_0, 0), (r_1, 1), (r_2, 2), \ldots, (r_{p-1}, p-1))$$
$$\mapsto \text{proj011} s = r_0 - \frac{1}{p-1} \sum_{1 \le i \le p-1} r_i = -\frac{1}{p-1} \sum_{1 \le i \le p-1} (r_i - r_0)$$

---

[17]The mapping $\iota \circ \text{proj0} \circ \iota^{-1}$ is the identity of $\mathcal{G}$.

This is the linear mapping. The projection onto the constant elements of $\mathcal{S}$ is the mapping [18]

$$s \mapsto \mathrm{Proj011}\, s := (\mathrm{proj011}\, s, 0)$$

The vector space $\mathcal{S}$ has two further structures. Firstly, the space becomes a Euclidean space with the inner product

$$(\mathrm{pairsinnerpairs}\ ((r_1, 1), (r_2, 2), \ldots, (r_{p-1}, p-1))$$
$$((s_1, 1), (s_2, 2), \ldots, (s_{p-1}, p-1)))$$
$$:= r_1 s_1 + r_2 s_2 + \cdots + r_{p-1} s_{p-1}$$

Moreover, the space becomes an algebra with the multiplication

$$(\mathrm{VII.5.2}) \qquad (\mathrm{pairsmalpairs}\ ((r_1, 1), (r_2, 2), \ldots, (r_{p-1}, p-1))$$
$$((s_1, 1), (s_2, 2), \ldots, (s_{p-1}, p-1)))$$
$$:= \sum_{1 \le i \le p-1 \ \mathrm{and}\ 1 \le k \le p-1} ((r_i\, s_k, (i+k) \mod p))$$

Sometimes I denote the algebra multiplication by the symbol $\star$. Given d orthogonal vectors $(b_1, b_2, \ldots, b_d)$ and any vector $s \in \mathcal{S}$, the orthogonal projection of s onto the span of $(b_1, b_2, \ldots, b_d)$ is given by the procedure

$$(\mathrm{proj}\ (b_1, b_2, \ldots, b_d) s) := \bigoplus_{1 \le i \le d} \frac{(\mathrm{pairsinnerpairs}\ b_i s)}{(\mathrm{pairsinnerpairs}\ b_i b_i)}\, b_i$$

Here the sum $\bigoplus$ refers to the operation "adtion" of addition in the space $\mathcal{S}$. The coefficients

$$c_i := \frac{(\mathrm{pairsinnerpairs}\ b_i s)}{(\mathrm{pairsinnerpairs}\ b_i b_i)} \quad \text{for } 1 \le i \le d$$

are given by the procedure "coordinates". Its output is the list

$$(\mathrm{coordinates}\ (b_1, b_2, \ldots, b_d)\, s) = (c_1 \ldots c_d)$$

If $s \ne \mathrm{proj}\, s$ the result of the procedure coordinates does not allow to reconstruct the vector s.

Next we have to express the operations and calculations in the way they appear directly in the complex plane. Let $\mathcal{G} \subset \mathbb{C}$ be the vector space with the elements

$$r_1 \exp \frac{2\pi i}{p} + r_2 \exp \frac{4\pi i}{p} + \cdots + r_{p-1} \exp \frac{(p-1)2\pi i}{p}$$

---

[18]The mapping $\iota \circ \mathrm{Proj011} \circ \iota^{-1}$ is the projection onto the constant elements of $\mathcal{G}$.

where the r_i are arbitrary rationals. The space $\mathcal{G}$ is a $p-1$-dimensional vector space over the rationals with the basis

$$(\text{VII.5.3}) \qquad \exp\frac{2\pi i}{p}, \ \exp\frac{4\pi i}{p}, \ldots, \exp\frac{(p-1)2\pi i}{p}$$

but now the common addition and multiplication of complex numbers are its operations of addition, scalar multiplication respectively algebra multiplication. Nevertheless, (sometimes) I denote the algebra multiplication by the symbol $\star$. The vector space $\mathcal{G}$ has further structures, too. The space becomes a Euclidean space with the inner product

$$(\text{VII.5.4}) \qquad \sum_{1\leq k\leq p-1} r_k \exp\frac{2k\pi i}{p} \cdot \sum_{1\leq l\leq p-1} s_l \exp\frac{2l\pi i}{p} := \sum_{1\leq k\leq p-1} r_k s_k$$

Moreover, the space becomes an algebra with the ordinary complex multiplication

$$\sum_{1\leq k\leq p-1} r_k \exp\frac{2k\pi i}{p} \star \sum_{1\leq l\leq p-1} s_l \exp\frac{2l\pi i}{p} := \sum_{1\leq l\leq p-1 \text{ and } 1\leq k\leq p-1} r_k\, s_l \exp\frac{(k+l)2\pi i}{p}$$

The procedure iota yields the isomorphism $\iota : \mathcal{S} \mapsto \mathcal{G}$ which fixes the coefficients r_i:

$$(\text{iota } ((r_0,0),(r_1,1),(r_2,2),\ldots,(r_{p-1},p-1))) := \sum_{0\leq k\leq p-1} r_k \exp\frac{2k\pi i}{p}$$

Indeed ι is an isomorphism of these vector spaces as well as Euclidean spaces and algebras. In other words for any $r, s \in \mathcal{S}$ and rational q there hold the compatibilities

$$\iota\, r + \iota\, s = \iota\,(\text{adtion}\, r s)$$
$$q\,\iota\, r = \iota\,(\text{scalarcation}\, q r)$$
$$\iota\, r \cdot \iota\, s = \iota\,(\text{pairsinnerpairs}\, r s)$$
$$\iota\, r \star \iota\, s = \iota\,(\text{pairsmalpairs}\, r s)$$

The redundant elements $((r_0,0))$ are included. In the space $\mathcal{S}$ has been defined

$$(1,0) = (-1,1) + (-1,2) + \cdots + (-1,p-1)$$

which is mapped by the above isomorphism ι to the complex number $1 \in \mathbb{C}$. The mapping $\iota \circ \text{proj0} \circ \iota^{-1}$ is indeed the identity of $\mathcal{G}$ since the sum of the $p-1$ non-positive p-th roots of unity from the basis (VII.5.3) equals 1. The mapping $\iota \circ \text{Proj011} \circ \iota^{-1}$

is the one-dimensional projection onto the constant elements of $\mathcal{G}$. These are the numbers

$$\sum_{1\leq k\leq p-1} r\exp\frac{2k\pi i}{p} = -r$$

with rational r. We see that the constants in $\mathcal{S}$ are isomorphic to the rational numbers in $\mathcal{G}$.

The computational implementation with mathematica of the above is as follows

```
nul = {{0,1}}; unit = {{1,0}};

Clear[simpli]; simpli[pairs_] := Which[
   pairs === {}, nul,
  Rest[pairs] === {} && First[pairs] == {}, nul,
  Rest[pairs] === {}, If[0 == First[First[pairs]], nul, pairs],
  First[pairs] == {}, simpli[Rest[pairs]],
   0 == First[First[pairs]], simpli[Rest[pairs]],
  pairs[[1]][[2]] == pairs[[2]][[2]],
  Prepend[
   simpli[Rest[Rest[pairs]]], {pairs[[1]][[1]] + pairs[[2]][[1]],
    pairs[[1]][[2]]} ],
  Rest[pairs] == nul  , Most[pairs],
  True, Prepend[simpli[Rest[pairs]] , First[pairs]  ]]

ordersimpli = Function[pairs,
   ordered = Sort[pairs, #1[[2]] < #2[[2]] &];
   simpli[simpli[ordered]]];

Clear[adtion]; adtion = Function[{pairs1, pairs2},
   ordersimpli[Join[pairs1, pairs2]]];

Clear[lstadtion]; lstadtion = Function[lstofs,
  all = Fold[Join, nul, lstofs]; ordersimpli[all]];

scalarcation = Function[{numb, pairs},
   Map[{numb*#[[1]], #[[2]]} &, pairs]];

proj0 = Function[{lstofpairs, p},
```

```
ordered = ordersimpli[lstofpairs];
firstpair1 = ordered[[1]][[1]];
firstpair2 = ordered[[1]][[2]];
pseudozero = Array[{-1*firstpair1, # - 1} &, p];
result =
 If[0 == firstpair2,
    adtion[pseudozero, lstofpairs], lstofpairs];
     ordersimpli[result]] ;


proj011 = Function[{lstofpairs, p},
  If[lstofpairs == {}, 0,
   proj0now = proj0[lstofpairs, p];
   comp11=Total[Map[#[[1]] &,
       proj0now]]/(p - 1); -1*comp11]];


Proj011 = Function[{lstofpairs, p},
              {{proj011[lstofpairs, p], 0}}] ;


pairsinnerpairs = Function[{pairlist1, pairlist2, p},
   pairinnercation = Function[{pair1, pair2},
     kroneckerdelta = Function[{k, l},
       If[Mod[k, p] == Mod[l, p], 1, 0]] ;
     delta = kroneckerdelta [pair1[[2]], pair2[[2]]];
     delta * pair1[[1]] *  pair2[[1]]] ;
   pairinnerpairs = Function[{anypair, pairlist},
 Total[Map[pairinnercation[anypair, #] &, pairlist]]];
 Total[Map[pairinnerpairs[#, pairlist2] &,pairlist1]]];


pairtimespairs = Function[{anypair, pairlist, p},
   pairvertexmultication2 = Function[{pair1, pair2},
     factor = pair1[[1]]*pair2[[1]];
     multica = Mod[pair1[[2]] + pair2[[2]], p];
     {factor, multica}] ;
   Map[pairvertexmultication2[anypair, #] &, pairlist]] ;


pairsmalpairs[pairlist1_, pairlist2_, p_] :=  Which[
  pairlist1 == {}, {},
  pairlist2 == {}, {},
```

```
   Rest[pairlist1] == {},
   ordersimpli[pairtimespairs[First[pairlist1], pairlist2, p]],
   True, rest2 = pairsmalpairs[Rest[pairlist1], pairlist2, p];
    first2 = pairtimespairs[First[pairlist1], pairlist2, p];
   adtion[rest2, first2]]


coordinates = Function[{anybasis, vector, p},
   compfactor = Function[bi,
      lbi = pairsinnerpairs[bi, bi, p];
     pairsinnerpairs[vector, bi, p]/lbi];
   Map[compfactor, anybasis]];


components = Function[{anybasis, vector, p},
   compfactor = Function[bi,
      lbi = pairsinnerpairs[bi, bi, p];
     pairsinnerpairs[vector, bi, p]/lbi];
     Map[ordersimpli[scalarcation[
           compfactor[#], #]] &, anybasis]];


proj = Function[{anybasis, vector, p},
   lstadtion[components[anybasis, vector, p]]];


idminusproj = Function[{anybasis, vector, p},
  adtion[vector,scalarcation[-1,proj[anybasis,vector,p]]]];


eulerterm = Function[{pair, p},
   eulers = Function[t,
     exponent = 2*Pi *I*t/p;
     Exp[exponent]] ;
   pair[[1]]* eulers[ pair[[2]]] ] ;


iota = Function[{pairlist, p},
        Total[Map[ eulerterm[#, p] &, pairlist]]];
```

VII.5.2   Polynomials and Elementary Symmetric Forms

Following the convention of mathematica, I write polynomials and their lists of coefficients both with ascending powers. Thus I give a polynomial $a_0 + a_1 z + \cdots + a_{d-1} z^{d-1} + a_d z^d$ by the list $(a_0, a_1, \ldots a_{d-1}, a_d)$.

The value of the above polynomial at z is given by the procedures "Hornernew" in the algebra $\mathcal{G}$ as well as "Horner" in the algebra $\mathcal{S}$.

$$(\text{Hornernew } (a_d, a_{d-1}, \ldots, a_1, a_0)z) := a_0 + z(a_1 + \cdots + z(a_{d-1} + za_d))$$

which of course may be expanded to $a_0 + a_1 z + \cdots + a_{d-1} z^{d-1} + a_d z^d$. Both procedures take advantage of the Horner scheme.

From the list of zeros $(z_1, z_2, \ldots, z_d)$ which may be given in any order, the procedure elemsyms calculates the list of coefficients of the monic polynomial:

$$(\text{elemsyms } (z_1, z_2, \ldots, z_d)) := (a_0, a_1, \ldots a_{d-1}, 1)$$

As well known, these coefficients are the signed elementary symmetric functions

$$a_0 = (-1)^d \prod_{1 \leq i \leq d} z_i; \ldots$$

$$a_{d-2} = \sum_{1 \leq i < j \leq d} z_i z_j$$

$$a_{d-1} = - \sum_{1 \leq i \leq d} z_i$$

Corresponding to elemsyms we define the procedure Selemsyms to produce the elementary symmetric forms in the algebra $\mathcal{S}$: Some implementation in mathematica follows.

```
(* Polynomials and elementary symmetric forms *)

Clear[Hornernew]; Hornernew[anypoly_, z_] := Which[
  Length[anypoly] == 1, First[anypoly],
  Length[anypoly] > 1,
  First[anypoly] + z*Hornernew[Rest[anypoly], z],
  True, Indeterminate]

Clear[Horner]; Horner[anypoly_, z_, p_] := Which[
  Length[anypoly] == 1, First[anypoly],
```

282

```
    Length[anypoly] > 1,
    adtion[First[anypoly],
     pairsmalpairs[z, Horner[Rest[anypoly], p], z, p]],
    True, Indeterminate]

showpoly = Function[coeff,
  l = Length[coeff];
  Do[Print["+ ", z^n ,
        "*(", coeff[[n + 1]], ")"], {n, 0, l - 1, 1}]];

syms12 = {z1*z2, (-1) (z1 + z2), 1};
syms123 = {-z1*z2*z3, z1*z3 + z2*z3, -z3};
syms1230 = {-z1*z2*z3, z1*z3 + z2*z3, -z3, 0};
syms012 = {0, z1*z2, (-1) (z1 + z2), 1};
syms012 + syms1230

{-z1 z2 z3, z1 z2 + z1 z3 + z2 z3, -z1 - z2 - z3, 1}

Clear[elemsyms]; elemsyms[zeros_] := Which[zeros == {}, {1},
  Rest[zeros] == {}, {-First[zeros], 1},
  True, syms12 = elemsyms[Most[zeros]];
  syms123 = Map[Times[#, -Last[zeros]] &, syms12];
  syms1230 = Append[syms123, 0];
  syms012 = Prepend[syms12, 0];
  syms012 + syms1230];

innerplus = Function[{plus, list1, list2},
  If[Length[list1] == Length[list2],
    Array[plus[list1[[#]], list2[[#]]] &,
      Length[list1]], "nonexistent"]];

Clear[Selemsyms]; unit = {{1, 0}};
  Selemsyms[zeros_, p_] :=
 Which[zeros == {}, {unit},
  Rest[zeros] == {}, {scalarcation[-1, First[zeros]], unit},
  True, syms12 = Selemsyms[Most[zeros], p];
  syms123 = Map[pairsmalpairs[scalarcation[
          -1, Last[zeros]], #, p] &, syms12];
```

```
syms1230 = Append[syms123, nul];
syms012 = Prepend[syms12, nul];
Map[proj0[#, p] &, innerplus[adtion, syms012, syms1230]]];
```

VII.5.3   Gaussian Periods

Let $p \geq 3$ be any odd prime. The construction of the regular p-gon by (real and complex) radicals involves the solution of the equation

$$\Phi_p(z) = 0 \tag{VII.5.5}$$

The Gauss periods allow to reduce this problem to a successive solution of corresponding equations and extraction of roots for lower primes. To begin with, Euler's formula $e^{it} = \cos t + i \sin t$ gives the analytic solutions

$$z_k = \exp \frac{2\pi i \cdot k}{p} \quad \text{with } k = 1, 2, \ldots, p. \tag{VII.5.6}$$

of equation (VII.5.5). By theorem 6 we know there exists a primitive root a for prime p. Hence all $k \in \mathcal{U}(p)$ in the unit group may be enumerated as the powers

$$k \equiv a^r \mod p \quad \text{with } 0 \leq r < p - 1.$$

Indeed we have obtained a bijection $r \in \mathbf{Z}/\mathbf{Z}(p-1) \mapsto k \in \mathcal{U}(p)$. One puts this formula into equation (VII.5.6) to obtain

$$d_r := \exp \frac{2\pi i \cdot a^r}{p} \quad \text{with } 0 \leq r < p - 1. \tag{VII.5.7}$$

which is indeed a rather clever formula for the solutions of equation (VII.5.5).

Gauss' method to solve equation (VII.5.5) proceeds by finding in several steps suitable sums of the d_r.

Definition 37. Let $d \mid p-1$ be any divisor of $p-1$. One defines the *Gaussian periods*

$$S[p, d, s] := \{a^{s+td} \in \mathcal{U}(p) : 0 \leq t < \tfrac{p-1}{d}\} \quad \text{for } 0 \leq s < d; \tag{VII.5.8}$$

and the *Gaussian sums*

$$G[p, d, s] := \sum_{0 \leq t < e} d_{s+td} = \sum_{0 \leq t < \frac{p-1}{d}} \exp \frac{2\pi i \cdot a^{s+td}}{p} \quad \text{for } 0 \leq s < d. \tag{VII.5.9}$$

Remark. Definition (VII.5.12) is indeed a generalization of equation (II.0.14) to arbitrary primes and primitive root. The special case involves Fermat prime $p = F_n$. its, primitve root $a = 3$, and the divisors $d = 2^q$ with $0 \le q \le 2^n$. With the the low digits, $s = l$, taking the values $0 \le l < 2^q$ holds

$$G(F_n, 2^q, s) = S(n, q, s) = \sum \{\exp \tfrac{2\pi i \cdot 3^{t2^q + s}}{F_n} : \text{ with } 0 \le t < 2^{2^n - q}\}$$

Remark. The Gaussian periods may also be considered as sums

$$(\text{VII.5.10}) \qquad S[p, d, s] := \bigoplus \{((1, a^{s+td})) : 0 \le t < \tfrac{p-1}{d}\}$$

with the operation of addition in the space $\mathcal{S}$.

Lemma 65. *The Gaussian sums $G[p, d, s]$ for $0 \le s < d$ are orthogonal for the inner product from equation (VII.5.4). A fortiori, they are linearly independent and all different from each other. Their sum is*

$$(\text{VII.5.11}) \qquad \sum_{0 \le s < d} G[p, d, s] = -1$$

Proof. At first we remark that

$$s := \sum_{0 \le r < p-1} d_r = -1$$

since the d_r and 1 are the roots of $z^p - 1 = 0$, and by Viëta's formula the roots of the latter equation have the sum zero. Hence

$$(\text{VII.5.12}) \qquad \sum_{0 \le s < d} G[p, d, s] := \sum_{0 \le s < d \text{ and } 0 \le t < \frac{p-1}{d}} d_{s+td} = \sum_{0 \le r < p-1} d_r = -1$$

Since the Gaussian sum $G[p, d, s]$ with $0 \le s < d$ are orthogonal for the inner product from equation (VII.5.4), they are linearly independent. $\square$

Any element $\sigma \in \mathcal{U}(p)$ acts on the elements of the vector space $\mathcal{S}$ by putting
$$(\text{VII.5.13})$$
$$\sigma((r_1, 1), (r_2, 2), \ldots, (r_{p-1}, p-1)) := ((r_1, \sigma 1), (r_2, \sigma 2), \ldots, (r_{p-1}, \sigma(p-1)))$$

Take two different basis elements $(1, k) \in \mathcal{S}$ and $(1, l) \in \mathcal{S}$. The image $\sigma(1, k)$ is $(1, \sigma k \mod p)$. with the product σk taken modulo p. The definition (VII.5.2) of the multiplication $\star$ and the distributive (!) law yield

$$\sigma\left(\exp\frac{2\pi i\cdot k}{p}\star\exp\frac{2\pi i\cdot l}{p}\right)=\sigma\exp\frac{2\pi i\cdot(k+l)}{p}$$

$$=\exp\frac{2\pi i\cdot\sigma(k+l)}{p}=\exp\frac{2\pi i\cdot(\sigma k+\sigma l)}{p}=\left(\exp\frac{2\pi i\cdot\sigma k}{p}\star\exp\frac{2\pi i\cdot\sigma l}{p}\right)$$

and similarly in the algebra $\mathcal{S}$.

(VII.5.14)
$$\sigma((1,k)\star(1,l))=\sigma(1,k+l)=(1,\sigma k+\sigma l)=(1,\sigma k)\star(1,\sigma l)$$
$$=\sigma(1,k)\star\sigma(1,l)$$

By linear extension σ becomes an automorphism of the algebra $\mathcal{S}$.

Remark. The automorphism $\sigma=a^v$ acts in on the algebra $\mathcal{S}$ not by the operation of algebra multiplication, but using the multiplication modulo p inside the denominator of the exponent of $\exp\frac{2\pi ik}{p}$. This situation leads to the equation (VII.5.14).

Fix the odd prime p. By $\mathcal{G}(p)\subset\mathcal{G}$ is denoted the subspace of unit roots with p fixed.

Lemma 66 (The field from the p-th unit roots). *Fix the odd prime p. The algebra $\mathcal{G}(p)$ is a field, too.*

Remark. Commonly, the field $\mathcal{G}(p)$ is denoted by $\mathbf{Q}(\zeta_p)$.

Proof. Fix the odd prime p. The polynomial $\Phi_p(z)$ is irreducible, as may be proved with the Eisenstein criterium 37. Let (Φ_p) denotes the <u>principal ideal</u> generated by the irreducible polynomial $\Phi_p(z)$. By Corollary 15 the quotient ring $\mathbf{Q}[x]/(\Phi_p)$ is a field.

Let $r_k\in\mathbf{Q}$ be rational numbers. There exists bijection

(VII.5.15)
$$\sum_{0\le k\le p-1}r_k\,z^k+(\Phi_p(z))\in\mathbf{Q}[x]/(\Phi_p)\mapsto\sum_{1\le k\le p-1}(r_k-r_0)\exp\frac{2\pi i\cdot k}{p}\in\mathcal{G}(p)$$

which is a ring isomorphism. Since the quotient ring $\mathbf{Q}[x]/(\Phi_p)$ is a field, we conclude. that $\mathcal{G}(p)$ is a field, too.

We need to exploit the special properties of the unit roots, to confirm that the formula

(VII.5.15) defines a ring isomorphism. At first note that the formula (VII.5.15) defines a mapping. Indeed

$$\sum_{0\leq k\leq p-1} (r_k - s_k)\, z^k \in (\Phi_p(z)) \;\Rightarrow\; (r_k - s_k) = r \text{ for all } k$$

$$\Rightarrow \sum_{1\leq k\leq p-1} (r_k - r_0) - (s_k - s_0)\, \exp\frac{2\pi i \cdot k}{p} = 0$$

Moreover the mapping is injective since

$$\sum_{1\leq k\leq p-1} (r_k - r_0)\, \exp\frac{2\pi i \cdot k}{p} = 0 \;\Rightarrow\; r_k = r_0 \text{ for all } k$$

$$\Rightarrow \sum_{0\leq k\leq p-1} r_k\, z^k = r_0\Phi_p(z) \;\Rightarrow\; \sum_{0\leq k\leq p-1} r_k\, z^k + (\Phi_p(z)) = 0 \in \mathbf{Q}[x]/(\Phi_p)$$

To show that the mapping preserves the addition and multiplication I may leave to the reader. As an example, I check that for $1 \leq k \leq p-1$ holds

$$z_k \mapsto z^k + (\Phi_p(z)) \ \text{ and } \ ]z_{p-k} \mapsto z^{p-k} + (\Phi_p(z)) \ \text{ and } \ 1 \mapsto = 1 + (\Phi_p(z))$$

but indeed $z_k \star z_{p-k} = 1$ and

$$(z^k + \Phi_p(z))(z^{p-k} + \Phi_p(z)) = (z^p - 1 + 1 + \Phi_p(z)) = 1 + (\Phi_p(z))$$

$\square$

Lemma 67 (Main Lemma). *Fix the odd prime p. The algebra $\mathcal{G}(p)$ is a field, too. Any automorphism of the algebra $\mathcal{G}(p)$ is of the form $\sigma = \Gamma(l)$ for some $l \in \mathcal{U}(p)$ and acts on the base elements by*

(VII.5.16)
$$\Gamma(l)\exp\frac{2\pi i \cdot k}{p} = \exp\frac{2\pi i \cdot l\, k}{p} \quad \text{for all } k \in \mathcal{U}(p)$$

With a primitive root a modulo p, we get (sloppily) $\sigma = a^v$ with some exponent modulo $p - 1$.

Proof. For any $k \in \mathcal{U}(p)$, the root

$$z_k := \exp\frac{2\pi i \cdot k}{p}$$

satisfies the equation

$$((\text{VII.5.5})) \qquad\qquad \Phi_p(z_k) = 0$$

Applying any automorphism σ of the algebra $\mathcal{G}(p)$, one gets

$$\Phi_p(\sigma z_k) = 0$$

hence σz_k is just another solution of equation (VII.5.5). Let $l \in \mathcal{U}(p)$ be given by $\sigma z_1 = z_l$. Hence for all k holds

$$\sigma z_k = \sigma(z_1)^k = (\sigma z_1)^k = (z_l)^k = z_{lk}$$

as to be shown $\qquad\qquad\qquad\qquad\qquad\qquad\qquad\qquad\qquad\qquad\qquad\qquad\square$

Problem 137. *Convince yourself that the algebra $\mathcal{S}(p)$ is a field, too.*

Problem 138. *Prove that any automorphism of the algebra $\mathcal{S}(p)$ is an element $\sigma = \gamma(l)$ with $l \in \mathcal{U}(p)$ and acts on the base elements by*

$$(\text{VII.5.17}) \qquad \gamma(l)(1,k) = (1, l\,k) \ \text{ for all } k \in \mathcal{U}(p)$$

With a primitive root a modulo p, we get (sloppily) $\sigma = a^v$ with some exponent modulo $p-1$.

Short solution. Let σ be an automorphism of $\mathcal{S}(p)$. Hence $\iota^{-1} \circ \sigma \circ \iota$ is an automorphism of $\mathcal{G}(p)$. By the previous main lemma 67 exists $l \in \mathcal{U}(p)$ such that $\iota^{-1} \circ \sigma \circ \iota = \Gamma(l)$ and hence

$$\iota^{-1} \circ \sigma \circ \iota = \Gamma(l)$$
$$\sigma = \iota \circ \Gamma(l) \circ \iota^{-1}$$
$$\sigma(1,k) = \iota \circ \Gamma(l) \exp \frac{2\pi i \cdot k}{p} = \iota \circ \exp \frac{2\pi i \cdot l\,k}{p}$$
$$= (1, lk) \ \text{ for all } k \in \mathcal{U}(p)$$

$$\qquad\qquad\qquad\qquad\qquad\qquad\qquad\qquad\qquad\qquad\qquad\qquad\square$$

A less dressed-up solution. Fix the odd prime p and assume $1 \le k \le p-1$. According to the definition the multiplication holds for the base element $(1,k) \in \mathcal{S}(p)$

$$(1,k)^p = (1, pk \ \mathrm{mod}\ p) = (1,0)$$

which is the unit element in $\mathcal{S}(p)$. Applying any automorphism $\sigma \in \mathcal{S}(p)$ yields.

$$(\sigma(1,k))^p = (1,0)$$

Put $k = 1$. There exists l such that $0 \le l \le p-1$ and $\sigma(1,1) = (1,l)$. The case $l = 0$ is ruled out since σ maps onto $\mathcal{S}(p)$ For all $k \in \mathcal{U}(p)$ holds

$$\sigma(1,k) = \sigma(1,1)^k = (\sigma(1,1))^k = (1,l)^k = (1,lk)$$

as to be shown. $\qquad\square$

Problem 139. *Rather obviously holds $S[p,d,v] = S[p,d,w] \Leftrightarrow d \mid v - w$. Hence $S[p,d,v] = S[p,d,v \bmod d]$. Prove these claims.*

Solution. Equivalent are because of definition (VII.5.8)

$$S[p,d,v] = S[p,d,w]$$
$$\{a^{v+td} \in \mathcal{U}(p) : 0 \le t < p-1\} = \{a^{w+td} \in \mathcal{U}(p) : 0 \le t < p-1\}$$
$$a^v \equiv a^{w+td} \pmod{p} \quad \text{holds for some integer } t$$
$$v \equiv w + td \pmod{p-1} \quad \text{holds for some integer } t$$
$$d \mid v - w$$

$\qquad\square$

Problem 140. *Take any Gaussian period $S[p,d,s]$ and any automorphism $\sigma = a^v$. Show that*

(VII.5.18) $$a^v S[p,d,s] = S[p,d,v+s]$$

.

Solution . For any automorphism $\sigma = a^v$ holds the transformation

$$\sigma S[p,d,s] = \bigoplus \{((1,\sigma a^{s+td})) : 0 \le t < p-1\}$$
$$= \bigoplus \{((1,a^{v+s+td})) : 0 \le t < p-1\} = S[p,d,v+s]$$

$\qquad\square$

Proposition 66. *Let $d \mid p-1$ be a divisor of $p-1$. The Gaussian sums $G[p,d,s]$ for $0 \le s < d$ are the zeros of a monic polynomial $P(p,d;x)$ of degree d with integer coefficients.*

Proof. The elementary symmetric forms from the zeros
$S[p, d, 0], S[p, d, 1], \ldots, S[p, d, d-1]$ transform under $\sigma = a^v$ by

$$\sigma(\text{Selemsyms}\,(S[p, d, 0], S[p, d, 1], \ldots, S[p, d, d-1]))$$
$$= (\text{Selemsyms}\,(\sigma S[p, d, 0]; \sigma S[p, d, 1], \ldots, \sigma S[p, d, d-1]))$$
$$= (\text{Selemsyms}\,(S[p, d, v], S[p, d, v+1], \ldots, S[p, d, v+d-1]))$$
$$= (\text{Selemsyms}\,(S[p, d, v \bmod d]; S[p, d, (v+1) \bmod d] \ldots S[p, d, (v+d-1) \bmod d]))$$
$$= (\text{Selemsyms}\,(S[p, d, 0], S[p, d, 1], \ldots, S[p, d, d-1]))$$

and hence they are invariant under all automorphisms $\sigma \in \mathcal{U}(p)$. On the other hand, only the rational constants (VII.5.1) are invariant under all automorphisms $\sigma \in \mathcal{U}(p)$. Hence the symmetric forms

$$(VII.5.19) \qquad\qquad (\text{Selemsyms}\,(S[p, d, 0], S[p, d, 1], \ldots, S[p, d, d-1]))$$

are rational constants. Too, there definition implies that they are combinations $((r_1, 1), (r_2, 2), \ldots, (r_{p-1}, p-1))$ where the r_i are integers. Hence the symmetric forms (VII.5.19) are integer constants. In other words, the Gaussian periods $S[p, d, s]$ for $0 \le s < d$ are the zeros of a monic polynomial of degree d with integer constant coefficients. Mapping by the isomorphism ι , we conclude that the Gaussian sums $G[p, d, s]$ for $0 \le s < d$ are the zeros of a monic polynomial of degree d with integer coefficients. $\qquad\square$

Lemma 68. *In the most simple case $d = 2$, one obtains the following result about the Gaussian sums*

$$G[p, 2, 0] = \sum_{0 \le t < \frac{p-1}{2}} \exp \frac{2\pi i \cdot a^{2t}}{p} = \sum \left\{ \exp \frac{2\pi i \cdot k}{p} : \ k \text{ is a quadratic residue} \right\}$$

$$G[p, 2, 1] = \sum_{0 \le t < \frac{p-1}{2}} \exp \frac{2\pi i \cdot a^{1+2t}}{p} = \sum \left\{ \exp \frac{2\pi i \cdot k}{p} : \ k \text{ is a non-residue} \right\}$$

The Gaussian sums $G[p, 2, 0]$ and $G[p, 2, 1]$ satisfy the quadratic equation

$$x^2 + x + \frac{1 - (-1)^{\frac{p-1}{2}} \cdot p}{4} = 0$$

and moreover holds

$$G[p, 2, 0] = \frac{-1 + \sqrt{(-1)^{\frac{p-1}{2}} \cdot p}}{2} \quad and \quad G[p, 2, 1] = \frac{-1 - \sqrt{(-1)^{\frac{p-1}{2}} \cdot p}}{2}$$

These quantities are real for $p \equiv 1 \pmod 4$ and conjugate complex for $p \equiv 3 \pmod 4$. The signs of the roots are harder to know. I have checked for all primes $p < 1\,000$ that the signs <u>in front</u> of the roots to be as stated.

Problem 141. *Prove that the Gaussian sums*

$$\text{(VII.5.20)} \qquad G[p, \tfrac{p-1}{2}, t] = 2\cos\frac{2\pi a^t}{p} \qquad \text{for } 0 \le t < \tfrac{p-1}{2}.$$

The roots of unity $\exp\pm\frac{2\pi a^t}{p}$ *are hence obtained as solutions of the quadratic equations*

$$z^2 - G[p, \tfrac{p-1}{2}, t]z + 1 = 0$$

Solution. The corresponding Gaussian periods are

$$\text{(VII.5.21)} \qquad S[p, \tfrac{p-1}{2}, t] = \{a^t,\, a^{t+\frac{p-1}{2}}\} \qquad \text{for } 0 \le t < \tfrac{p-1}{2}.$$

Since a is assumed to be a primitive root holds

$$a^{\frac{p-1}{2}} \equiv -1 \pmod p$$

By definition the corresponding Gaussian sums are

$$G[p, \tfrac{p-1}{2}, t] = 2\cos\frac{2\pi a^t}{p} = \exp\frac{2\pi i \cdot a^t}{p} + \exp\frac{2\pi i \cdot (-a^t)}{p} = 2\cos\frac{2\pi \cdot a^t}{p}$$

as claimed. $\qquad\qquad\qquad\qquad\qquad\qquad\qquad\qquad\qquad\qquad\qquad\qquad\qquad\square$

VII.5.4 Computations for Proposition 66

The procedure Gauss[p,d,s] calculates the Gaussian sums $G[p, d, s]$ symbolically. The procedure zeros[p,d] outputs the list of Gaussian sums $G[p, d, s]$ with $s = 0, 1, \ldots, d-1$. One may check numerically that these are the zeros of an integer polynomial.

```
Gauss = Function[{p,d,s},
   e = (p - 1)/d;
   If[IntegerQ[e] && PrimeQ[p],
    proot = PrimitiveRoot[p];
    ars = PowerMod[proot, Array[s + (# - 1)*d &, e ], p];
    eulersp = Function[t,
      exponent = 2*Pi *I*t/p;
```

```
      Exp[exponent]] ;
    Total[Map[eulersp, ars]],
    {"Nonprime or Noninteger", p, e} ]];

zeros = Function[{p, d}, Array[Gauss[ p, d, # - 1]&,  d]] ;

showpoly[Map[N[Expand[#]]&, elemsyms[zeros[31, 5]]]];
+ 1*(5. +0. I)
+ z*(1. +0. I)
+ z^2*(-21.+0. I)
+ z^3*(-12.+0. I)
+ z^4*(1. +0. I)
+ z^5*(1.)
```

More to the point, one may calculate in the algebra $\mathcal{S}$ and thus avoid all numerics. The procedure S[p,d,s] calculates the Gaussian period $S[p, d, s]$ symbolically. The procedure Szeros[p,d] outputs the list of Gaussian periods $S[p, d, s]$ with $s = 0, 1, \ldots, d-1$.

```
S = Function[{p,d,s},
    e = (p - 1)/d;
    If[IntegerQ[e],
      proot = PrimitiveRoot[p];
      ars = PowerMod[proot, Array[s + (# - 1)*d &, e ], p];
      Map[{1, #} &, ars],
      {"Noninteger", e} ]];

Szeros = Function[{p,d},Array[Function[s,S[ p,d,s - 1]],d]];
```

As an example, I choose one more

```
p = 31; d = 5; Szeros[p, d];
Selemsyms[Szeros[p, d], p];
```

The output is long, but one may check that this is a list of constant integer coefficients. Moreover, the result of

```
Array[idminusproj[{proj0[{{1, 0}}, p]},
        Selemsyms[Szeros[p,d],p][[#]],p] &, d]
```

```
{{{0, 1}}, {{0, 1}}, {{0, 1}}, {{0, 1}}, {{0, 1}}}
```

confirms once more that the coefficients are integers. Hence we may obtain the polynomial $P(p, d; z)$ by the commands

```
n759=Hornernew[Map[proj011[#,p] &,Selemsyms[Szeros[p,d],p]],z];
Expand[n759]
```

```
5 + z - 21 z^2 - 12 z^3 + z^4 + z^5
```

and may even check once more that its zeros are the Gaussian sums $G[p, d, s]$ with $s = 0, 1, \ldots, d - 1$.

```
n126 = Expand[n759] /. z -> iota[#, p] &[Szeros[p,d][[1]]];
FullSimplify[n126]
0
```

Note that all is done without need of any decimal number computations. In both ways one gets the polynomial

$$P(31, 5, z) = +5z^0 + 1z^1 - 21z^2 - 12z^3 + 1z^4 + 1z^5$$

Too, I have checked and proved that its zeros are indeed the Gaussian sums $G(31, 5, r)$ for $0 \leq r < 5$.

VII.5.5   Building a Tree of Gaussian Sums

Fix the odd prime p. By $\mathcal{S}(p) \subset \mathcal{S}$ is denoted the subspace of unit root exponents with p fixed.

Problem 142. *Fix the prime p and a divisor $c \mid p-1$. Prove the following statement: Any $x \in \mathcal{S}(p)$ is invariant under all automorphisms $\sigma = a^{cw}$ with arbitrary w if and only if x lies in the span the Gaussian periods $S[p, c, r]$ with $0 \leq r < c$.*

Proof. Assume that x lies in the span the Gaussian periods $S[p, c, r]$ with $0 \leq r < c$. In other words, x is a rational combinations of the Gaussian periods $S[p, c, r]$ with $0 \leq r < c$:

$$x = \oplus_{0 \leq r < c} \, q_r S[p, c, r]$$

with rational coefficients q_r. Under the automorphism a^{cw} one gets the transformation

$$a^{cw} x = \oplus_{0 \leq r < c} q_r a^{cw} S[p, c, r] = \oplus_{0 \leq r < c} q_r \, S[p, c, r + cw] = \oplus_{0 \leq r < c} q_r S[p, c, r] = x$$

confirming the invariance. Conversely take any $x \in \mathcal{S}(p)$ and assume $a^{cw} x = x$ to hold for all w. The standard basis in $\mathcal{S}(p)$ may be gives as powers of a primitive root. Hence there exist rational numbers q_t such that

$$x = \oplus_{0 \leq t < p-1} \left(q_t . a^t \right) = \oplus_{0 \leq t < p-1} q_t S[p, p-1, t]$$

and is transformed under an arbitrarily automorphism a^v to

$$a^v x = \oplus_{0 \leq t < p-1} \left(q_t . a^{t+v} \right) = \oplus_{0 \leq t < p-1} q_t S[p, p-1, t+v]$$

Use division with remainder to get $t = r + cs$ with $0 \leq r < c$ and $0 \leq s < \frac{p-1}{c}$. The assumed invariance implies

$$a^{-cw} x = x \quad \text{for all } w$$
$$\oplus_{0 \leq t < p-1} \left(q_t . a^{t-cw} \right) = \oplus_{0 \leq t < p-1} \left(q_t . a^t \right)$$
$$\oplus_{0 \leq r < c} \oplus_{0 \leq s < (p-1)/c} \left(q_{r+cs+cw} . a^{r+cs} \right) = \oplus_{0 \leq r < c} \oplus_{0 \leq s < (p-1)/c} \left(q_{r+cs} . a^{r+cs} \right)$$

Linear independence of the basis implies

$$q_{r+cs+cw} = q_{r+cs} \quad \text{for all } 0 \leq r < c, \, 0 \leq s < (p-1)/c \text{ and all } w.$$

Hence $q_{r+cs} = q_r$ for all $0 \leq r < c$ and $0 \leq s < (p-1)/c$. Since $S[p, c, r] = \oplus_{0 \leq s < (p-1)/c} \left(1 . a^{r+cs} \right)$ one gets

$$x = \oplus_{0 \leq r < c} \oplus_{0 \leq s < (p-1)/c} \left(q_{r+cs} . a^{r+cs} \right) = \oplus_{0 \leq r < c} \oplus_{0 \leq s < (p-1)/c} \left(q_r . a^{r+cs} \right)$$
$$= \oplus_{0 \leq r < c} q_r S[p, c, r]$$

In other words, the element x lies in the span the Gaussian periods $S[p, c, r]$ with $0 \leq r < c$, as to be shown. $\qquad \square$

We now explain how the Gaussian sums are used for the calculation of the p-th roots of unity. Let $c, d \mid p-1$ and $cd \mid p-1$ be divisors of $p-1$. The Gaussian periods are elements of the algebra $\mathcal{S}$ and

$$S[p, c, r] = \{ a^r, a^{r+c}, \ldots, a^{r+(p-1-c)} \}$$
$$S[p, cd, r+cs] = \{ a^{r+cs}, a^{r+cs+cd}, \ldots, a^{r+cs+(p-1-cd)} \}$$

The latter is a list of l elements where $(p-1-cd) = (l-1)cd$ hence $l = \frac{p-1}{cd}$. We get

$$S[p, c, r] = \oplus_{0 \leq s < d} S[p, cd, r+cs]$$

As illustration take the example $p = 31$, $c = 5$, $r = 1$, $d = 3$

$$S[31, 5, 1] = \{a^1, a^6, \ldots, a^{21}, a^{26}\} \quad \text{and} \quad S[31, 15, t] = \{a^t, a^{t+15}\}$$
$$S[31, 5, 1] = \oplus_{t=1,6,11} S[31, 15, t] = \oplus_{0 \leq s < 3} S[31, 15, r + 5s]$$

One may build a rooted tree with three levels. One puts $\mathcal{U}(p)$ at the root, which has c children counted by index $0 \leq r < c$. At each of these child vertices is put one Gaussian period $S[p, c, r]$. Each one of them has in turn d children counted by a second index $0 \leq s < d$. At each of these grandchildren is put one Gaussian period $S[p, cd, r + cs]$. Since

$$\mathcal{U}(p) = \oplus_{0 \leq r < c} S[p, c, r] \quad \text{and}$$
$$S[p, c, r] = \oplus_{0 \leq s < d} S[p, cd, r + cs] \quad \text{for } 0 \leq r < c,$$

we see that the Gaussian periods are just partitioned as one proceeds down one level, from parent to its children. We see that, as it should be, holds

$$\oplus_{0 \leq r < c} S[p, c, r] = \oplus_{0 \leq r < c} \oplus_{0 \leq s < d} S[p, cd, r + cs] = \oplus_{0 \leq t < cd} S[p, cd, t] = \mathcal{U}(p)$$

Proposition 67. *Let $c, d \mid p - 1$ and their product $cd \mid p - 1$ be fixed divisors of $p - 1$ and fix any child's index $0 \leq r < c$. The grandchildren's Gaussian sums $G[p, cd, r + cs]$ for $0 \leq s < d$ are the zeros of monic polynomial $P(p, c, d, r; x)$ of degree d. The coefficients of this polynomial are integer combinations of the Gaussian sums $G[p, c, r]$ with $0 \leq r < c$. Moreover we get from examples the conjecture that*

$$(\text{VII}.5.22) \quad P(p, c, d, r; x) = \left[\sum_{0 \leq i < d} \sum_{0 \leq t < c} q[p, c, d, (r - t) \bmod c, i] G[p, c, t] x^i \right] + x^d$$

with integer coefficients $q[p, c, d, (r - t) \bmod c, i]$. This conjecture has been proved below, too.

Proof. The Gaussian periods $S[p, cd, r + cs]$ for $0 \leq s < d$ are the zeros of the monic polynomial $P(p, c, d, r; x)$ the coefficients of which are the elementary symmetric forms listed as

$$(\text{VII}.5.23) \qquad \text{Selemsyms}\,(S[p, cd, r], S[p, cd, r + c], \ldots, S[p, cd, r + (d-1)c])$$

These forms transform under any automorphism $\sigma = a^v$ by

$$\sigma\,[\text{Selemsyms}\,(S[p, cd, r], S[p, cd, r + c], \ldots, S[p, cd, r + (d-1)c])]$$
$$= \text{Selemsyms}\,(\sigma S[p, cd, r], \sigma S[p, cd, r + c], \ldots, \sigma S[p, cd, r + (d-1)c])$$
$$= \text{Selemsyms}\,(S[p, cd, v + r], S[p, cd, v + r + c], \ldots, S[p, cd, v + r + (d-1)c])$$

By problem 142 the span for the Gaussian periods $S[p, c, r]$ for $0 \leq r < c$ consists of the $x \in \mathcal{S}$ which are invariant under all automorphisms $\sigma = a^{cw}$, with arbitrary $w..$ To confirm that the coefficients of polynomial $P(p, c, d, r; x)$ are rational combinations of the Gaussian sums $G[p, c, r]$ with $0 \leq r < c$, we need to check that the relevant elementary symmetric forms (VII.5.23) are invariant under all such automorphisms $\sigma = a^{cw}$. From the above calculation we see that

$$a^{cw}\left[\text{Selemsyms}\left(S[p, cd, r], S[p, cd, r + c], \ldots, S[p, cd, r + (d - 1)c]\right)\right]$$
$$= \text{Selemsyms}\left(\sigma S[p, cd, r + cw], \sigma S[p, cd, r + c(1 + w)], \ldots, \sigma S[p, cd, r + (d - 1 + w)c]\right)$$
$$= \text{Selemsyms}\left(\sigma S[p, cd, r], \sigma S[p, cd, r + c], \ldots, \sigma S[p, cd, r + (d - 1)c]\right)$$

By problem 139 the third indexes $r + cw, r + c(1 + w), \ldots, r + c(d - 1 + w)$ in the second line only matter modulo cd. Use division with remainder to get $w = u + qd$ with $0 \leq u < d$. One may replace in these indexes w by u to produce the sequence
$$r + cu, r + c(1 + u), \ldots, r + c(d - 1 + u)$$
Finally use a cyclic permutation and the result of problem 139 once more and get.the sequence of indexes $r, \ldots, r + c(u - 1), r + cu, r + c(1 + u), \ldots, r + c(d - 1)$ as in the last line. Thus the elementary symmetric forms (VII.5.23) are invariant under all automorphisms $\sigma = a^{cw}$. The coefficients of polynomial $P(p, c, d, r; x)$ have the following two properties:

(i) They are rational combinations of the Gaussian periods $S[p, c, r]$ with $0 \leq r < c$.

(ii) Too, they are integer combinations to Gaussian periods $S[p, cd, t]$ with $0 \leq t < cd$.

The latter claim follows from the definition of the elementary symmetric forms (VII.5.23). Because of the identity

$$S[p, c, r] = \oplus_{0 \leq s < d} S[p, cd, r + cs] \quad \text{for } 0 \leq r < c,$$

both statements (i) and (ii) together imply that elementary symmetric forms (VII.5.23) are integer combinations of the Gaussian periods $S[p, c, r]$ with $0 \leq r < c$. In other words, the Gaussian periods $S[p, cd, r + cs]$ for $0 \leq s < d$ are the zeros of a monic polynomial of degree d with integer constant coefficients.

Mapping by the isomorphism ι, we conclude that the Gaussian sums $G[p, cd, r + cs]$ for $0 \leq s < d$ are the zeros of a monic polynomial of degree d with integer coefficients. $\square$

Problem 143. *Prove the last claim equation* (VII.5.22).

Solution. From the above proof we get

$$(VII.5.24) \qquad P(p,c,d,r;x) = \left[\sum_{0 \le i < d} \sum_{0 \le t < c} q[p,c,d,r,t,i] G[p,c,t] x^i \right] + x^d$$

with separate indexes r, t which are both counted modulo c. We transform both sides under the automorphism $\sigma = a^v$. The right-hand side transforms as $G[p,c,t] \mapsto G[p,c,t+v]$. The monic polynomial $P(p,c,d,r;x)$ of degree d has as its zeros the grandchildren's Gaussian sums $G[p,cd,r+cs]$ for $0 \le s < d$. The latter are transformed under the automorphism a^v to the Gaussian sums $G[p,cd,r+v+cs]$ for $0 \le s < d$. Hence the polynomial on the left-hand side transforms as $P(p,c,d,r;x) \mapsto P(p,c,d,r+v;x)$. Thus the transformed equation is

$$(VII.5.25) \qquad P(p,c,d,r+v;x) = \left[\sum_{0 \le i < d} \sum_{0 \le t < c} q[p,c,d,r,t,i] \, G[p,c,t+v] \, x^i \right] + x^d$$

Hence $q[p,c,d,r,t,i] = q[p,c,d,r+v,t+v,i]$. We may put $v := -t$ and get $q[p,c,d,r,t,i] = q[p,c,d,r-t,0,i]$. The index 0 may be omitted The separate indexes r, t and hence their difference are counted modulo c. $\qquad \square$

VII.5.6   Computational Implementation of Proposition 67

The list of Gaussian periods $S[p,cd,r+cs]$ for $0 \le s < d$ are output by the procedure Szerosbranch[p,c,r,d]. The Gaussian periods $S[p,c,r]$ with $0 \le r < c$ yield the basis, which list is output by basis[p, c]. The mathematica code for these procedures follows

```
Szerosbranch = Function[{p,c,d,r},
   Array[ordersimpli[S[p,c*d,r + c*#]] & , d]];
```

```
basis = Function[{p, c},Array[ordersimpli[S[p,c,# - 1]] &, c]];
```

Since the grandchildren Szerosbranch[p,c,d,r] are the Gaussian periods $S[p,cd,r+cs]$ for $0 \le s < d$. it is clear that the replacement $r \mapsto r + c$ maps them cyclically. One may check directly for $0 \le r < c$ the equalities

```
Szerosbranch[p,c,d,r + c] ==
 Array[Szerosbranch[p,c,d,r][[Mod[# + 1, d, 1]]] &, d]
```

to be true. Hence the monic polynomial $P(p,c,d,r;x)$ depends only on $r \bmod c$.

But by Proposition 67 we know much more. The isomorphism $\iota^{-1} : \mathcal{G} \mapsto \mathcal{S}$ implies the following: The list of coefficients Selemsyms[Szerosbranch[p, c, d, r], p] of the monic polynomial $\iota^{-1} P(p, c, d, r; x)$ consists of integer combinations of the Gaussian periods $S[p, c, r]$ with $0 \leq r < c$ and hence all of them lie in the span of basis[p,c].

This statement is checked directly in the following way. We assemble the coordinates with respect to basis[p,c] by the procedure qmatrix[p,c,d,r,t,i]. The entire vector is produced by the procedure backqmatrix[p,c,d,r].

```
qmatrix = Function[{p,c,d,r,t,i}, coordinates[basis[p,c],
  Selemsyms[Szerosbranch[p,c,d,r], p][[1 + d - i]],p][[t+1]] ];


backqmatrix =Function[{p,c,d,r},Array[proj[basis[p,c],
      Selemsyms[Szerosbranch[p,c,d,r],p][[#]],p] &,d+1]];
```

We check that indeed holds

```
backqmatrix[p,c,d,r] == Selemsyms[Szerosbranch[p,c,d,r],p]
True
```

for any $0 \leq r < c$, as claimed above.

VII.5.7   Examples

The coefficients of this polynomial $P(p, c, d, r; x)$ may be computated explicitly as are integer combinations of the Gaussian sums $G[p, c, r]$ with $0 \leq r < c$. The implementation chooses the example $p = 31, c = 3, d = 5,$ for whhich we output all three values $0 \leq r < 3$. The monic polynomial with the zeros $G(31, 3, r)$ turns out to be

$$P(31, 3, z) = -8z^0 - 10z^1 + 1z^2 + 1z^3$$

More interestingly, we now get the "grandchildren polynomials" $P(p, c, d, r; x)$ output by means of procedure

```
coeffPr66 = Function[{p,c,d,r}, Append[Array[
      (Sum[qmatrix[p,c,d,r,t, # - 1]
            *G[p,c,t], {t,0,c - 1}]) &, d],1]];


showpoly[coeffPr66[p, c, d, 0], z];
showpoly[coeffPr66[p, c, d, 1], z];
showpoly[coeffPr66[p, c, d, 2], z];
```

We get from this example

$$P(31,3,5,0;z) = \begin{array}{l} (\quad 1 \quad G[31,3,0]+ \ 1G[31,3,1]+ \ 1G[31,3,2])z^0 \\ +(\quad 2 \quad G[31,3,0]+ \ 2G[31,3,1]+ \ 4G[31,3,2])z^1 \\ +(\ (-1)\ G[31,3,0]- \ 3G[31,3,1]- \ 4G[31,3,2])z^2 \\ +(\quad 1 \quad G[31,3,0]+ \ 2G[31,3,1]+ \ 1G[31,3,2])z^3 \\ +(\ (-1)\ G[31,3,0]+ \ 0G[31,3,1]+ \ 0G[31,3,2])z^4 \end{array} \ + z^5$$

$$P(31,3,5,1;z) = \begin{array}{l} (1 \quad G[31,3,0]+1 \ \ G[31,3,1]+1 \ \ G[31,3,2])z^0 \\ +(4 \quad G[31,3,0]+2 \ \ G[31,3,1]+2 \ \ G[31,3,2])z^1 \\ +((-4)\ G[31,3,0]-1 \ \ G[31,3,1]-3 \ \ G[31,3,2])z^2 \\ +(1 \quad G[31,3,0]+1 \ \ G[31,3,1]+2 \ \ G[31,3,2])z^3 \\ +(0 \quad G[31,3,0]-1 \ \ G[31,3,1]+0 \ \ G[31,3,2])z^4 \end{array} \ + z^5$$

$$P(31,3,5,2;z) = \begin{array}{l} (1 \quad G[31,3,0]+1 \ \ G[31,3,1]+1 \ \ G[31,3,2])z^0 \\ +(2 \quad G[31,3,0]+4 \ \ G[31,3,1]+2 \ \ G[31,3,2])z^1 \\ +((-3)\ G[31,3,0]-4 \ \ G[31,3,1]-1 \ \ G[31,3,2])z^2 \\ +(2 \quad G[31,3,0]+1 \ \ G[31,3,1]+1 \ \ G[31,3,2])z^3 \\ +(0 \quad G[31,3,0]+0 \ \ G[31,3,1]-1 \ \ G[31,3,2])z^4 \end{array} \ + z^5$$

Thus we confirm the last statement (VII.5.22) from Proposition 67.

VII.6   Roots of Unity

VII.6.1   Subgroups of $\mathcal{U}(p)$ and Fixed Fields

Let p be an odd prime and $d \mid p - 1$ be any divisor. We define the subgroup of the Euler group

$$(\text{VII.6.1}) \qquad\qquad H_d := \{g \in \mathcal{U}(p) : g^{\frac{p-1}{d}} = e\}$$

Problem 144. *Let a be a primitive root modulo p. Show that the sets $a^u \cdot H_d$ with $0 \le u < d$ are a complete system of cosets of the subgroup $H_d \subseteq \mathcal{U}(p)$.*

Solution. Take any $g \in \mathcal{U}(p)$. Since a is a primitive root, there exists v such that $0 \le v < p - 1$ and $g = a^v$. Division by d with remainder yields $v = u + q \cdot d$ with $0 \le u < d$. Hence

$$g = a^v = a^u \cdot (a^q)^d$$

Little Fermat implies that $a^d \in H_d$ and $(a^q)^d \in H_d$. Hence g lies in the coset $a^u \cdot H_d$, as claimed. $\qquad\square$

Hence the subgroup has the index

$$(\text{VII.6.2}) \qquad\qquad \text{index}[\mathcal{U}(p) : H_d] = d$$

and its order is $\frac{p-1}{d}$.

Remark. The subgroup $H_2 \subseteq \mathcal{U}(p)$ turns out to consist of all quadratic residues. The second coset $a \cdot H_2$ consists of the quadratic non-residues.

Problem 145. *Let a be a primitive root modulo p. Convince yourself that $a^v \in H_d \Leftrightarrow d \mid v$. Hence*

$$(\text{VII.6.3}) \qquad\qquad H_d = \{a^{dw} : 0 \le w < \tfrac{p-1}{d}\}$$

In other words, for all $g \in H_d$ holds $g = a^v$ with a unique v such that $0 \le v < p - 1$ and $d \mid v$.

Solution. By definition $a^v \in H_d \Leftrightarrow e = (a^v)^{\frac{p-1}{d}} = a^{v\frac{p-1}{d}}$. By the properties of a primitive root holds $a^w = e \Leftrightarrow p - 1 \mid w$. Hence $a^v \in H_d \Leftrightarrow p - 1 \mid v \cdot \frac{p-1}{d} \Leftrightarrow d \mid v$.

Take any $g \in \mathcal{U}(p)$. Assuming $0 \le v < p - 1$ there exists a unique v such that $g = a^v$. For all other values v' with $g = a^{v'}$ holds $p - 1 \mid v - v'$. Since $d \mid p - 1$, the equivalence $a^v \in H_d \Leftrightarrow d \mid v$ remains true after introducing the restriction $0 \le v < p - 1$. $\qquad\square$

Definition 38 (Fixed field). Let H be any subgroup of the automorphisms of the field $\mathcal{G}$. The *fixed field* $\text{Fix}(H)$ is the subfield consisting of those $x \in \mathcal{G}$ for which $hx = x$ holds for all $h \in H$. In other words

$$\text{Fix}(H) = \{x \in \mathcal{G} : hx = x \text{ holds for all } h \in H\}$$

Problem 146. *Convince yourself that $\text{Fix}(H)$ is indeed a subfield of $\mathcal{G}(p)$.*

Proof. As shown in the main Lemma VII.6.25 the group of the automorphisms of the field $\mathcal{G}(p)$ is the Euler group $\mathcal{U}(p)$. In other words, in the study of roots of unity holds $H \subseteq \mathcal{U}(p)$.

Take any such subgroup H and any elements $x, y \in \text{Fix}(H) \subseteq \mathcal{G}(p)$. For any element $\sigma \in H \subseteq \mathcal{U}(p)$ hold by definition $\sigma x = x$ and $\sigma y = y$. Since σ is extended distributively beyond unit roots addition yields $\sigma(x + y) = \sigma x + \sigma y = x + y$. Since $\sigma \in H$ is chosen arbitrarily, we get $x + y \in \text{Fix}(H)$ as claimed.

300

Take two unit roots z_k and z_l in $\mathcal{G}(p)$. The definition (VII.5.2) of the multiplication $\star$ and the distributive (!) law in the exponent yield

$$\sigma\left(\exp\frac{2\pi i \cdot k}{p} \star \exp\frac{2\pi i \cdot l}{p}\right) = \sigma\exp\frac{2\pi i \cdot (k+l)}{p}$$

$$= \exp\frac{2\pi i \cdot \sigma(k+l)}{p} = \exp\frac{2\pi i \cdot (\sigma k + \sigma l)}{p} = \left(\exp\frac{2\pi i \cdot \sigma k}{p} \star \exp\frac{2\pi i \cdot \sigma l}{p}\right)$$

in short-hand we get $\sigma(z_k \star z_l) = \sigma z_k \star \sigma z_l$.

Take any such subgroup H and any elements $x, y \in \mathrm{Fix}(H) \subseteq \mathcal{G}(p)$. By definition of the algebra $\mathcal{G}(p)$, its elements are sums

$$x = \sum_{1 \leq k \leq p-1} r_k \exp\frac{2k\pi i}{p} \quad \text{and} \quad y = \sum_{1 \leq l \leq p-1} s_l \exp\frac{2l\pi i}{p}$$

with rational coefficients r_k and s_l. The distributive law on the base level gives

$$x \star y = \sum_{1 \leq l \leq p-1 \text{ and } 1 \leq k \leq p-1} r_k\, s_l \exp\frac{(k+l)2\pi i}{p}$$

Since σ is extended distributively beyond unit roots we get

$$\sigma x = \sum_{1 \leq k \leq p-1} r_k \exp\frac{(\sigma k)2\pi i}{p} \quad \text{and} \quad \sigma y = \sum_{1 \leq l \leq p-1} s_l \exp\frac{(\sigma l)2\pi i}{p}$$

$$\sigma(x \star y) = \sum_{1 \leq l \leq p-1 \text{ and } 1 \leq k \leq p-1} r_k\, s_l\, \sigma \exp\frac{(k+l)2\pi i}{p}$$

$$= \sum_{1 \leq l \leq p-1 \text{ and } 1 \leq k \leq p-1} r_k\, s_l \exp\frac{\sigma(k+l)2\pi i}{p}$$

$$= \sum_{1 \leq l \leq p-1 \text{ and } 1 \leq k \leq p-1} r_k\, s_l \exp\frac{(\sigma k)2\pi i}{p} \exp\frac{(\sigma l)2\pi i}{p}$$

$$= \left(\sum_{1 \leq k \leq p-1} r_k \exp\frac{(\sigma k)2\pi i}{p}\right)\left(\sum_{1 \leq l \leq p-1} s_l \exp\frac{(\sigma l)2\pi i}{p}\right)$$

$$= (\sigma x) \star (\sigma y)$$

The above semi-obvious but not really easy handling of terms implies $\sigma(x \star y) = (\sigma x) \star (\sigma y)$. For any element $\sigma \in H \subseteq \mathcal{U}(p)$ hold by definition $\sigma x = x$ and $\sigma y = y$. We now get finally $\sigma(x \star y) = (\sigma x) \star (\sigma y) = x \star y$. Since $\sigma \in H$ is chosen arbitrarily, we get $x \star y \in \mathrm{Fix}(H)$ as claimed. $\qquad\square$

Proposition 68. *Let $d \mid p - 1$ be any divisor. The following subsets of $\mathcal{S}$ are equal*

(i) *The linear span of the Gaussian periods $S[p, d, s]$ for $0 \le s < d$;*

(ii) *the set of $x \in \mathcal{S}$ which are invariant under all automorphisms $\sigma = a^v$ with $d \mid v$;*

(iii) *the fixed field $\mathrm{Fix}(H_d)$ of the subgroup H_d from the automorphisms of $\mathcal{S}$.*

Proof. By the result of problem 142 the sets defined by items (i) and (ii) are equal. Any element $\sigma \in \mathcal{U}(p)$ may be written as a power $\sigma = a^v$ where the (later double!) exponent v is determined modulo $p - 1$. Equivalent are the following lines

$$\sigma\, S[p, d, s] = S[p, d, s] \quad \text{for all } \sigma \in H_d$$
$$\Leftrightarrow a^v S[p, d, s] = S[p, d, s] \quad \text{for all } a^v \in H_d$$
$$\Leftrightarrow S[p, d, v + s] = S[p, d, s] \quad \text{for all } v \text{ such that } d \mid v$$

But the last line is true by Problem 139. We conclude that any Gaussian periods $S[p, d, s]$, and hence the span of the Gaussian periods $S[p, d, s]$ for $0 \le s < d$ are a subset of the fixed field of the subgroup H_d. Hence the inclusion (i) $\subseteq$ (iii) holds.

Conversely take any $x \in \mathcal{S}$ lying in the fixed field of the subgroup H_d . This subgroup consists of the automorphisms $\sigma = a^v$ with $d \mid v$. Hence holds $a^v x = x$ for all v such that $d \mid v$. Thus x lies in the set specified by item (ii). Hence the inclusion (iii) $\subseteq$ (ii) holds. Hence the sets defined by items (i) and (iii) are equal. $\qquad\square$

Lemma 69 (Subgroups-lemma)**.** *Any subgroup $H \subseteq \mathcal{U}(p)$ is equal to $H = H_d$ for some divisor $d \mid p - 1$.*
Hence the group $\mathcal{U}(p)$ has exactly one subgroup for each divisor of its order.

Proof. Let $H \subseteq \mathcal{U}(p)$ be any subgroup. We define

$$d := \frac{p - 1}{|H|} \quad \text{and} \quad D := \max\{e : H \subseteq H_e\}$$

Take any element $g \in H$. By Lagrange's Theorem holds $e = g^{|H|} = g^{\frac{p-1}{d}}$. Hence $g \in H_d$. Thus we have obtained the inclusion $H \subseteq H_d$. From the maximality in the definition of D, we get $d \le D$.

The inclusion $H \subseteq H_D$ implies $|H| \le |H_D|$ and from equation (VII.6.2) we get

$$D = \mathrm{index}[\mathcal{U}(p) : H_D] = \frac{p - 1}{|H_D|}$$

$$d \le D = \frac{p - 1}{|H_D|} \le \frac{p - 1}{|H|} = d$$

Hence $d = D$ and $H = H_D$. $\qquad\square$

Remark. The last statement holds more generally for each cyclic group. Hence it holds especially for each group $\mathcal{U}(m)$ where a primitive root modulo m exists. On the other hand, it does not hold for all other cases. For example, the group $\mathbf{Z}_2 \times \mathbf{Z}_2 \times \mathbf{Z}_4 \simeq \mathcal{U}(60)$ has even two non-isomorphic subgroups of order 4.

Definition 39. Let p be an odd prime and let $d \mid p - 1$ be any divisor. In the study of the p-th unit roots occur the fields

$$F_d := \operatorname{span}\{G[p, d, s] : 0 \le s < d\}$$

generated by Gaussian sums.

Lemma 70. *Let p be an odd prime and let $d \mid p - 1$ be any divisor. The sets F_d are indeed fields and*

(VII.6.4) $$\operatorname{Fix}(H_d) = F_d$$

Moreover, the group H_d is cyclic and has the generator a^d.

Proof. By proposition 68, the fixed field $\operatorname{Fix}(H_d)$ is F_d, the span of the Gaussian periods $G[p, d, s]$ for $0 \le s < d$. Hence the set F_d is indeed a field. $\qquad\square$

Proposition 69. *For any subgroup $H \subseteq \mathcal{U}(p)$ of the automorphisms of the field $\mathcal{G}(p)$ holds*

$$|H| = \dim[\mathcal{G}(p) : Fix(H)] = \frac{p-1}{d}$$

(VII.6.5) $$\operatorname{index}[\mathcal{U}(p) : H] = \dim[Fix(H) : \mathbf{Q}] = d$$

Moreover, any such subgroup is equal to $H = H_d$ for the divisor $d \mid p - 1$ which appears in the equality (VII.6.5), too

Proof. By Lemma 69 we know that $H = H_d$ holds for some divisor $d \mid p - 1$. By equation (VII.6.2) holds

((VII.6.2)) $$\operatorname{index}[\mathcal{U}(p) : H_d] = d$$

By proposition 68, the fixed field $\operatorname{Fix}(H_d)$ is F_d, the span of the Gaussian periods $G[p, d, s]$ for $0 \le s < d$. As shown in Lemma 65, these Gaussian periods are linearly independent over the rationals. Hence

$$\operatorname{Fix}(H_d) = F_d$$

(VII.6.6) $$\dim[\operatorname{Fix}(H_d) : \mathbf{Q}] = \dim F_d = d$$

Finally equations (VII.6.2) and (VII.6.6) yield the claimed result (VII.6.5). $\qquad\square$

Problem 147. *Prove that under the assumption of proposition 69 holds*

$$\text{(VII.6.7)} \qquad\qquad |H| = \dim[\mathcal{G}(p) : Fix(H)] = \frac{p-1}{d}$$

Lemma 71. *Let p be any odd prime, a a primitive root and assume $d \mid p-1$. The polynomial $P[p, d; z]$ may be reduced by substituting $y = dz + 1$. Let $H(y) = d^d P[p, d; \frac{y-1}{d}]$ be the resulting monic polynomial. The polynomial $H(y)$ is irreducible.*

Problem 148. *Prove lemma 71.*

The following conjecture I have checked for some low primes.

Conjecture 3. *Under the assumption of lemma 71, the polynomial $H(y)$ satisfies with prime p the assumptions of the Eisenstein criterium 37.*

VII.6.2   The Groups of Automorphisms

Definition 40 (Group of automorphisms with a fixed field). Let $F \supseteq E \supseteq \mathbf{Q}$ be any fields. The group of automorphisms of the field F that fix all elements of the subfield E is denoted by $\text{Aut}(F/E)$.

Lemma 72. *Let p be an odd prime and let $d \mid p-1$ be any divisor.*

$$\text{Aut}(\mathcal{G}(p)/F_d) = H_d \quad\text{and}\quad \text{Fix}(H_d) = F_d$$

Proof.

Problem 149. *Assume $F \supseteq E$ and $F \supseteq \text{Fix}(H)$. Convince yourself that the inclusions*

$$\text{(VII.6.8)} \qquad\qquad \text{Fix}(\text{Aut}(F/E)) \supseteq K \quad\text{and}$$
$$\text{(VII.6.9)} \qquad\qquad \text{Aut}(F/\text{Fix}(H)) \supseteq H$$

follow directly from the definitions 38 and 40.

Check of inclusion (VII.6.8). Take any automorphism $\sigma \in \text{Aut}(F/E)$. By definition 40 holds $\sigma k = k$ for all $k \in E$.

   Hence $\sigma k = k$ holds for all automorphisms $\sigma \in \text{Aut}(F/E)$ and for all $k \in E$.

   Take any field element $k \in E$. By definition 38 we see that $k \in \text{Fix}(\text{Aut}(F/E))$.

$\square$

Check of inclusion (VII.6.9). For all $k \in \text{Fix}(H)$ holds by definition 38 $\sigma k = k$ for all $\sigma \in H$..

Hence $\sigma k = k$ holds for all $k \in \text{Fix}(H)$ and for all automorphisms $\sigma \in H$..

Take any $\sigma \in H$. By definition 40 we see that $\sigma \in \text{Aut}(F/\text{Fix}(H))$. $\qquad\square$

We have shown above that

$$\text{(VII.6.10)} \qquad\qquad\qquad \text{Fix}(H_d) = F_d$$

and the main Lemma VII.6.25 tells that

$$\text{Aut}(\mathcal{G}(p)/\mathbf{Q}) = \mathcal{U}(p)$$

Hence

$$\mathcal{U}(p) \supseteq \text{Aut}(\mathcal{G}(p)/F_d) = \text{Aut}(\mathcal{G}(p)/\text{Fix}(H_d)) \supseteq H_d$$

Clearly $H \supseteq H' \Rightarrow \text{Fix}(H) \subseteq \text{Fix}(H')$ and hence

$$\text{Fix}(\text{Aut}(\mathcal{G}(p)/F_d)) \subseteq \text{Fix}(H_d)$$

On the other hand inclusion (VII.6.8) yields

$$\text{Fix}(\text{Aut}(\mathcal{G}(p)/F_d)) \supseteq F_d$$

which together with equation (VII.6.10) imply

$$\text{Fix}(\text{Aut}(\mathcal{G}(p)/F_d)) = \text{Fix}(H_d) = F_d$$

By the subgroups-lemma 69 there exists some divisor $D \mid p-1$ such that $\text{Aut}(\mathcal{G}(p)/F_d) = H_D$. Put together this yields

$$\begin{aligned}
S_D &= \text{Fix}(H_D) = \text{Fix}(H_d) = F_d \\
D &= \dim S_D = \dim F_d = d \\
\text{Aut}(\mathcal{G}(p)/F_d) &= H_D = H_d
\end{aligned}$$

$$\square$$

Problem 150. *Let p be an odd prime and let $d \mid p - 1$ be any divisor. Convince yourself that any automorphism a^v with $d \mid v$ leaves any space F_d invariant.*

Solution. By definition 39

$$F_d := \mathrm{span}\{G[p, d, s] : 0 \le s < d\}$$

and by problem 140 holds

$$a^v G[p, d, s] = G[p, d, v + s]$$

Hence

$$a^v F_d = \mathrm{span}\{G[p, d, v + s] : 0 \le s < d\}$$

But $G[p, d, s]$ depends only on s modulo d. Hence

$$a^v F_d = \mathrm{span}\{G[p, d, v + s] : 0 \le s < d\} = \mathrm{span}\{G[p, d, s] : 0 \le s < d\} = F_d$$

as claimed. $\qquad\square$

Lemma 73. *Let p be an odd prime and let $e \mid d \mid p - 1$ be any divisors and assume $e < d$. The order of the automorphism group is*

$$(\text{VII.6.11}) \qquad\qquad |\mathrm{Aut}(F_d/F_e)| = \frac{d}{e}$$

Moreover we conclude the mapping a^{ew} with $0 \le w < \frac{d}{e}$ already give the entire group $\mathrm{Aut}(F_d/F_e)$. This group is cyclic and a^e is a generator.

Its generator a^e permutes the Gaussian sums spanning F_d as follows

$$a^e G[p, d, s] = G[p, d, e + s \mod d]$$

We see that the basis of the vector space F_d is permuted in e cycles each of which has length $\frac{d}{e}$.

Proof. Because of lemma 72 holds

$$H_d \cdot \mathrm{Aut}(F_d/F_e) = \mathrm{Aut}(\mathcal{G}(p)/F_d) \cdot \mathrm{Aut}(F_d/F_e) \subseteq \mathrm{Aut}(\mathcal{G}(p)/F_e) = H_e$$

and the left-hand side is even a subgroup. Hence equation (VII.6.5) yields

$$|H_d| \cdot |\mathrm{Aut}(F_d/F_e)| \mid |H_e|$$

$$|\mathrm{Aut}(F_d/F_e)| \mid \frac{|H_e|}{|H_d|} = \frac{d}{e}$$

giving one inequality.

We now prove the converse inequality. From equations (VII.6.3) and (VII.5.18) one gets

$$\mathrm{Aut}(\mathcal{G}(p)/F_e) = H_e = \{a^{ew} : 0 \le w < \tfrac{p-1}{e}\}$$

On the other hand

$$\mathrm{Aut}(\mathcal{G}(p)/F_d) = H_d = \{a^{e\cdot(\frac{d}{e}u)} : 0 \le u < \tfrac{p-1}{d}\}$$
$$= \{a^{ew} : w = \tfrac{d}{e}u \text{ and } 0 \le u < \tfrac{p-1}{d}\}$$
$$= \{a^{ew} : \tfrac{d}{e} \mid w \text{ and } 0 \le w < \tfrac{p-1}{e}\}$$

which is for $e < d$ a properly smaller group. By problem 150 any automorphism a^v leaves any space F_d invariant. For all $0 \le w < \tfrac{d}{e}$. the mappings a^{ew} may be restricted to an automorphism of F_d that fixes $F_e \subset F_d$. But these restrictions are all different automorphisms. Indeed, by problem 140 holds

$$a^{ew} S[p, d, s] = S[p, d, ew + s]$$

for any Gaussian period $S[p, d, s]$. We conclude the inequality

$$|\mathrm{Aut}(F_d/F_e)| \ge \frac{d}{e}$$

Together we get equality (VII.6.11).

Moreover we conclude the mapping a^{ew} with $0 \le w < \tfrac{d}{e}$ already give the entire group $\mathrm{Aut}(F_d/F_e)$. This group is cyclic and a^e is a generator. $\qquad \square$

VII.6.3   The 7-th and 13-th Roots of Unity

Problem 151. *Calculate the polynomials in $\cos^k t$ for $\cos 3t$ and $\cos 5t$.*

Proof.

$$\cos 3x = \cos(x + 2x) = \cos x \cos 2x - \sin x \sin 2x$$
$$= \cos x(2\cos 2x - 1) - 2\sin 2x \cos x = \cos x(4\cos 2x - 3)$$
$$\sin 3x = \sin(x + 2x) = \sin x \cos 2x + \cos x \sin 2x$$
$$= \sin x(2\cos 2x - 1) + 2\sin x \cos 2x = \sin x(4\cos 2x - 1)$$

$$\cos 5x = \cos(3x + 2x) = \cos 3x \cos 2x - \sin 3x \sin 2x$$
$$= \cos x(4\cos 2x - 3)(2\cos 2x - 1) - 2\sin 2x(4\cos 2x - 1)\cos x$$
$$= \cos x(8\cos 4x - 10\cos 2x) + 3 + (2\cos 2x - 2)(4\cos 2x - 1))$$
$$= \cos x(8\cos 4x - 10\cos 2x) + 3 + 8\cos 4x - 10\cos 2x + 2)$$
$$= \cos x(16\cos 4x - 20\cos 2x + 5)$$

$\square$

Problem 152. *Take the very simple example $p = 7$. Find the six solutions of $\Phi_7(z) = 0$ by first obtaining the three roots of $P[7, 3; z] = 0$.*

Solution. Since $P[7, 3; z] = -1 - 2z + z^2 + z^3$, we put $y = 3z + 1$ and get the reduced polynomial $H(y) = 33G[7, 3, \frac{y-1}{3}]$ to be

$$H(y) = -7 - 3 \cdot 7y + y^3$$

Use $\cos 3t = \cos t(4\cos 2t - 3)$ and try the Ansatz $y = a\cos t$

$$H(y) = -7 - 21a\cos t + a^3 \cos 3t! = -7 - 3b\cos t + 4b\cos 3t = -7 + b\cos 3t$$

The equality marked by $! =$ holds if $b = 7a$, $4b = a^3$. Hence $28a = a^3$ and we put $b = 14\sqrt{7}$ to get

$$H(y) = 7(-1 + 2\sqrt{7}\cos 3t)$$

By solving $H(y) = 0$ we get $\cos 3t = \sqrt{7}/14$, and may calculate next $3t$ and the angle t. Finally there are three solutions with $r = \{0, 1, 2\}$ and $a^r = 3^r = \{1, 3, 2\} \mod 7$ which turn out to be:

$$\{t_r\} = \frac{1}{3}\arccos\frac{\sqrt{7}}{14} + \frac{2\pi r}{3} \qquad \text{and}$$

$$2\cos\frac{3^r\,2\pi}{7} = \{z_r\} := \frac{y-1}{3} = \frac{2\sqrt{7}\cos t_r - 1}{3}$$

$$2\cos\frac{\{1,3,2\}\,2\pi}{7} = \frac{2}{3}\sqrt{7}\cos\left(\frac{1}{3}\arccos\frac{\sqrt{7}}{14} + \frac{2\pi\{0,1,2\}}{3}\right) - \frac{1}{3}$$

To check how these three solutions are matched, I see no possibility than numerics. It turns out that the zeros match in the order indicated in the last line. From here one gets the complex zeros of $\Phi_7(z) = 0$ easily by problem 141. $\square$

Remark. Let

$$\omega = \frac{-1 + i\sqrt{3}}{2} \qquad \text{and} \qquad c = \sqrt[3]{\frac{7}{2}(1 + 3i\sqrt{3})}$$

The zeros of polynomial $P[7,3;w]$ are indeed

$$w_0 = \frac{1}{3}\left(-1 + c + \frac{7}{c}\right)$$

$$w_1 = \frac{1}{3}\left(-1 + \omega c + \omega^2\frac{7}{c}\right)$$

$$w_2 = \frac{1}{3}\left(-1 + \omega^2 c + \omega\frac{7}{c}\right)$$

$$2\cos\frac{2\pi \cdot 3^k}{7} = w_k = \frac{1}{3}\left(-\omega^0 + \omega^k c + \omega^{2k}\frac{7}{c}\right) \quad \text{for } k = 0,1,2 \mod 3$$

The automorphism σ mapping $c \mapsto \sigma c := \omega c$ maps the zeros cyclically according to $w_l \mapsto w_{\lfloor k + l \mod 3\rfloor}$.

Problem 153. *Take the very simple example $p = 7$. Find the six solutions of $\Phi_7(z) = 0$ by first obtaining the two roots of $P[7,2;z] = 0$.*

Solution. I still stay with the case $p = 7$ but switch the numbers c and d. One gets the alternative $c = 2$. Since $P[7,2;z] = 2 + z + z2$, one gets the already complex solutions

$$G[7,2;\{0,1\}] = z_{1,2} = \frac{-1 \pm \sqrt{-7}}{2}$$

The signs in front of the root match as claimed by Lemma 68. In the next and last level of the Gauss-sum tree, one has to solve two cubic equations, both of which now have complex coefficients.

$$P(7,2,3,0;z) = -1 + G[7,2,1]z - G[7,2,0]z^2 + z^3 = 0$$
$$P(7,2,3,1;z) = -1 + G[7,2,0]z - G[7,2,1]z^2 + z^3 = 0$$

As claimed by Proposition 67 the three solutions of the first equation are the grandchildren's Gaussian sums $G[p,cd,r+cs] = G[7,6,2s]$ for $0 \le s < 3$ hence

$$G[7,6,\{0,2,4\}] = \exp\frac{2\pi i\{3^0,3^2,3^4\}}{7} = \exp\frac{2\pi i\{1,2,-3\}}{7}$$

The three solutions of the second equation are the grandchildren's Gaussian sums $G[p,cd,r+cs] = G[7,6,1+2s]$ for $0 \le s < 3$ hence

$$G[7,6,\{1,3,5\}] = \exp\frac{2\pi i\{3^1,3^3,3^5\}}{7} = \exp\frac{2\pi i\{3,-1,-2\}}{7}$$

Once more, I have to check numerically how these solutions match. $\square$

Problem 154. *For the next example I take $p = 13$. Since $p - 1 = 12 = 3 \cdot 2 \cdot 2$, the solutions of $\Phi_{13}(z) = 0$ may be obtained in three steps. It is convenient to put the first factor $c = 3$. Thus only one cubic equation occurs. In this way get all twelve solutions of $\Phi_{13}(z) = 0$.*

Solution. From the DrRacket program one gets $P[13, 3; z] = 1 - 4z + z^2 + z^3$ We have to obtain the three roots of $P[13, 3; z] = 0$. Put $y = 3z + 1$ and get the reduced polynomial $H(y) = 33P[13, 3; \frac{y-1}{3}]$ to be

$$H(y) = +5 \cdot 13 - 3 \cdot 13y + y^3$$

For third degree polynomials with real zeros we have the Ansatz $y = a \cos t$

$$H(y) = 65 - 39a \cos t + a^3 \cos^3 t! = 65 - 3b \cos t + 4b \cos^3 t = 65 + b \cos(3t)$$

The equality marked by ! $=$ holds with $b = 13a$, $4b = a^3$. Hence $52a = a^3$ and we put $a = 2\sqrt{13}$, $b = 26\sqrt{13}$ to get

$$H(y) = 13(5 + 2\sqrt{13} \cos(3t))$$

By solving $H(y) = 0$ we get $\cos 3t = -5\sqrt{13}/26$, and calculate $3t$ and the angle t. Finally there are three solutions with $l = \{0, 1, 2\}$:

$$t_l = \frac{1}{3} \arccos \frac{-5\sqrt{13}}{26} + \frac{2\pi l}{3}$$

$$z_l = \frac{y-1}{3} = \frac{-1 + 2\sqrt{13} \cos t_l}{3}$$

$$z_l = -\frac{1}{3} + \frac{2}{3}\sqrt{13} \cos \left(\frac{1}{3} \arccos \frac{-5\sqrt{13}}{26} + \frac{2\pi l}{3} \right)$$

From proposition 66 it is known that the Gauss sums are $G[13, 3; r] = z\sigma(r)$ with the correct permutation σ. To find this permutation, one uses that

$$G[13, 3, r] = 2 \cos \frac{2\pi \cdot 2r}{13} + 2 \cos \frac{2\pi \cdot 2r + 3}{13}$$

Here we have used the definition of the Gaussian sums. Moreover that $a = 2$ is the primitive root used by the DrRacket program, and that the third step is given by problem 141. Numerically these values are

$$G[13, 3, \{0, 1, 2\}] = \{0.273891; \ 1.3772, -2.65109\}$$

310

The correct permutation turns out to be $\sigma(0) = 2$, $\sigma(1) = 0$, $\sigma(2) = 1$. Thus we have obtained from the solutions of the third order equation $P[13, 3; z] = 0$ to be

$$G[13, 3, \{0, 1, 2\}] = -\frac{1}{3} + \frac{2}{3}\sqrt{13}\cos\left(\frac{1}{3}\arccos\frac{-5\sqrt{13}}{26} + \frac{2\pi\{2, 0, 1\}}{3}\right)$$

In the next level of the Gauss-sum tree, one has to solve the following three quadratic equations:

$$P(13, 3, 2, 0; z) = G[13, 3, 2] - G[13, 3, 0]z + z^2 = 0$$
$$P(13, 3, 2, 1; z) = G[13, 3, 0] - G[13, 3, 1]z + z^2 = 0$$
$$P(13, 3, 2, 2; z) = G[13, 3, 1] - G[13, 3, 2]z + z^2 = 0$$

The extra symmetry from equation (VII.5.22) has been used. The solutions are for $r = 0, 1, 2$

$$(\text{VII.6.12}) \quad G[13, 6, \{r, r+3\}] = \frac{G[13, 3, r]}{2} \pm \frac{\sqrt{G[13, 3, r]^2 - 4G[13, 3, r + 2 \mod 3]}}{2}$$

Once more, one has to check numerically how these pairs of solutions match. Indeed, as claimed by Proposition 67 the six solutions for the grandchildren's Gaussian sums $G[p, cd, r + cs] = G[13, 6, r + 3s]$ for $0 \leq r < 3$ and $0 \leq s < 2$:

$$G[13, 6, \{r, r + 3\}] = 2\cos\frac{2\pi\{2^r, 2^{r+3}\}}{13} \quad \text{for } r = 0, 1, 2.$$
$$G[13, 6, \{0, 3\}] = 2\cos\frac{2\pi\{1, 8\}}{13} = \{1.77091, -1 : 49702\}$$
$$G[13, 6, \{1, 4\}] = 2\cos\frac{2\pi\{2, 16\}}{13} = \{1.13613, 0.241073\}$$
$$G[13, 6, \{2, 5\}] = 2\cos\frac{2\pi\{4, 32\}}{13} = \{-0.70921, -1.94188\}$$

With these numerical values one has to check how the pairs of solutions obtained from the quadratic formula (VII.6.12) match. $\qquad\square$

VII.6.4   The 11-th Roots of Unity

For the next example I have tried $p = 11$, but could finish the work only after considerable effort. From the DrRacket program one gets the polynomial

$$P[11, 5; z] = +1 + 3z - 3z^2 - 4z^3 + z^4 + z^5$$

One knows by proposition 66 and primitive root $a = 2$ that its zeros are the Gauss sums

$$G[11, 5, r] = 2\cos\frac{2\pi \cdot 2^r}{11} \quad \text{with } 0 \le r < 5.$$

To find the zeros algebraically, we put $y = 5z + 1$ and get the reduced polynomial $H(y) = 5^5 P[11, 5, \frac{y-1}{5}]$ to be

$$H(y) = 11 \cdot 89 + 2 \cdot 3 \cdot 5 \cdot 7 \cdot 11 y - 5 \cdot 11 y^2 - 2 \cdot 5 \cdot 11 y^3 + y^5$$

We call its zeros $tx[j]$ for $j = 0, 1, 2, 3, 4$. To find them I use the discrete Fourier transformation and put

$$(\text{VII.6.13}) \qquad tx[j] = \sum_{0 \le k \le 4} A[k] \exp\frac{2\pi i \cdot kj}{5} \quad \text{with } j \in \mathbf{Z}_5.$$

The reader may note that already Cardano's formula contains a similar structure, but with the third roots of unity in place of the fifths. To take advantage of the inversion of the discrete Fourier transformation I put

$$(\text{VII.6.14}) \qquad B[j] = \frac{1}{5} \sum_{0 \le k \le 4} t[k] \exp\frac{-2\pi i \cdot kj}{5} \quad \text{with } j \in \mathbf{Z}_5.$$

By the inversion of the discrete Fourier transformation holds $B[j] = A[j]$. The fact that the roots are real, and Viëta's formula $\sum tx[j] = 0$ imply

$$A[0] = 0, \ A[1] = a + ib, \ A[2] = c + id, \ A[3] = c - id, \ A[4] = a - ib$$

with real a, b, c, d. Hence

$$(\text{VII.6.15}) \qquad tx[j] = (a + ib)\exp\frac{2\pi i \cdot j}{5} + (c + id)\exp\frac{4\pi i \cdot j}{5}$$
$$+ (a - ib)\exp-\frac{2\pi i \cdot j}{5} + (c - id)\exp-\frac{4\pi i \cdot j}{5}$$

Since the roots $tx[r]$ in question are some permutation of the Gaussian sums $G[11, 5, r]$, there exists an automorphism σ of period five which leaves the fifth unit roots fixed and transforms the roots cyclically:

$$(\text{VII.6.16}) \qquad \sigma\, tx[k] = tx[k + 1 \mod 5]$$

One checks by formula (VII.6.13) that

$$\sigma\, B[j] = \exp\frac{-2\pi i\, j}{5} B[j]$$

Hence the fifth powers $B[j]^5$ are members of the fixed field of the automorphism σ. Moreover, $B[0] = 0$ is obtained from Viëta's formula. I set up a forth order equation $R[u]$ with the roots $B[j]^5$ for $j = 1, 2, 3, 4$. Its coefficients are elementary symmetric functions of the $B[j]^5$ and hence lie the fixed field of the automorphism σ, too.

For the actual computation, we need to find the permutations of the roots $tx[j]$ for which the automorphism σ acts in the simple way given by equation (VII.6.16). I was able to achieve this goal by going through the 120 permutations of the roots and checking for each one if the polynomial $R(u)$ has even <u>integer</u> coefficients. It turns out that there exist 20 such permutations, and the polynomial $R(u)$ is the same for all of them. Moreover, I may number the zeros $tx[j]$ in such a way that the identical permutation is among these twenty permutations. It turns out that they form a group $\mathcal{G}$ and contain the cyclical group C_5 as a normal divisor. The quotient group $\mathcal{G}/C_5$ turns out to be isomorphic to the cyclic group C_4. Finally the resolvent polynomial turn out to be

$$\text{(VII.6.17)} \qquad R(u) = 25937424601 + 157668929u + 467181u^2 + 979u^3 + u^4$$

The zeros of this polynomial can be calculated symbolically with mathematica. Moreover, I permute them in such a way that

$$A[1]^5 = B(1)^5 = \frac{11}{4}\left(-89 + 25\sqrt{5} - 5i\sqrt{410 + 178\sqrt{5}}\right)$$

$$A[2]^5 = B(2)^5 = -\frac{11}{4}\left(89 + 25\sqrt{5} - 5i\sqrt{410 - 178\sqrt{5}}\right)$$

$$A[3]^5 = B(3)^5 = -\frac{11}{4}\left(89 + 25\sqrt{5} + 5i\sqrt{410 - 178\sqrt{5}}\right)$$

$$A[4]^5 = B(4)^5 = \frac{11}{4}\left(-89 + 25\sqrt{5} + 5i\sqrt{410 + 178\sqrt{5}}\right)$$

which result has been checked numerically to hold.

We can now calculate the roots from formula (VII.6.13) since $A = B$. To this end, one needs to get the fifth root of the complex $A[j]^5$. For $j = 1, 2, 3, 4$, one needs to check out, which one of the root actually is $A[j]$, Hence there occur in the symbolic calculation the integers $l_1, l_2, l_3, l_4 \in \mathbf{Z}_5$ The complex symbolic calculation uses polar coordinate. This is again similar to the situation with Cardano's formula in the case of real roots. The absolute values turn out to be easy, and one gets

$$A(1) = \sqrt{11}\exp\left(\frac{1}{5}i\left(2\pi l_1 - \pi - \arctan\left[\frac{5\sqrt{410 + 178\sqrt{5}}}{-89 + 25\sqrt{5}}\right]\right)\right)$$

$$A(2) = \sqrt{11}\exp\left(\frac{1}{5}i\left(2\pi l_2 + \pi + \arctan\left[\frac{5\sqrt{410 - 178\sqrt{5}}}{-89 - 25\sqrt{5}}\right]\right)\right)$$

$$A(3) = \sqrt{11}\exp\left(\frac{1}{5}i\left(2\pi l_3 - \pi - \arctan\left[\frac{5\sqrt{410 - 178\sqrt{5}}}{-89 - 25\sqrt{5}}\right]\right)\right)$$

$$A(4) = \sqrt{11}\exp\left(\frac{1}{5}i\left(2\pi l_4 + \pi + \arctan\left[\frac{5\sqrt{410 + 178\sqrt{5}}}{-89 + 25\sqrt{5}}\right]\right)\right)$$

$$A(5) = 0$$

I now check numerically that $l_1 = -l_4 = 2$ and $l_2 = -l_3 = 1$ By construction there exists a permutation ρ of the indexes $j \in \mathbf{Z}_5$ such that

$$2\cos\left(\frac{2\pi 3^j}{11}\right) = \frac{tx[\rho\, j] - 1}{5}$$

and one checks that indeed

$$2\cos\left(\frac{2\pi 3^j}{11}\right) = \frac{tx[4 - j] - 1}{5} \quad \text{for all } j \in \mathbf{Z}_5.$$

By means of equation (VII.6.13) we obtain finally

$$2\cos\left(\frac{2\pi 3^j}{11}\right) = -\frac{1}{5} + \frac{2\sqrt{11}}{5}\cos\left(\frac{1 + 4\pi j}{5} + \frac{1}{5}\arctan\left[\frac{5\sqrt{410 - 178\sqrt{5}}}{89 + 25\sqrt{5}}\right]\right)$$

$$+\frac{2\sqrt{11}}{5}\cos\left(\frac{1 - 2\pi j}{5} + \frac{1}{5}\arctan\left[\frac{410 + 5\sqrt{178\sqrt{5}}}{89 - 25\sqrt{5}}\right]\right) \quad \text{for all } j \in \mathbf{Z}_5.$$

VII.6.5   The Area of the Regular 11-gon

The area of the regular 11-gon with side length a is

$$\frac{11a^2}{4}\cot\frac{\pi}{11}$$

One gets from mathematica the irreducible polynomial with zeros

$$\sqrt{11}\cot\frac{\pi\cdot 3^r}{11}\quad\text{for } r = 1,2,3,4,5$$

to be

$$Pcot(z) = 121 - 363z + 242z^2 - 22z^3 - 11z^4 + z^5$$

To get rid of the z^4- term , we put $z = (2y+11)/5$ and get the reduced polynomial to be

$$H(y) = \frac{5^5}{2^5}Pcot\left(\frac{2y+11}{5}\right) = 5203 + 1210y - 1815y^2 - 440y^3 + y^5$$

We call its zeros $tx[j]$ for $j = 0,1,2,3,4$. To find them I use the discrete Fourier transformation given by formula (VII.6.13) To take advantage of the inversion of the discrete Fourier transformation I use formula (VII.6.14). By the inversion of the discrete Fourier transformation holds $B[j] = A[j]$. The fact that the roots are real, and Viëta's formula $\sum tx[j] = 0$ imply

$$A[0] = 0,\ A[1] = a + ib,\ A[2] = c + id,\ A[3] = c - id,\ A[4] = a - ib$$

with real a,b,c,d. Hence the roots are given once more by formula (VII.6.15).

It is now natural to go on with the assumption that similarly as in the previous section, the roots $tx[r]$ in question allow for some permutation an automorphism σ of period five which leaves the fifth unit roots fixed and transforms the roots cyclically:

(VII.6.18) $$\sigma\, tx[k] = tx[k+1 \mod 5]$$

One checks by formula (VII.6.13) that under these assumptions holds

$$\sigma\, B[j] = \exp\frac{-2\pi i\, j}{5}B[j]$$

Hence the fifth powers $B[j]^5$ are members of the fixed field of the automorphism σ. Moreover, $B[0] = 0$ is obtained from Viëta's formula. I set up a forth order equation $R(u)$ with the roots $B[j]^5$ for $j = 1,2,3,4$. Its coefficients are elementary symmetric functions of the $B[j]^5$ and hence lie the fixed field of the automorphism σ, too.

For the actual computation, we need to find the permutations of the roots $tx[j]$ for which the automorphism σ acts in the simple way given by equation (VII.6.18). I was able to achieve this goal by going through the 120 permutations of the roots and checking for each one if the polynomial $R(u)$ has even <u>integer</u> coefficients. Finally the resolvent polynomial turn out to be

(VII.6.19)
$$R(u) = 4177248169415651 + 3577006647247u + 1363940919u^2 - 61347u^3 + u^4$$

The actual calculation I had to do with the polynomial

$$(VII.6.20) \quad S(v) = 11^{-8}R(121v) = 19487171 + 2019127v + 93159v^2 - 507v^3 + v^4$$

It turns out that there exist 20 such permutations, and the polynomial $R(u)$ is the same for all of them. Moreover, I may number the zeros $tx[j]$ in such a way that the identical permutation is among these twenty permutations. It turns out that they form a group $\mathcal{G}$ and contain the cyclical group C_5 as a normal divisor. The quotient group $\mathcal{G}/C_5$ turns out to be isomorphic to the cyclic group C_4. The zeros of polynomial $R(u)$ can be calculated symbolically with mathematica. Moreover, I permute them in such a way that

$$A[1]^5 = B(1)^5 = \frac{121}{4}(507 + 245\sqrt{5} - 5i\sqrt{10970 + 4882\sqrt{5}})$$

$$A[2]^5 = B(2)^5 = \frac{121}{4}(507 - 245\sqrt{5} - 5i\sqrt{10970 - 4882\sqrt{5}})$$

$$A[3]^5 = B(3)^5 = \frac{121}{4}(507 - 245\sqrt{5} + 5i\sqrt{10970 - 4882\sqrt{5}})$$

$$A[4]^5 = B(4)^5 = \frac{121}{4}(507 + 245\sqrt{5} + 5i\sqrt{10970 + 4882\sqrt{5}})$$

which result has been checked numerically to hold.

We can now calculate the roots from formula (VII.6.13) since $A = B$. To this end, one needs to get the fifth root of the complex $A[j]^5$. For $j = 1, 2, 3, 4$, one needs to check out, which one of the root actually is $A[j]$, Hence there occur in the symbolic calculation the integers $l_1, l_2, l_3, l_4 \in \mathbf{Z}_5$

$$A[1] = \sqrt{11(4 + \sqrt{5})}\exp\frac{i}{5}\left(2l_1\pi - \arctan\left[\frac{5}{89}\sqrt{205 - 22\sqrt{5}}\right]\right)$$

$$A[2] = \sqrt{11(4 - \sqrt{5})}\exp\frac{i}{5}\left((2l_2 - 1)\pi + \arctan\left[\frac{5}{89}\sqrt{205 + 22\sqrt{5}}\right]\right)$$

$$A[3] = \sqrt{11(4 - \sqrt{5})}\exp\frac{i}{5}\left((2l_3 + 1)\pi - \arctan\left[\frac{5}{89}\sqrt{205 + 22\sqrt{5}}\right]\right)$$

$$A[4] = \sqrt{11(4 + \sqrt{5})}\exp\frac{i}{5}\left(2l_4\pi + \arctan\left[\frac{5}{89}\sqrt{205 - 22\sqrt{5}}\right]\right)$$

$$A(5) = 0$$

I now check numerically that $l_1 = -l_4 = 0$ and $l_2 = -l_3 = 1$ One gets By construction there exists a permutation ρ of the indexes $j \in \mathbf{Z}_5$ such that

$$\cot\left(\frac{\pi 3^j}{11}\right) = \frac{\sqrt{11}}{5} + \frac{2\sqrt{11}}{11} tx[\rho\, j]$$

and one checks that indeed $\rho j = 5 - j$. The complex symbolic calculation uses polar coordinate. This is again similar to the situation with Cardano's formula in the case of real roots. The absolute values turn out to be done by mathematica. By means of equation (VII.6.13) we obtain finally

$$\cot\frac{3^r \cdot \pi}{11} = \frac{\sqrt{11}}{5} + 4\sqrt{4 + \sqrt{5}}\cos\left(\frac{2\pi r}{5} + \frac{1}{5}\arctan\left[\frac{5}{89}\sqrt{205 - 22\sqrt{5}}\right]\right) +$$
$$4\sqrt{4 - \sqrt{5}}\cos\left(\frac{(1 - 4r)\pi}{5} + \frac{1}{5}\arctan\left[\frac{5}{89}\sqrt{205 + 22\sqrt{5}}\right]\right) \quad \text{for all } R \in \mathbf{Z}_5.$$

These formulas have been checked numerically.

VII.6.6   The Entire Tree of Gaussian Periods

Theorem 23. *Let p be an odd prime. Let*

$$p - 1 = \prod_{1 \leq s \leq u} q_s$$

be the prime decomposition. The prime factors q_s may be arranged in any order and clearly repetition has to be allowed. The solutions of the equation

(VII.5.5) $$\Phi_p(z) = 0$$

are obtained by successively for $1 \leq t < u$ solving polynomial equations of order q_t which all can be solved by radicals and unit roots of order q_t.

Corollary 37. *The p-th unit roots can be expressed by radicals all of order less that p.*

Proof. Since all prime factors $q_t \mid p - 1$ are less than p, we may use Theorem 23 and induction. $\qquad\square$

Proof of Theorem 23. Gaussian sums $G[p, d, s]$ for $d \mid p - 1$ are put at the nodes of a rooted tree with $u + 1$ levels. At the root one puts the sum $G[p, 1, 0] = -1$ of all non-positive p-th roots of unity. The root gets q_1 children. At each of the child vertices is put one Gaussian sum $G[p, q_1, r_1]$ with running index $0 \leq r_1 < q_1$. Each child has in turn q_2 children counted by a second index $0 \leq r_2 < q_2$. At each of these grandchildren is put a Gaussian sum $G[p, q_1 q_2, r_1 + q_1 r_2]$. The grandchildren get the two indices $0 \leq r_1 < q_1, 0 \leq r_2 < q_2$.

At the t-th generation one gets altogether $\prod_{1 \leq s \leq t} q_s$ successors. These are enumerated by the indices $0 \leq r_1 < q_1, 0 \leq r_2 < q_2, \ldots, 0 \leq r_t < q_t$. I put

$$n_t := r_1 + q_1 r_2 + q_1 q_2 r_3 + \cdots + q_1 \cdots q_{t-1} r_t$$

which takes the values $0 \leq n_t < q_1 \cdots q_t$ each exactly once. At each vertex of the t-th generation is put the Gaussian sum

(VII.6.21) $$G[p, q_1 \cdot q_2 \cdots q_t, n_t]$$

Finally at the u-th generation occur the Gaussian sums with only one term

$$G[p, p - 1, n] = \exp \frac{2\pi i \cdot a^n}{p}$$

since $p - 1 = q_1 \cdots q_u$. The third variable

$$n = r_1 + q_1 r_2 + q_1 q_2 r_3 + \cdots + q_1 \cdots q_{u-1} r_u$$
$$\text{with } 0 \leq r_1 < q_1, 0 \leq r_2 < q_2, \ldots 0 \leq r_u < q_u$$

takes all values $0 \leq n < p - 1$ exactly one time. Hence the formulas for $G[p, p - 1, n]$ yield all the solutions of the equation

(VII.5.5) $$\Phi_p(z) = 0$$

Lemma 74. *Let $1 \leq t \leq u$ and fix any $0 \leq n_{t-1} < q_1 \cdots q_{t-1}$. The zeros of monic polynomial $P(p, q_1 \cdots q_{t-1}, q_t, n_{t-1}; x)$ of degree q_t are the Gaussian sums $G[p, q_1 \cdots q_t, n_{t-1} + q_1 \cdots q_{t-1} r_t]$ with running index $0 \leq r_t < q_t$. Hence they lie in the field*

$$E_t := S_{q_1 \cdots q_t}$$

But the coefficients of these polynomials lie in the <u>smaller</u> field E_{t-1}. We have put $E_0 = \mathbf{Q}$.

Proof. In case $t = 1$, we use Proposition 66 with $d := q_1$. Equivalently, we may put $c = 1, r = 0$ into Proposition 67. For $t \geq 2$, we use Proposition 67 with $c := q_1 \cdots q_{t-1}$

318

and $d := q_t$. We put $r = n_{t-1}$ and $s = r_t$. Hence the Gaussian sums $G[p, q_1 \cdots q_t, n_t]$ with

$$n_t = r + cs = n_{t-1} + q_1 \cdots q_{t-1} r_t$$
$$= r_1 + q_1 r_2 + q_1 q_2 r_3 + \cdots + q_1 \cdots q_{t-2} r_{t-1} + q_1 \cdots q_{t-1} r_t$$

for $0 \leq r_t < q_t$ are the zeros of monic polynomial $P(p, q_1 \cdots q_{t-1}, q_t, n_{t-1}; x)$ of degree q_t. The coefficients of this polynomial are integer combinations of the Gaussian sums $G[p, q_1 \cdots q_{t-1}, n_{t-1}]$ with $0 \leq n_{t-1} < q_1 \cdots q_{t-1}$ and hence lie in the field E_{t-1} spanned by them. $\qquad\square$

Lemma 75. *A basis of E_t is $G[p, q_1 \cdots q_t, n_t]$ with $0 \leq n_t < q_1 \cdots q_t$. Hence*

$$\dim[E_t : \mathbf{Q}] = q_1 \cdots q_t$$

Proof. The assertion follows directly from Lemma 65. $\qquad\square$

Lemma 76. *Moreover, the polynomial equations from lemma 74 can be solved by radicals and unit roots of order q_t.*

Proof. I fix any n_{t-1} in the range $0 \leq n_{t-1} < q_1 \cdots q_{t-1}$. That is the parent whose children I may get as zeros of equation $P(p, q_1 \cdots q_{t-1}, q_t, n_{t-1}; x) = 0$. I put $q := q_t$ and $j := r_t$. Of course, we know that these zeros are

$$tx[j] := G[p, q_1 \cdots q_t, n_{t-1} + q_1 \cdots q_{t-1} j]$$

for $0 \leq j < q$. But this expression already uses the p-th roots of unity.

I use as Ansatz the discrete Fourier transformation and put

$$\text{(VII.6.22)} \qquad tx[j] = \sum_{0 \leq k < q} A[k] \exp \frac{2\pi i \cdot kj}{q} \quad \text{with } j \in \mathbf{Z}_q.$$

There is an easy inversion formula

$$\text{(VII.6.23)} \qquad A[k] = \frac{1}{q} \sum_{0 \leq j < q} tx[j] \exp \frac{-2\pi i \cdot kj}{q} \quad \text{with } k \in \mathbf{Z}_q.$$

We use use lemma 73 with $e := q_1 \cdots q_{t-1}$ and $d := q_1 \cdots q_t$. The group $\text{Aut}(E_t/E_{t-1})$ of automorphisms of the field E_t that fixes all elements of the subfield E_{t-1} is a group of (prime) order q_t. Its generator $\sigma = a^e$ permutes the Gaussian sums spanning $E_t = F_d$ as follows:

$$a^e G[p, d, s] = G[p, d, e + s \mod d]$$

Put $s = n_{t-1} + q_1 \cdots q_{t-1} j$ and hence $e + s = n_{t-1} + q_1 \cdots q_{t-1}(j+1)$. Among the zeros $tx[j]$ of <u>one</u> parent we get

$$a^e tx[j] = tx[(j+1) \mod q]$$

hence a q-cyclic permutation.By lemma 73 we know that $a^e \in \mathrm{Aut}(E_t/E_{t-1})$. Moreover, it is a generator of this <u>cyclic</u> group.

By problem 145 the automorphism a^e fixes all elements of the space $F_e = E_{t-1}$ and hence the coefficients of polynomial $P(p, q_1 \cdots q_{t-1}, q_t, n_{t-1}; x)$.

Without proof I claim that the automorphism a^e can be extended to an automorphism of $a^e \in \mathrm{Aut}(E_t(\zeta_q)/E_{t-1}(\zeta_q))$, at least for q prime.

Problem 155. *Prove this claim.*

Under this additional assumption, how does a^e transform the Fourier coefficients $A[k]$? The gentleman calculates:

$$
\begin{aligned}
a^e A[k] &= \frac{1}{q} \sum_{0 \leq j < q} tx[j+1 \mod q] \exp \frac{-2\pi i \cdot kj}{q} \\
&= \frac{1}{q} \sum_{0 \leq j < q} tx[j] \exp \frac{-2\pi i \cdot k(j-1)}{q} \\
&= \exp \frac{2\pi i \cdot k}{q} A[k]
\end{aligned}
$$

for all $k \in \mathbf{Z}_q$. Hence the powers $A[k]^q$ are fixed by all automorphisms in $\mathrm{Aut}(E_t(\zeta_q)/E_{t-1}(\zeta_q))$

Problem 156. *Is it true or not that the powers $A[k]^q$ are fixed by all automorphisms in* $\mathrm{Aut}(E_t/E_{t-1})$, *too.*

Luckily, it is not necessary to know the answer. In any case, by Problem 145 we may conclude that $A[k]^q \in E_{t-1}(\zeta_q)$.

Since a basis $G[p, e, n_{t-1}]$ with $0 \leq n_{t-1} < e$ of the space E_{t-1} has been obtained in the $t-1$-th level,of the tree, we are able to obtain expressions for the $A[k]^q$ in terms of the Gaussian sums $G[p, e, u]$ for $0 \leq u < e$ and powers the unit root ζ_q. Hence we may now obtain the Fourier coefficients $A[k]$ by extraction of q-th roots after having, too, adjoined the unit root ζ_q. $\qquad\square$

$$\square$$

Remark. Have we really obtained a constructive proof of theorem 23 and the last claim. Yes, but we seriously are missing useful formulas, which are needed to write an algorithm.

The drawbacks are obvious. Neither do we know how to obtain the expansion of the $A[k]^q$ in the basis of E_{t-1} and the unit root ζ_q. Nor do we know which unit roots $(\zeta_q)^l$ are additionally involved in the calculation of the $A[k]$.

Above I could get only the algorithm for the 11-th root, by making some additional guesses about these points and checking them numerically. In the style of Leibniz I say:

> *this proof is an amphibion of concrete and abstract.*

VII.6.7   Abstract Constructibility of Fermat Polygons

Take especially the case that $p = F_n$ is a Fermat prime, and confirm the constructibility of the p-th unit roots and the numbers $2 \pm 2 \cos \frac{2\pi 3^r}{p}$ and $\sin^2 \frac{2\pi l}{p}$. We have already been covering in earlier sections the construction with straightedge and compass, and with Hilbert tools, only. My point was to aim at explicit solution of the problem whenever possible. Now we are taking a more abstract approach, but get as I want to stress,— only more modest results.

Definition (VII.5.12) is indeed a generalization of equation (II.0.14) to arbitrary primes and primitive root. The special case involves Fermat prime $p = F_n$, its primitive root $a = 3$, and the divisors $d = 2^q$ with $0 \leq q \leq 2^n$. With the the low digits $s = l$ taking the values $0 \leq l < 2^q$ holds

$$\text{(VII.6.24)} \quad G(F_n, 2^q, s) = S(n, q, s) = \sum \left\{ \exp \frac{2\pi i \cdot 3^{(t2^q + s)}}{F_n} \ : \ \text{with } 0 \leq t < 2^{2^n - q} \right\}$$

Since the dimension agrees with the grad these Gaussian sums span the vector space of the field

$$\mathbf{F}_q = \operatorname{span}\{ G[F_n, 2^q, s] \ : \ 0 \leq s < 2^q \}$$

[19] The dimension of the field extension is $[\mathbf{F}_q : \mathbf{Q}] = 2^q$.

Theorem 24 (Abstract constructibility). *For all Fermat prime F_n, a construction of the regular F_n-gon with straightedge exists in principle.*

---

[19]The corresponding fields are given for any prime p and divisor $d \mid p - 1$ by definition 39, and there named F_d.

Proof. Finally, with $0 \leq q \leq 2^n$ we have obtained an explicit example for the tower

$$\mathbf{Q} = \mathbf{F}_0 \subset \mathbf{F}_1 \subset \mathbf{F}_2 \subset \cdots \subset \mathbf{F}_{2^n}$$

$$= \mathcal{G}(F_n) \ni \text{ all conjugates of } \exp\frac{2\pi i \cdot k}{F_n} \text{ with } 1 \leq k < F_n$$

of finitely many two dimensional extensions. Such a tower has in Corollary 51 been shown to consist of subfields of the constructible field and give a criterium for constructibility. In the present example, the tower is ending with splitting field for the cyclotomic polynomial

$$\Phi_{F_n}(x) = P(F_n, F_n - 1; x) \quad \text{ with } F_n - 1 = 2^q = 2^{2^n}.$$

Thus we obtain a quite abstract proof that real parts of the F_n-th unit roots lie in the constructible field. $\qquad\square$

Galois Theorem deals with the automorphisms of field extensions. We have the reader exposed in the present section only to a baby version of this theory,— restricting ourselves to the field extensions $[\mathcal{G}(p)/\mathbf{Q}]$ generated by p-th unit roots where p is any odd prime.

In the main lemma 67 is shown that any automorphism of the algebra $\mathcal{G}(p)$ is a representation $\sigma = \Gamma(l)$ for some element $l \in \mathcal{U}(p)$ from the Euler group. The action on the base elements is

$$(\text{VII.6.25}) \qquad \Gamma(l)\exp\frac{2\pi i \cdot k}{p} = \exp\frac{2\pi i \cdot l\,k}{p} \text{ for all } k \in \mathcal{U}(p)$$

With a primitive root a modulo p, we write (sloppily) $\sigma = a^v$ with some exponent modulo $p - 1$.

We go back to the specific example with Fermat prime $p = F_n$ and choose as its primitve root $a = 3$. In this case the group of automorphisms of the extension $[\mathcal{G}(p)/\mathbf{Q}]$ is cyclic and has as a single generator the automorphism $\sigma = \Gamma(3)$ in equation (VII.6.25). From lemma 70 we conclude

Lemma 77.

$$\mathbf{F}_q = \text{span}\{G[F_n, 2^q, s] : 0 \leq s < 2^q\}$$

consists of all elements of $\mathcal{G}(p)$ which are fixed by the automorphism σ^d.

Remark. We may even get a bid more concrete in the theorem 24 about abstract constructibility, and reproduce at least part of equation (II.0.30) from lemma 23. Let

$0 \leq q < 2^n - 1$ and $0 \leq l < 2^q$. Because of the invariance.

$$\sigma^{2^q} G(F_n, 2^{q+1}, l) \cdot G(F_n, 2^{q+1}, l + 2^q)$$
$$= G(F_n, 2^{q+1}, l + 2^q) \cdot G(F_n, 2^{q+1}, l + 2^q + 2^q)$$
$$= G(F_n, 2^{q+1}, l) \cdot G(F_n, 2^{q+1}, l + 2^q)$$

the product of Gaussian sums $G(F_n, 2^{q+1}, l) \cdot G(F_n, 2^{q+1}, l + 2^q)$ lies in the (smaller) field F_q. Hence there exist rational coefficients c such that

$$G(F_n, 2^{q+1}, l) \cdot G(F_n, 2^{q+1}, l + 2^q) = \sum_{0 \leq k < 2^q} c_{[q,l,k]} G(F_n, 2^q, k)$$

Remark. Above in the subsection on Gauss' polygons II.0.2 we had obtained several additional results:

- $c \geq 0$ are integers and depend only on n, q and $(l - k) \mod 2^q$, as shown by equation (II.0.38).

- We have the formulas (II.0.32) and (II.0.34) for an effective calculation.

By my opinion, only these additional more definite results allow one to claim to have obtained a <u>construction</u>. Therefore I have called theorem 24 more carefully a "result about abstract constructibility".

Lemma 78. F_q *is the splitting field of integer polynomial*

$$P(F_n, 2^q; x) = \prod_{0 \leq s < 2^q} (x - G[F_n, 2^q, s])$$

and this polynomial is irreducible. For $q = 2^n - 1$ holds additionally

(VII.6.26)
$$P(F_n, 2^q; x) = U_{F_n - 1}\left(\frac{\sqrt{x + 2}}{2}\right)$$

Proof. Proposition 66 yields the integer polynomial

$$P(F_n, 2^q; x) = \prod_{0 \leq s < 2^q} (x - G[F_n, 2^q, s])$$

The roots of polynomial $P(F_n, 2^q; x)$ span the field F_q.

Hence F_q is the splitting field of $P(F_n, 2^q; x)$. Moreover, the action of the group of automorphisms on the extension $[\mathcal{G}(p)/\mathbf{Q}]$ is determined by

(VII.6.27)
$$\sigma G[F_n, 2^q, s] = G[F_n, 2^q, s + 1 \mod 2^q]$$

since there is a single generator σ. Indeed the, group action is a cyclic permutation of those Gaussian sums which provide all the roots.

Assume towards a contradiction the polynomial $P(F_n, 2^q; x) = A(x)B(x)$ is reducible. Any automorphism permutates the roots of factor A among themselves. Hence σ may contain only cycles of length at most $\deg A$. Hence $\deg A = 0$ or $\deg A = 2^q$, and the polynomial $P(F_n, 2^q; x)$ is irreducible.

In other words, because of the transitive action of the permutation σ on the roots, the polynomial is irreducible.

Let $q = 2^n - 1$. Because of equations (VII.6.24) and (IV.4.16) and since F_n is a prime holds additionally

$$P(F_n, 2^q; x) = \prod_{0 \le s < 2^q} \left(x - 2\cos \frac{2\pi \cdot 3^s}{F_n} \right) = \Psi_{F_n - 1}(x) = U_{F_n - 1}\left(\frac{\sqrt{x+2}}{2} \right)$$

$\square$

Remark. We now use the case $n = 3$, $F_n = 257$ for a further check of the numerical result from subsection II.0.3. There we had obtained numerical values for the x-coordinates of the vertices of the 257-gon, via the straightedge and compass construction. Using this numeric result, we may calculate the integer polynomial $\Psi_{F_n - 1}(x)$ by multiplying out its linear factors.

On the other hand, we may use the recursion formula from problem 81 for the associated Chebyshev polynomials to calculate $U_{F_n - 1}\left(\frac{\sqrt{x+2}}{2} \right)$. It may be astonishing, the accuracy of 40 digits is needed to check that the two results agree!

Next we define the polynomial

$$D(F_n, 2^q; x) = \prod_{0 \le s < 2^q} \left(x - \frac{\Delta(n, q, s)}{4} \right)$$

$$= \prod_{0 \le s < 2^q} \left(x - (G[F_n, 2^{q+1}, s] - G[F_n, 2^{q+1}, s + 2^q])^2 \right)$$

which has as its zeros the discriminants appearing in the pair of equation (II.0.40) and (II.0.41) under the root.

Lemma 79. *The polynomial $D(F_n, 2^q; x) \in \mathbf{Z}[x]$ is an irreducible integer polynomial. All its roots are conjugates and totally positive.*

Proof. The action of the single generator σ on the roots is given because of equation (VII.6.27) to be

$$\sigma\,\Delta(n,q,s) = (G[F_n, 2^{q+1}, s+1 \mod 2^{q+1}] - G[F_n, 2^{q+1}, s+1+2^q \mod 2^{q+1}])^2$$
$$= \Delta(n,q,s+1 \mod 2^{q+1})$$

but even more holds because of the square.

$$[\sigma^{2^q}\,\Delta(n,q,s) = (G[F_n, 2^{q+1}, s+2^q \mod 2^{q+1}] - G[F_n, 2^{q+1}, s \mod 2^{q+1}])^2$$
$$= \Delta(n,q,s \mod 2^{q+1}) = \Delta(n,q,s)$$

We see that the, group action is a cyclic permutation of (shorter) length 2^q. This is a cyclic permutation of the roots $\Delta(n,q,s)$ of polynomial $D(F_n, 2^q; x)$. Hence the coefficients of this polynomial are invariant under σ and hence they are rational.

Because of the transitive action of the permutation σ on the roots, the polynomial is irreducible. Hence all $\Delta(n,q,s)$ for $0 \leq s < 2^q$ are conjugates. Since they are a complete set of conjugates, they are totally positive, too. $\square$

Problem 157. *Use the formula*

$$\Delta(n,q,l) = S(n,q,l)^2 - 4 \sum_{0\leq d<2^q} e[n,q,d]S(n,q,l+d)$$

to show that the coefficients of polynomial $D(F_n, 2^q; x)$ are even integers.

Problem 158 ("For the tombstone of Eisenstein"). *With the accurate data from remark VII.6.7, computate the polynomials $D(257, 2^q; x)$ for $0 \leq q \leq 6$. Check directly with the Eisenstein criterium 37 that they are irreducible.*

Part of he solution.

$$\left[\frac{1}{257}\,\mathrm{polydelta}\,(2^n - 2)\right] =$$

$$\frac{x^{64}}{257} - x^{63} + 124x^{62} - 9885x^{61} + 569527x^{60} - 25278457x^{59}$$

$$+899686731x^{58} - 26389891931x^{57} + 650728407923x^{56} - 13691814584350x^{55}$$

$$+248684031900652x^{54} - 3934991897498482x^{53} + 54645943475707811x^{52}$$

$$-670049974890935582x^{51} + 7290284049666775904x^{50} - 70672501386574608587x^{49}$$

$$+612494260654567516875x^{48} - 4759081696681424337564x^{47}$$

$$+33229539004069359993495x^{46} - 208897242772273424506942x^{45}$$

$$+1184175558949464296609887x^{44} - 6060403327252270807566737x^{43}$$

$$+28027878790604511514558865x^{42} - 117211272496236299890598797x^{41}$$

$$+443426308158962829407510006x^{40} - 1517859443550440014543907772x^{39}$$

$$+4701030431506561352374531034x^{38} - 13170791205794078311485134073x^{37}$$

$$+33366402459488084708983105774x^{36} - 76387887978588110961787399636x^{35}$$

$$+157911784924798602512483725471x^{34} - 294479701680750574780095757199x^{33}$$

$$+494812509409524708495995934636x^{32} - 748136037571164768174208233175x^{31}$$

$$+1016247549400243280924417488568x^{30} - 1238036779466417946320501484307x^{29}$$

$$+1349979568356893375082075428321x^{28} - 1314709576858988450783613187713x^{27}$$

$$+1140753433276487033210249640929x^{26} - 879549613575631368416055600082x^{25}$$

$$+600863164475934503857994309220x^{24} - 362548603293126247409812511758x^{23}$$

$$+192550836981402198466920359817x^{22} - 89682106908337617050676449650x^{21}$$

$$+36484947861349464458183194600x^{20} - 12909239374294946681586962321x^{19}$$

$$+3954123131388340039659507793x^{18} - 1043224398756670112309607428x^{17}$$

$$+235775666812510778130894637x^{16} - 45371450780759416777789746x^{15}$$

$$+7383837266462634841528749x^{14} - 1008414043476504484709083x^{13}$$

$$+114538031903544605183627x^{12} - 10704937566421065367623x^{11}$$

$$+812704719667823576634x^{10} - 49323472889820297236x^9 + 2345093784721654886x^8$$

$$-85084549780617843x^7 + 2274495835208802x^6 - 42667636106420x^5 + 522918278716x^4$$

$$-3733060620x^3 + 12527480x^2 - 12176x + 1$$

$$\square$$

Part VIII

Excursion to Galois Theory

VIII.0.1   The Splitting Field

We depose of the restriction to use the rational field as the only ground field. Instead, we call any two fields $F \subseteq E$ a *field extension* and denote this pair by E/F or $[E : F]$. The smaller field is called the *ground field*. With scalars defined to be from the ground field, the larger field E becomes a vector space over the smaller one. The dimension of this vector space is denoted $\dim[E : F]$.

Definition 41 (Splitting field). Let $f \in F[x]$ be a nonconstant polynomial with degree $n = \deg f \geq 1$. The field extension E/F is called a *splitting extension* and E is called a *splitting field* iff the following two conditions hold:

(a) The polynomials f splits in E, in other words, there exist $a, x_1, \ldots, x_n \in E$ such that
$$f(x) = a \prod_{1 \leq i \leq n} (x - x_i)$$

(b) E is the <u>minimal</u> field for which condition (a) holds. In other words, for any field extension L/F with $L \subseteq E$ in which f splits, holds $L = E$.

Proposition 70 (Existence and uniqueness properties of the splitting field). *Let $f \in F[x]$ be a nonconstant polynomial. About existence and uniqueness of the splitting field hold*

(a) *For each polynomial $f \in F[x]$, there exists a splitting extension E/F. Its degree has the upper bound $\dim[E : F] \leq n\,!$*

(b) *The splitting field is unique up to isomorphism. In other words: suppose that E/F and L/F are two splitting extensions for the polynomial $f \in F[x]$, then there exists an isomorphism $\psi : E \mapsto L$ such that $\psi \uparrow F = \mathrm{id}_F$ and the zeros of f in E are mapped bijectively to the zeros of f in L.*

(c) *Moreover, for any irreducible factor $g \in F[x]$ of the polynomial f and any to zeros α and β of g, there exists isomorphism $\psi : E \mapsto L$ such that $\psi \upharpoonright F = \mathcal{I}d$ and $\psi(\alpha) = \beta$.*

A detailed proof is given by Gerd Fischer [8], p. 320.

Proposition 71 (Three characterizations of the splitting field). *For any finite dimensional field extension E/F the following three requirements are equivalent:*

(i) *E is the splitting field for some polynomial $f \in F[x]$.*

(ii) *For each extension L/E and each monomorphism $\phi : E \mapsto L$ such that $\phi \uparrow F = \mathrm{id}_F$ holds $\phi(E) \subseteq E$.*

(iii) *If any irreducible polynomial $g \in F[x]$ has one zero $\alpha \in E$, then it splits in E.*

(i) $\Rightarrow$ (ii). We may exclude the case $n = 0$, $E = F$ and assume $n = \deg f \geq 1$. Let $x_1, \ldots, x_n$ be the zeros of the polynomial $f \in F[x]$. The splitting field is the smallest field containing F and the zeros: hence $E = F(x_1, \ldots, x_n)$. The given monomorphism ϕ maps zeros to zeros since

$$f(\phi(x_i)) = \phi(f(x_i)) = \phi(0) = 0$$

Hence $\phi(x_i) \in E$ for all $1 \leq i \leq n$. Finally

$$\phi(E) = \phi(F(x_1, \ldots, x_n)) \subseteq F(\phi(x_1), \ldots, \phi(x_n)) = F(x_1, \ldots, x_n) = E$$

$\square$

(ii) $\Rightarrow$ (iii). Given is any irreducible polynomial $g \in F[x]$ which has one zero $\alpha \in E$. We may assume that g is monic. Because of its irreducibility, g is the minimal polynomial of α. Since the extension E/F is assumed to be finite dimensional, there exists further $z_1, \ldots, z_m$ such that

$$E = F(\alpha, z_1, \ldots, z_m)$$

The case $m = 0$ is possible to occur. Let $h_1, \ldots h_m \in F[x]$ be the minimal polynomials of $z_1, \ldots, z_m$ and define

$$f = g \cdot \prod_{1 \leq i \leq m} h_i$$

which has some splitting field $L \supseteq E$.

For any zero α_2 of polynomial g, we have to confirm $\alpha_2 \in E$. According to part (b) of proposition 70 about the existence and uniqueness of the splitting field, there exists an isomorphism (even automorphism) $\psi : L \mapsto L$ such that $\psi \uparrow F = \mathrm{id}_F$ and $\psi(\alpha) = \alpha_2$.

To the restriction $\phi = \psi \uparrow E$ is now applied part (ii). Hence $\phi(K) \subseteq K$ and $\phi(\alpha) = \alpha_2 \in K$, as to be shown. $\square$

(iii) $\Rightarrow$ (i). Since the extension E/F is assumed to be finite dimensional, there exists a basis $z_1, \ldots, z_m$. For the basis holds

$$E = F(z_1, \ldots, z_m)$$

Let $h_1, \ldots h_m \in F[x]$ be the minimal polynomials of $z_1, \ldots, z_m$ and define

$$f = \prod_{1 \leq i \leq m} h_i$$

Because of the assumption from part (iii), the polynomials h_i all split in E. Hence f splits in E, too.

Moreover f does not split in a smaller field than E since an entire basis of extension $[E : F]$ are zeros of f. $\qquad\square$

Remark. For the degree of the polynomial $f \in F[x]$ and the dimension of its splitting extension, we have obtained only the very rough estimates

$$\dim[E : F] \leq \deg f! \ \text{ and } \ \deg f \leq \dim[E : F]^{\dim[E:F]}$$

Definition 42. Let $\alpha \in K$ and let m be its minimal polynomial over K. The roots of m are called the *Galois conjugates* of α over the field K. Equivalently, the Galois conjugates of are those elements of the algebraic closure $\overline{K}$ which have the same minimal polynomial as α.

In the most common case that $K = \mathbf{Q}$, we call them the *algebraic conjugates*.

The Galois conjugates of $\alpha \in K$ need not be members of the base field K, except for α itself, but only members of the algebraic closure $\overline{K}$.

Definition 43. Let L/K be a Galois extension and $\alpha \in L$. The *norm* of α over K, denoted $N_{L/K}(\alpha)$, is the product of the Galois conjugates of α.

In the most common case that $K = \mathbf{Q}$, we write simply $N(\alpha)$. No dependence of L occurs.

Lemma 80. *Let L/K be a Galois extension. All Galois conjugates of any $\alpha \in L$ are members of L.*

In general $N(\alpha)$ is an element of L, but not necessarily of K. α and all its Galois conjugates have the same norm.

If the minimal polynomial m splits in the one-element extension $K(\alpha)$, then holds $N(\alpha) \in K$.

Otherwise, $N(\alpha) \in K$ may or may not hold.

Reason. Since L/K is a Galois extension, by item (iii) of the proposition 71 about the three characterizations of the splitting field, any polynomial with one root in L, splits in L. Hence the minimal polynomial m splits in L. Hence all roots of m and their product $N(\alpha)$ are members of L. $\qquad\square$

Theorem 25. *The norm of an algebraic integer,—over the field $\mathbf{Q}$,—is an integer.*

Proof. Let $m \in \mathbf{Z}[z]$ be a minimal polynomial of α. Since α is an algebraic integer, its minimal polynomial $m \in \mathbf{Z}[z]$ is a *monic* polynomial. The norm $N(\alpha)$ is the product of the roots of m, which by Vi eta's formula for a monic polynomial is equal to $\pm$ the constant term of m. Hence $N(\alpha) \in \mathbf{Z}$ as claimed. $\qquad\square$

Take the number $\sqrt{\sqrt{2}-1}$, which is an algebraic integer. Over the rational number field, the norm is -1. The minimal polynomial over the extensions field $\mathbf{Q}(\sqrt{2})$ is $1 - \sqrt{2} + x^2$. Now we get a different norm:

$$N_{L/\mathbf{Q}(\sqrt{2})}(\sqrt{\sqrt{2}-1}) = 1 - \sqrt{2}$$

Lemma 81. *Let L/K be a Galois extensions, and let $\alpha \in L$ be an algebraic integer over K. Then $N_{L/K}(\alpha)$ is a member of K.*

Proof. Let $m \in K[z]$ be a minimal polynomial of α. Since α is an algebraic integer over K, its minimal polynomial $m \in K[z]$ is a *monic* polynomial. The norm $N_{L/K}(\alpha)$ is the product of the roots of m, which by Vi eta's formula for a monic polynomial is equal to $\pm$ the constant term of m. Hence $N_{L/K}(\alpha) \in K$ as claimed. $\qquad\square$

VIII.0.2   The Main Theorem about Symmetric Polynomials

Let

$$\text{(VIII.0.1)} \qquad s_r = \sum_{1 \le i_1 < i_2 < \cdots < i_r \le n} X_{i_1} \cdots X_{i_r} \quad \text{for } 0 \le r \le n$$

be the elementary symmetric polynomials in n variables. One puts $s_0 = 1$.

Theorem 26 (Main Theorem about symmetric polynomials). *Let $\mathcal{R}$ be any ring with unit. The mapping*

$$\Phi : R[S_1, \ldots, S_n] \mapsto R[X_1, \ldots, X_n]$$

obtained by substituting for S_i the elementary symmetric polynomial s_i is a monomorphism of rings mapping onto the subring $R_{sym}[X_1, \ldots, X_n]$ of symmetric polynomials in the variables $X_1, \ldots X_n$.

In other words, every symmetric polynomial $f \in R_{sym}[X_1, \ldots, X_n]$ becomes in a unique way a polynomial $g \in R[S_1, \ldots, S_n]$ of the elementary symmetric polynomials:

$$f(x_1, \ldots, x_n) = g(s_1(x_1, \ldots, x_n), \ldots, s_n(x_1, \ldots, x_n))$$

I find it really more convincing to give an independent proof,—-independent from these overly abstract isomorphism constructions.

Independent proof of (i) $\Rightarrow$ (iii). We may exclude the case $n = 0$, $E = F$ and assume $n = \deg f \geq 1$. Let $x_1, \ldots, x_n$ be the zeros of the polynomial $f \in F[x]$. The splitting field is the smallest field containing F and the zeros: hence $E = F(x_1, \ldots, x_n)$. Given is any irreducible polynomial $g \in F[x]$ which has one zero $\alpha \in E$. We may assume that g is monic. Because of its irreducibility, g is the minimal polynomial of α. Every element in E and especially α may be written as

$$\alpha = P_1(x_1, \ldots, x_n)$$

with some polynomial $P_1(u_1, \ldots, u_n) \in F[u_1, \ldots, u_n]$. Let $P_1, \ldots P_r$ be the polynomials obtained by permutating the variables of P_1, in other words the orbit of the action of the symmetry group $\mathcal{S}_n$ on P_1. Let r be the length of the orbit and

$$\alpha_k = P_k(x_1, \ldots, x_n) \ \text{ for } 1 \leq k \leq r$$
$$R(x) = \prod_{1 \leq k \leq r} (x - P_k(x_1, \ldots, x_n)) = \sum_{0 \leq k \leq r} r_k(x_1, \ldots, x_n)x^k$$

The coefficients r_k are symmetric in the variables $x_1, \ldots, x_n$. Hence, by the Symmetric Functions Theorem 26l, they are polynomials of the elementary symmetric functions of the variables $x_1, \ldots, x_n$. These are by Viëta's formulas in turn the coefficients of the polynomial $f \in F[x]$. Hence $r_k \in F$ for $0 \leq k \leq r$ and hence $R \in F[x]$. By construction R has the zeros α_k for $1 \leq k \leq r$.

Since $\alpha = \alpha_1$ is a common root of g and R and $\deg R \geq 1$, and the polynomial g is irreducible, we conclude that g is a divisor of R. Hence the roots of g are a subset of the α_k for $1 \leq k \leq r$. Thus polynomial g is splitting in the extension E/F, as to be shown. $\square$

VIII.0.3 Multiple Zeros

Definition 44 (Characteristic of a field). A field F is defined to have *characteristic zero* if and only if $F \supseteq \mathbf{Q}$. In this case the smallest subfield of F is $\mathbf{Q}$. For any prime p, the field F is defined to have *characteristic $p > 0$* if and only if $F \supseteq \mathbf{Z}_p$. In this case the smallest subfield of F is $\mathbf{Z}_p$.

Explanation. Given any field F, we define the mapping $\phi : \mathbf{N} \mapsto F$ by setting $\phi(0) = 0$ and successively $\phi(n + 1) = \phi(n) + 1$. The mapping is extended to a mapping $\phi : \mathbf{Z} \mapsto F$ by requiring $\phi(-n) = -\phi(n)$ for all natural n. We now distinguish the cases:

ϕ **is injective.** The mapping is once more extended to a mapping $\phi : \mathbf{Q} \mapsto F$ by requiring $\phi(\frac{p}{q}) = \frac{\phi(p)}{\phi(q)}$ for all integers p and integers $q > 0$. This extension is still injective. We may identify $\mathbf{Q}$ with $\phi(\mathbf{Q})$. In this sense, $\mathbf{Q}$ is the smallest field contained in F.

ϕ **is not injective.** Since the mapping ϕ is a ring homomorphism, the kernel $\{a \in F : \phi(a) = 0\}$ is an ideal. Hence there exists m such that $\phi^{-1}(0) = m\mathbf{Z}$. Since $\phi(1) = 1$, we know that $m \geq 2$. Moreover there is the canonical isomorphism $\mathbf{Z}_m = \mathbf{Z}/m\mathbf{Z} \simeq \phi(\mathbf{Z})$. Since the image $\phi(\mathbf{Z})$ has no null divisors, the same holds for $\mathbf{Z}_m$. Hence $m = p$ is a prime number. From the Euclidean property of primes (or the extended Euclidean algorithm), we have deducted that $\mathbf{Z}_p$ is even a field. We may identify $\mathbf{Z}_p$ with $\phi(\mathbf{Z}_p)$. In this sense, $\mathbf{Z}_p$ is the smallest field contained in F.

$\square$

Definition 45 (Separable polynomial). A polynomial $Q \in F[x]$ is called *separable* iff its irreducible factors have no multiple roots. In more detail

$$Q(x) = c(x - \alpha_1) \cdots (x - \alpha_r) P_1(x) \cdots P_s(x)$$

where $r \geq 0, s \geq 0$ and $c, \alpha_1, \ldots, \alpha_r \in F$, and moreover, the irreducible polynomials $P_t(x) \in F[x]$ have degree at least two, and have no multiple roots.

Remark. The polynomials P_i may be chosen to be monic. In this case, each two of them are either relatively prime or equal.

Lemma 82. *Let the polynomial $f \in F[x]$ be irreducible and $\deg f \geq 1$. f is separable if and only if it has only simple zeros, and in turn if and only if its formal derivative $f' \neq 0$ is not identically zero.*

Proof. We show that an irreducible polynomial has multiple zeros if and only if $f' = 0$.

Assume towards a contradiction that f is irreducible, has a multiple zero $x_1 \in E$, but nevertheless $f' \neq 0$. Clearly this is only possible for $\deg f \geq 2$. Since $f' \neq 0$, the greatest common divisor $\gcd(f, f') \in F[x]$ exists. This is a polynomial in $F[x]$ of degree $\deg \gcd(f, f') \leq \deg f' \leq \deg f - 1$. From the extended Euclidean algorithm, there exist polynomials p and q in $F[x]$ such that $h = pf + qf'$.

In the splitting field E/F holds the factoring $f(x) = (x - x_1)^2 g(x)$ and hence $f'(x) = 2(x - x_1)g(x) + (x - x_1)^2 g'(x)$. In the field extension the polynomial $x - x_1$ divides both f and f' and hence divides $\gcd(f, f') \in F[x]$, Hence $\deg \gcd(f, f') \geq 1$, and

$$\deg(f) > \deg(f') \geq \deg \gcd(f, f') \geq 1$$

But now we see that polynomial f is reducible. Here is it important to know that $\gcd(f, f') \in F[x]$, and therefore the extended Euclidean algorithm is essential. But we have assumed from the beginning that $f \in F[x]$ is irreducible. From that contradiction, we conclude that existence of the multiple root x_1 of the irreducible polynomial f implies $f' = 0$ identically.

Conversely, the assumption $f' = 0$ implies $f'(x_i) = 0$ at all roots. Hence all roots are multiple. $\qquad\qquad\square$

Remark. We need not even assume prime characteristic. But in characteristic zero we always get $\deg \gcd(f, f') \geq 1$ for any polynomial with a multiple root.

Lemma 83. *For an irreducible polynomial with simple roots, the number of roots equals the degree: $r = n = \deg(f)$. This simpler situation occurs always in a field of characteristic zero, and as well in a field of prime characteristic p under the additional assumption that $f' \neq 0$.*

Corollary 38. *In a field $F \supseteq \mathbf{Q}$ of characteristic zero, each nonconstant polynomial is separable. Each irreducible polynomial (of $\deg f \geq 1$) has simple zeros, in its splitting field.*

Proof. For a nonconstant polynomial holds

$$f = cx^n + \text{ lower terms} \quad \text{and} \quad f' = cnx^{n-1} + \text{ lower terms}$$

with $c \neq 0$ and $n \geq 1$. Hence $cn \neq 0$ and $f' \neq 0$.

For an irreducible polynomial of $\deg f \geq 2$ holds $\deg f' \geq 1$ and $f' \neq 0$. By the lemma 82 it has simple zeros. $\qquad\qquad\square$

VIII.0.4　　Galois Extensions

The Galois Theory deals with the interrelation of field extensions and groups of field automorphisms. For any field E, the set of automorphisms

$$\operatorname{Aut}(E) = \{\sigma : E \mapsto E : \sigma \text{ is an automorphism}\}$$

with composition as group operation becomes a group.

Definition 46 (Group of relative automorphisms). Let E/F be any field extension. The set of automorphisms

$$\operatorname{Aut}(E : F) = \{\sigma \in \operatorname{Aut}(E) : \sigma \upharpoonright F = \operatorname{id}_F\}$$

is a subgroup of $\operatorname{Aut}(E)$. It is called the group of *relative automorphisms* of the field extension E/F.

Definition 47 (Galois group). For any polynomial $f \in F[x]$, let E/F be the splitting extension. The group

$$\mathrm{Gal}(f : F) = \mathrm{Aut}(E : F)$$

is called the *Galois group* of the polynomial f over the ground field F.

Proposition 72 (Fundamental Lemma about the Galois-group). *For any polynomial $f \in F[x]$ of degree $n = \deg f$, let E/F be the splitting extension. Let $N = \{x_1, \ldots, x_r\} \subseteq E$ with $r \leq n$ be the set of (pairwise different) zeros of f.*

(a) *The Galois group does not depend on the choice of the splitting field.*

(b) *For any automorphism $\sigma \in \mathrm{Gal}(f : F)$ the restriction $\sigma \uparrow N$ is a permutation of the zeros, and $\sigma \uparrow N$ determine σ uniquely. In other words, the restriction induces a monomorphism $\mathrm{Gal}(f : F) \mapsto \mathcal{S}_r$ and hence an action of the Galois group $\mathrm{Gal}(f : F)$ on the set of zeros N. Especially, the Galois group is finite and its order is a divisor of r!*

Proof. Because of the uniqueness part of proposition 70 about the existence and uniqueness of the splitting field, all splitting extension are isomorphic with the ground field fixed. Hence the group of relative automorphisms does not depend on the choice of the extension field. Any relative automorphism $\sigma \in \mathrm{Aut}(E : F)$ maps any zero $\alpha \in E$ of polynomial $f \in F[x]$ to a zero of f. Indeed

$$f(\sigma(\alpha)) = \sigma(f(\alpha)) = \sigma(0) = 0$$

Since σ is a bijection and N is finite, the restriction $\sigma \uparrow N$ is a bijection $N \mapsto N$, in other words a permutation of the zeros.

Assume that $\sigma \uparrow N = \mathrm{id}_N$. Hence σ is the identity on the field $E = F(\alpha_1, \ldots, \alpha_r)$ which is defined to be a splitting field. Thus the operation of restriction to N defines a monomorphism. By Lagrange's Theorem, the order of the subgroup $\mathrm{Gal}(f : F) \leq \mathcal{S}_r$ is a divisor of the order $r! = |\mathcal{S}_r|$ of the permutation group. $\qquad\square$

VIII.0.5 About Groups

In the following, let G be a finite group that acts on a set X. By definition that means that a mapping $G \times X \mapsto X$ is given, usually denoted simply by a dot, for which holds

$$g(h \cdot x) = (gh) \cdot x$$

for all $g, h \in G$ and $x \in X$. For each $g \in G$ let $X^g = \{x \in X : g \cdot x = x\}$ denote the set of elements in X that are fixed by g.

For each $x \in X$, the *orbit of x* is defined to be $G \cdot x = \{g \cdot x : g \in G\} \subseteq X$. The set of all orbits is denoted by X/G. Notice that X is the disjoint union of all its orbits in X/G.

Let $G_x = \{g \in G : g \cdot x = x\}$ be the subgroup of G that fixes the point $x \in X$, which is also called the *stabilizer subgroup*.

Problem 159 (Bahn Lemma). *Prove the orbit-stabilizer theorem,*[20]

For each $x \in X$ there is a natural bijection between its orbit $G \cdot x = \{g \cdot x | g \in G\} \subseteq X$, and the set of left cosets G/G_x of its stabilizer subgroup G_x. Hence the length of any group orbit is a divisor of the group order.
Prove that even for an infinity group, in cardinal arithmetic holds

$$|G \cdot x||G_x| = |G|$$

Definition 48 (transitive action, simply transitive action). Let the group G act on the elements of the set X. We say that the action of the group is *transitive* if and only if for all $x, y \in X$ there exists $g \in G$ such that $g \cdot x = y$.

We say that the action of the group is *simply transitive* if and only if for all $x, y \in X$ there exists a *unique* $g \in G$ such that $g \cdot x = y$.

Problem 160. *Assume that the group G acts transitively on the set X. Show that the cardinality $|X|$ is a divisor of the group order $|G|$.*

Solution. For a transitive group action, the group orbit equals $G \cdot x = X$ for all $x \in X$. Hence the Bahn lemma implies

$$|X||G_x| = |G \cdot x||G_x| = |G|$$

and $|X|$ is a divisor of the group order $|G|$. $\square$

Problem 161. *Assume that the group G acts on the set X. Show that the following statements are equivalent.*

(i) *The action is simply transitive*

(ii) *For all $x \in X$ the stabilizer subgroup G_x has only one element.*

(iii) *There exists $x \in X$ for which the stabilizer subgroup G_x has only one element.*

(iv) $|X| = |G|$

---

[20] Also called "Bahn Lemma"

Lemma 84 (Burnside's Lemma). *The following formula holds for the number of orbits $|X/G|$*

$$|X/G| = \frac{1}{|G|} \sum_{g \in G} |X^g|$$

Thus the number of orbits (a natural number or $+\infty$) is equal to the average number of points fixed by an element of G (which is also a natural number or infinity). If G is infinite, the division by $|G|$ may not be well-defined; in this case the following statement in cardinal arithmetic holds:

$$|G||X/G| = \sum_{g \in G} |X^g|$$

Proof. The first step in the proof of the lemma is to re-express the sum over the group elements $g \in G$ as an equivalent sum over the set of elements $x \in X$:

$$\sum_{g \in G} |X^g| = |\{(g, x) \in G \times X : g \cdot x = x\}| = \sum_{x \in X} |G_x|$$

Together with Lagrange's theorem one gets

$$|G \cdot x| = [G/G_x] = \frac{|G|}{|G_x|}$$

Our sum over the set X may therefore be rewritten as

$$\sum_{x \in X} |G_x| = \sum_{x \in X} \frac{|G|}{|G \cdot x|} = |G| \sum_{x \in X} \frac{1}{|G \cdot x|}$$

Finally, notice that X is the disjoint union of all its orbits in X/G, which means the sum over X may be broken up into separate sums over each individual orbit.

$$\sum_{x \in X} \frac{1}{|G \cdot x|} = \sum_{A \in X/G} \sum_{x \in A} \frac{1}{|A|} = \sum_{A \in X/G} 1 = |X/G|$$

Putting everything together gives the desired result:

$$\sum_{g \in G} |X^g| = |G| \cdot |X/G|$$

$\square$

Remark. This proof is essentially also the proof of the class equation formula, simply by taking the action of G on itself $X = G$ to be by conjugation, $g \cdot x = gxg^{-1}$. Obviously holds $g \cdot x = x \Leftrightarrow gx = xg..$ In this case the set of fixed points G_x instantiates to the centralizer of x in G.

$$C_x = \{g \in G \ : \ gx = xg\}$$

But also

$$X^g = \{x \in X \ : \ g \cdot x = x\} = C_g$$

and the first step of the proof is tautological. The group orbits become the conjugacy classes of the group. The following formula holds for the number conjugacy classes

$$r = \frac{1}{|G|} \sum_{g \in G} |C_g|$$

Remark (History). Burnside's lemma, sometimes also called Burnside's counting theorem, the Cauchy–Frobenius lemma, orbit-counting theorem, or..."The Lemma that is not Burnside's". In any case, it is a result in group theory which is often useful in taking account of symmetry when counting mathematical objects. Its various eponyms are based on William Burnside, George Pólya, Augustin Louis Cauchy, and Ferdinand Georg Frobenius. The result is not due to Burnside himself, who merely quotes it in his 1897 book 'On the Theory of Groups of Finite Order', attributing it instead to Frobenius (1887).

But, even prior to Frobenius, the formula was known to Cauchy in 1845. In fact, the lemma was apparently so well known that Burnside simply omitted to attribute it to Cauchy. Consequently, this lemma is sometimes referred to as the lemma that is not Burnside's (see also Stigler's law of eponymy). This is less ambiguous than it may seem: Burnside contributed many lemmas to this field.

Problem 162. *Let a and b be relatively prime. Let $\mathcal{G}$ be a commutative group. Take the element $x \in \mathcal{G}$. Show that there exist $y \in \mathcal{G}$ such that $y^a = e$ and $z \in \mathcal{G}$ with $z^b = e$ and natural number m and n such that $x = y^m z^n$.*

Moreover if the element $x \in \mathcal{G}$ has the order $a \cdot b$, then $y \in \mathcal{G}$ has the order a and $z \in \mathcal{G}$ has the order b.

Solution. By the extended Euclidean algorithm there exist positive integers m and n such $\pm 1 = na - mb$. I may assume that $+1 = na - mb$ Hence because of the commutativity

$$x = (x^a)^n (x^{-b})^m = z^n \cdot y^m \qquad \text{with } z := x^a \text{ and } y := x^{-b}$$

But $y^a = x^{-ab} = 1$ and $z^b = x^{ab} = 1$. Let $a' \mid a$ be the order of y and $b' \mid b$ be the order of z. Hence

$$x^{a'b'} = (z^{b'})^{a'n} \cdot (y^{a'})^{b'm} = 1$$

If $a' < a$ or $b' < b$, the order of x would be smaller than ab. Hence $a' = a$ is the order of y and $b' = b$ is the order of z. $\qquad\square$

Problem 163. *Let the commutative group $\mathcal{G}$ have the order $a \cdot b$ with a and b relatively prime. We define the subgroups $H_a = \{x \in \mathcal{G} : x^a = e\}$ and $H_b = \{x \in \mathcal{G} : x^b = e\}$. Show that $\mathcal{G} = H_a \times H_b$. In other words, any element $x \in \mathcal{G}$ has a unique decomposition $x = yz$ with $y \in H_a$ and $z \in H_b$.*

Proof. Take any element $x \in \mathcal{G}$. By the previous problem there exist $y, z \in \mathcal{G}$ such that $x = y^m z^n$. Moreover $y^a = e$, hence $y, y^m \in H_a$. Similarly $z^b = e$, hence $z, z^n \in H_b$.

Finally, the decomposition is unique. Assume that $z = yz = y'z'$ with $y, y' \in H_a$ and $z, z' \in H_b$. Hence $u := yy'-1 = z'z^{-1} \in H_a \cap H_b$. The order t of u is both a divisor of a and of b. Since they are assumed to be relatively prime, we conclude that $t = 1$, hence $u = e$ and $y = y', z = z'$, as to be shown. $\qquad\square$

Lemma 85. *Let $H \subset \mathcal{G}$ be a subgroup that does not contain any elements of which the order is a prime power of any prime dividing the group order. Then $H = \{e\}$,*

Proof. We have assumed the $\mathcal{G}$ is commutative, finite and has the order

$$|\mathcal{G}| = \prod_{1 \leq i \leq r} p_i^{\mu_i} = \prod_{1 \leq i \leq r} q_i$$

For simpler notation I introduce the prime powers $q_i := p_i^{\mu_i}$. Let

$$a = q_1 = p_{\mu_1}^1 \quad \text{and} \quad b = \prod_{2 \leq i \leq r} q_i$$

Take any $x \in H$. It has a unique decomposition $x = yz$ with $y^a = e$ and $z^b = e$. From the assumption we know that $y = e$. Hence $x = z$ For $2 \leq i < r$, we continue with .

$$a_i = q_i \quad \text{and} \quad b_i = \prod_{i < j \leq r} q_j$$

and successively get z_i as follows: We begin with $z_1 := z$ There is a decomposition $z_i = u_i v_i$ with $u_i^{a_i} = e$ and $v_i^{b_i} = e$. From the assumption we know that $u_i = e$. Hence $z_{i+1} := v_i = z_i$.

Finally for $i = r - 1$, the second factor b_i is a prime power, too. From the assumption we conclude that $v_i^{b_i} = e$. implies $v_i = e$. Hence $z_r = v_i = e$. After finitely many steps one obtains $x = z_1 = z_2 = \cdots = z_r = e$. $\qquad\square$

Proposition 73. *Assume $\mathcal{G}$ is commutative, finite and has the order*

$$|\mathcal{G}| = \prod_{1 \leq i \leq r} q_i$$

Then the group is isomorphic to a direct product.with the multiplicities $n_{i,k} \geq 0$

(VIII.0.2)
$$\mathcal{G} \simeq \prod_{1 \leq i \leq r} \prod_{1 \leq k \leq \mu_i} (\mathbf{Z}_{p_i^k})^{n_{i,k}}$$
$$|\mathcal{G}| = \prod_{1 \leq i \leq r} \prod_{1 \leq k \leq \mu_i} p_i^{n_{i,k} \cdot k}$$

Proof. The group is deconstructed algorithmically. We look whether the group $\mathcal{G}$ contains a cyclic subgroup of order p_1. If yes, let its generator be c. Now we consider the quotient group $\mathcal{G}/(c)$. Repeat until no cyclic subgroup of order p_1 exists. One goes on to the orders $p_1^2, \ldots, p_1^{\mu_1}$. Because of Lagrange's Theorem any subgroup cannot have any higher powers of p_1 as order. After finitely many steps, the remaining quotient group $\mathcal{G}/(\prod c_{1,k})$ contains no more cyclic subgroup for any prime power of p_1. Let .

$$a = p_{\mu_1}^1 \text{ and } b = \prod_{2 \leq i \leq r} q_i \text{ and}$$

By problem 163, any element $x \in G_2 := \mathcal{G}/(\prod c_{1,k})$ has a unique decomposition $x = yz$ with $y^a = e$ and $z^b = e$. From the construction above, we know that $y = e$. [21]

With the group G_2, and the power of prime p_2, one proceeds similarly, finally with all prime factors p_i for $1 \leq i \leq r$ of the group order. In this way, all cyclic prime power subgroups are exhausted. The process stops after finitely many steps since in the beginning the group was assumed to be finite. Moreover the remaining quotient group $H = G_r$ does not contain any elements of which the order is a prime power. By the lemma holds $H = \{e\}$. The cycles we have been diving out constitute the entire given group. $\square$

Proposition 74. *If prime p divides the order of finite commutative group $\mathcal{G}$, there exists an element $g \in \mathcal{G}$ which has the order p, in other words $g \neq e$ and $g^p = e$.*

Proposition 75. *Every finite multiplicative subgroup $\mathcal{G}$ of a field is cyclic.*
For all the prime divisors $p_i \mid |\mathcal{G}|$ with $1 \leq i \leq r$, let c_i be an element with maximal prime power order $p_i^{k_i}$. The group is generated by an element $x = \prod_{1 \leq i \leq r} c_i$ with the order $\prod_{1 \leq i \leq r} p_i^{k_i}$.

---

[21]This conclusion is obtained without have at this point knowledge about the order of $\mathcal{G}/(\prod c_{1,k})$, but only from having exhausted the cyclic subgroup for any prime power of p_1.

Proof. Assume $\mathcal{G} \subseteq F \setminus \{0\}$ is finite and has the order

$$|\mathcal{G}| = \prod_{1 \le i \le r} q_i$$

Take any prime p dividing the group order. Let k be the maximal number such that an element c of order p^k exists in $\mathcal{G}$. By Lagrange's Theorem holds $p = p_i$ for some $1 \le i \le r$ and $0 \le k_i \le \mu_i$.

Fix i . Let

$$H = \{x \in F : x^{p^k} = 1\}$$

Because the polynomial $x^{p^k} - 1$ has degree p^k, it has in the field F at most p^k zeros. Hence $|H| \le p^k$. On the other hand $\{c^j : 0 \le j < p^k\} \subseteq H$ has already p^k elements. Hence

$$H = \{c^j : 0 \le j < p^k\} \text{ and } |H| = p^k$$

Allowing all $1 \le i \le r$, we may form the product group

$$G' := \{ \prod_{1 \le i \le r} c_i^{j_i} : \text{ with } 0 \le j_i < k_i \text{ for } 1 \le i \le r\}$$

Clearly $G' \subseteq \mathcal{G}$. From the problem 163 one gets by an easy induction that the element $x = \prod_{1 \le i \le r} c_i$ has the order $\prod_{1 \le i \le r} p_i^{k_i}$. Hence

$$\prod_{1 \le i \le r} p_i^{k_i} \le |G'| \le \prod_{1 \le i \le r} p_i^{k_i}$$

holds with equalities. Hence G' is cyclic.

I claim that $\mathcal{G} \subseteq G'$. Take any $x \in \mathcal{G}$. By Lagrange's Theorem it has as the order t which is a divisor of the group order and hence

$$t = \prod_{1 \le i \le r} p_i^{\nu_i} \text{ with } 0 \le \nu_i \le \mu_i \text{ for } 1 \le i \le r$$

By the problem, we obtain further group elements with orders $p_i^{\nu_i}$. Hence $0 \le \nu_i \le k_i$ for $1 \le i \le r$ and hence $t \in G'$. Hence $\mathcal{G} = G'$ and $k_i = \mu_i$ for $1 \le i \le r$. $\qquad\square$

Proposition 76 (Cauchy's Theorem). *If prime p divides the order of group $\mathcal{G}$, there exists an element $g \in \mathcal{G}$ which has the order p, in other words $g \ne e$ and $g^p = e$.*

Proof. Above the proposition is already proved for a commutative group,—the proof for that case is different but simpler. The proof for the noncommutative case is done by induction on the order of the group. So we assume the assertion to hold for any group of order smaller than n. Let a group of order $n = |\mathcal{G}|$ be given, and assume prime p divides the group order $|\mathcal{G}|$. Let

$$Z = \{g \in \mathcal{G} : gh = hg \text{ for all } h \in \mathcal{G}\}$$

be the center of $\mathcal{G}$.

- In the case that $p \mid |Z|$ we use that Z is commutative. Thus there exists $g \in Z$ of order p,—the assertion holds.

- In the case that $p \nmid |Z|$ and $|Z| > 1$ we consider the quotient group $H = G/Z$. Since $|H| < |\mathcal{G}|$ and $p \mid |H|$, we may use the induction assumption. There exists an element $hZ \in H$ of order p. Hence $h^p = z \in Z$ and $h \notin Z$.

  I claim there exists $u \in Z$ such that $u^p = z$. Indeed, since $|Z|$ and p are relatively prime, there exist integers a and b such that $ap - b|Z| = 1$. Hence $z = z^{ap}$ and $u = z^a$.

  In the end one puts $g := hu^{-1}$. Since $u \in Z$ we may use commutativity to get

  $$g^p = h^p u^{-p} = z z^{-1} = e$$

  Since $h \notin Z$ holds $g \neq e$ and hence g has the order p.

- In the case that $|Z| = 1$ we use the equivalence classes of group $\mathcal{G}$. Z consists of the equivalence classes with a single element. The remaining equivalence classes all contain at least two elements. Since the order of all equivalence classes add up to $|\mathcal{G}|$, there exists an equivalence class $[g]$ the order of which is at least two and not divisible by prime p. Let

  $$Z_g = \{h \in \mathcal{G} : hgh^{-1} = g\}$$

  be the centralizer of g. By the Bahn Lemma 159 one gets $|Z_g| \cdot |[g]| = |\mathcal{G}|$. Hence $p \mid |Z_g|$ and $|Z_g| < |\mathcal{G}|$. We may use the induction assumption. There exists an element $h \in Z_g$ of order p.

$\square$

VIII.0.6   About Existence of a Primitive Element

Definition 49 (Primitive element). Let E/F be a finite dimensional field extension. The element $\alpha \in E$ is called *primitive* if and only if $E = F(\alpha)$, in other words the extension field E is generated by the ground field F together with the single element α.

Theorem 27 (Steinitz' Theorem about the primitive element). *Let E/F be a finite dimensional field extension. A primitive element exists for the field extension E/F if and only if there exist only finitely many intermediate fields $E \supseteq L \subseteq F$.*

Necessity. We assume that the extension field E is generated by the single element α. Let $f \in F[x]$ be the minimal polynomial of α. By the assumption holds $n = \deg f = \dim[E : F]$. Take any intermediate field $L \subsetneq E$. A fortiori holds $E = L(\alpha)$. The minimal polynomial for the extension E/L is a divisor f_L of f with degree $\deg f_L = \dim[E : L]$.

Take a second smaller intermediate field $K \subseteq L$. Again holds $E = K(\alpha)$. The minimal polynomial f_K for the extension E/K is a divisor of f with degree $\deg f_K = \dim[E : K]$, and moreover f_L is a divisor of f_K. Hence

$$(\text{VIII.0.3}) \qquad \deg f_L = \dim[E : L] \leq \dim[E : K] = \deg f_K$$

The coup de grâce is to take for K the field obtained by adjoining to the field F only the coefficients of the polynomial f_L. By construction, f_L is irreducible for the extension E/L. Hence $f_L \in K[x]$ and this polynomial is a fortiori irreducible for the extension E/K and $f_L(\alpha) = 0$ and $E = K(\alpha)$. Hence f_L is the minimal polynomial for the extension E/K and hence $f_L = f_K$. After all that pain one gets equalities in (VIII.0.3) and hence $L = K$.

To each intermediate field L corresponds a divisor f_L of f, such that $f_L \in L[x]$[22] is the minimal polynomial of the extension E/L. Since there exist only finitely many divisors, there exist only finitely many intermediate fields, Indeed there exist at most 2^n intermediate fields. $\qquad \square$

Remark. Take any intermediate field L. We have obtained in the field L the splitting $f = \prod (f_L)^j \in L[x]$ into irreducible factors. By the proof above, each factor is equal to the unique minimal polynomial of the extension E/L. Hence

$$f = (f_L)^{\dim[L:F]}$$

---

[22] The painful part was actually to confirm $f_L \in L[x]$.

Sufficiency. We assume that the extension field E/F is finite dimensional and there exist only finitely many intermediate fields. The proof has to distinguish two cases.

The field F is finite. If F contains q elements and $\dim[E : F] = n$, then E contains q^n elements. We have shown in proposition 75 that the multiplicative group $E \setminus \{0\}$ of a finite field is cyclic. Let α be a generator. In that case holds $E = F(\alpha)$.

The field F contains infinitely many elements. Since $\dim[E : F] = n$ is assumed to be finite, the vector space E is generated by any basis β_i for $1 \leq i \leq n$. Hence it is enough to prove that the extension by adjunction of two elements α, β can be achieved by adjunction of one element γ.

There exist infinitely many elements $\alpha + c\beta$ with $c \in F$ but only finitely many field extensions $F(\alpha + c\beta)$. Hence there exist $c \neq d \in F$ such that $F(\alpha + c\beta) = F(\alpha + d\beta)$. Hence both

$$\alpha + c\beta,\ \alpha + d\beta \in F(\alpha + c\beta)$$
$$(c - d)\beta \in F(\alpha + c\beta)$$
$$\beta,\ \alpha \in F(\alpha + c\beta)$$
$$F(\alpha, \beta) \subseteq F(\alpha + c\beta) \subseteq F(\alpha, \beta)$$
$$F(\alpha, \beta) = F(\alpha + c\beta)$$

proves that the extension may be done by adjoining one less element, hence indeed by a single element.

$\square$

Theorem 28 (The common theorem about the primitive element). *Let E/F be a finite dimensional field extension. If either F is finite or $F \supseteq \mathbf{Q}$ (F has characteristic zero) a primitive element exists for the field extension E/F.*

Proof. The proof for the case that F is finite has been done above. Assume that $F \supseteq \mathbf{Q}$ has the characteristic zero. Hence F has infinitely many elements. As above, to prove that the extension by adjunction of two elements α, β can be achieved by adjunction of one element γ.

Let $P \in F[x]$ be the minimal polynomial of α and $Q \in F[x]$ be the minimal polynomial of β. Let $L \supseteq E \cup F(\alpha) \cup F(\beta)$ be a finite dimensional field extension in which both P and Q split completely. Let α_i and β_j be the respective conjugates.

There exist infinitely many elements $c \in F$ such that all $\alpha_i + c\beta_j$ for $1 \leq i \leq \deg P$ and $1 \leq j \leq \deg Q$ are different. We choose any such c and let $\alpha = \alpha_1$, $\beta = \beta_1$ and

$$h(x) := P(\alpha + c\beta - cx) \in F(\alpha + c\beta)[x]$$

By construction $h(\beta) = P(\alpha) = 0$. Moreover, $h(x) = 0 \Leftrightarrow \alpha + c\beta = \alpha_i + cx$ and hence $h(\beta_j) = 0 \Leftrightarrow (i = 1 \text{ and } j = 1)$.

The greatest common divisor $\gcd(Q, h)$ has only one zero, namely β_1 since any irreducible polynomial has simple zeros in characteristic zero. Hence $\gcd(Q, h) = x - \beta$ By extended Euclidean algorithm holds $\gcd(Q, h) = AQ - Bh$ with $A, B \in F(\alpha + c\beta)$. Hence

$$-\beta = A(0)Q(0) - B(0)h(0) \in F(\alpha + c\beta)$$
$$\beta, \alpha \in F(\alpha + c\beta)$$
$$F(\alpha, \beta) \subseteq F(\alpha + c\beta) \subseteq F(\alpha, \beta)$$
$$F(\alpha, \beta) = F(\alpha + c\beta)$$

proves that the extension may be done by adjoining one less element, hence indeed by a single element. $\qquad\square$

The following example for a finite dimensional extension E/F which has no primitive element is given by Artin [1]. By Theorems 27 and 28 we see that the ground field F needs to be an infinite field with characteristic $p > 0$.

The following field is odd enough to produce the counterexample. Let characteristic $p = 2$ and let $E = \mathbf{Z}_2(x, y)$ be the field of rational functions with two variables x, y over $\mathbf{Z}_2$. We agree to write down only the non-vanishing terms. Hence these terms all have the coefficient 1. Let $F \subset E$ be the subfield of rational functions depending only on the variables x^2, y^2.

Problem 164. *Prove that indeed F is a field. Determine the dimension $\dim[E : F]$ and give a basis.*

Proof. Addition and multiplication with the usual rules for rational fractions show that F is a field. Take any element in the field E. The following trick is used to obtain a denominator from the ring $\mathbf{Z}_2[x^2, y^2]$, in other words containing only the squares x^2 and y^2:

$$\alpha = \frac{p(x, y)}{q(x, y)} = \frac{p(x, y)q(x, y)}{(q(x, y))^2} = \frac{p(x, y)q(x, y)}{q(x^2, y^2)}$$

The numerator is from the ring $\mathbf{Z}_2[x, y]$. We may separate even and odd powers of x and y to obtain

$$\frac{p(x, y)}{q(x, y)} = \frac{\sum_{i=0,1 \text{ and } k=0,1} a_{ik}\, x^i y^k}{q(x^2, y^2)} = \sum_{i=0,1 \text{ and } k=0,1} b_{ik}\, x^i y^k$$

with $a_{ik} \in \mathbf{Z}_2[x^2, y^2]$,—the ring,— and

$$b_{ik} = \frac{a_{ik}}{q(x^2, y^2)} \in \mathbf{Z}_2(x^2, y^2) = F$$

We have obtained the spanning set $1, x, y, xy$ for the extension E/F. Too, these elements are linearly independent over F. Indeed, any dependence relation

$$\sum_{i=0,1 \text{ and } k=0,1} b_{ik}\, x^i y^k = 0$$

with coefficients $b_{ik} \in F$ may be put on a common denominator in $\mathbf{Z}_2[x^2, y^2]$. Finally one obtains from the numerator $b_{ik} = 0$ to confirm the linear independence.

We have obtained the basis $1, x, y, xy$ for the extension E/F. Hence the dimension of the extension is $\dim[E : F] = 4$. $\qquad\qquad\square$

Remark. On the other hand any one element extension $F(\alpha)$ has at most the dimension 2. Indeed $\alpha \in E$ implies $c := \alpha^2 \in F$. Hence α is a zero of the quadratic polynomial $X^2 - c \in F[X]$. Hence there does not exist a primitive element.

Remark. Too, we may confirm directly that there exist infinitely many intermediate fields $F \subset L \subset E$. Let

$$\alpha = \sum_{i=0,1 \text{ and } k=0,1} a_{ik}\, x^i y^k$$

with $a_{ik} \in F$. One gets a two-dimensional extension if and only if $\alpha \notin F \Leftrightarrow (a_{10}, a_{01}, a_{11}) \neq (0, 0, 0)$. In that case we get the extension

$$F(\alpha) = \{A + B\alpha : A, B \in F\} = \{C + Ba_{10}x + Ba_{01}y + Ba_{11}xy : C, B \in F\}$$

with $C = A + Ba_{00}$. The ratios of the last three coefficients are determined by the choice of α. All extensions $F(x + y^{2n-1})$ with $n \geq 1$ turn out to be different:

$$F(x + y^{2n-1}) = \{C + Bx + By^{2n-1} : C, B \in F\}$$

We see that we get indeed infinitely many intermediate fields, as predicted by Steinitz' Theorem 27 in the case of nonexistence of a primitive element.

VIII.0.7   Fixed Fields and Groups of Automorphisms

Definition 50 (Fixed field). Let E be a field and $G \leq \mathrm{Aut}(E)$ be any subgroup of automorphisms. The field

$$\text{(VIII.0.4)} \qquad \mathrm{Fix}(E : G) = \{x \in E \,:\, \sigma(x) = x \text{ for all } \sigma \in G\}$$

is called the *fixed field* of group G in the field E.

Problem 165. *Let E be any field and F be any subfield. Convince yourself that from the definitions 50 and 46 one gets for the fixed field $L := \mathrm{Fix}(E : \mathrm{Aut}(E : F))$ at least $E \supseteq L \supseteq F$.*

Proposition 77. *Let $E = F(x)$ be a one-element extension and $f \in F[x]$ be the minimal polynomial of x. For this case holds $n := \deg(f) = \dim[E : F]$, and the polynomial f is irreducible.*

Moreover, the automorphism group acts even simply transitively *on the roots of the minimal polynomial f. Hence the number of its roots $r = |G|$ equals the order $|G|$ of the Galois group $G = \mathrm{Aut}(E, F)$.*

Note that it is still possible that $r < n$. Only under the additional assumption that the minimal polynomial has simple zeros (or, equivalently, is separable) holds $r = n$, and hence $|G| = \dim[E : F]$.

Corollary 39. *Especially for any one-element extension $E = F(x)$ of a field of characteristic zero holds $\dim[E : F] = |\mathrm{Aut}(E, F)|$: the order of the group of relative automorphisms is equal to the dimension of the field extension.*

Proof. The minimal polynomial f is irreducible. It is known that $\deg(f) = \dim[E : F]$, a basic fact about any one-element extension. By Theorem 16.6.6 in Michael Artin's book [2] , irreducibility implies that the automorphism group $G = \mathrm{Aut}(E, F)$ acts *transitively* on roots of the minimal polynomial. [23]

Any automorphism $\sigma \in \mathrm{Aut}(E, F)$ is uniquely determined by its value $\sigma(x)$. Hence the automorphism group acts even *simply transitively* on the roots of the minimal polynomial f. Now we conclude (see problem 161) that all the stabilizer groups $|G_x|$ have only one element, and $r = |X| = |G|$. $\qquad\square$

We want to stress the importance of the equality $\dim[E : F] = |\mathrm{Aut}(E, F)|$: the order of the group of relative automorphisms is equal to the dimension of the field extension. One does not need to restrict oneself to one-dimensional extensions, and neither to fields with characteristic zero. Before I dig myself into Artin's genial approach, let me proceed by more elementary ideas available at this point.

---

[23] see also theorem 1, p.316 in Gerd Fisher [8].

A preliminary remark about Galois extensions

Take any polynomial $f \in F[x]$ of degree $\deg(f) \geq 1$, and let E be the splitting field of f. Such a splitting field may be constructed by successively adjoining zeros of f to the field. One obtains the splitting extension E/F after at most $n - 1$ steps. The single root extension has a dimension at most n. By induction is proved that the degree of the extension satisfies $\dim[E : F] \leq n!$.

Let $G = \mathrm{Aut}(E, F)$ be the group of automorphisms. Any $g \in G$ fixes all elements of the base field F while permuting the roots $X = \{x_1, \ldots, x_r\}$. We may have $r < n$ for the case of multiple zeros, otherwise $r = n$.

The automorphisms $\sigma \in G$ are in one-to-one correspondence of a subgroup of the permutations of the roots. Hence the order $|G|$ is a divisor of $r!$.

One is forced to restrict attention to irreducible polynomials, to get any further results. Since the order in which the zeros are adjoined to the field extension is arbitrary, one can prove that for an irreducible polynomial, and any two of its roots x_i and x_j, there exists an automorphism $\sigma \in \mathrm{Aut}(E, F)$ such that $\sigma(x_i) = x_j$. In that case, one says that the group G *acts transitively* on the roots. For complete details see Gerd Fisher [8], p.320. This is the theorem 70 about existence and uniqueness properties of the splitting field from above.

By the Bahn lemma 159, also called the orbit-stabilizer theorem, there is

> for each $x \in X$ a natural bijection between its orbit $G \cdot x = \{g \cdot x | g \in G\} \subseteq X$, and the set of left cosets G/G_x of its stabilizer subgroup $G_x = \{g \in G : g \cdot x = x\}$.

"Transitivity" means that for any root $x \in X = \{x_1, \ldots, x_r\}$, its orbit $G \cdot x = X$ is the entire set of roots. From the Bahn-lemma one gets

$$r = |X| = |G|/|G_x|$$

and hence the number of roots r is a divisor of the group order $|G|$.

Lemma 86. *For any irreducible polynomial, the order of the automorphism group is divisible by the number of roots r, and it is in turn a divisor of $r!$:*

$$r \mid |G| \mid r!$$

Under the additional assumption has simple zeros holds $r = n$ and $|G| = \dim[E : F]$. This case always occurs in a field of characteristic zero.

It is a main contribution of Artin to Galois theory to stress the equality $\dim[E : F] = |\mathrm{Aut}(E, F)|$, and prove it in a general setup. One does not need to restrict oneself to one-dimensional extensions, and neither to fields with characteristic zero and irreducible polynomials, as done above. Indeed the equality $\dim[E : F] = |\mathrm{Aut}(E, F)|$ for the finite dimensional case gets the *definition* of a Galois extension.

Theorem 29 (Artin's approach to the Galois Theory). *Let E be a field and $\mathcal{G} \leq \mathrm{Aut}(E)$ be a finite group of n automorphisms $\sigma_i : E \mapsto E$. Let $F = \mathrm{Fix}(E : G) \subseteq E$ be the fixed field of the group $\mathcal{G}$. Then $\dim[E : F] = |\mathcal{G}|$.*

We prove $\dim[E : F] \leq |\mathcal{G}|$. Define the group-averaging operator $P : E \mapsto F$

$$\text{(VIII.0.5)} \qquad\qquad Px = \frac{1}{n} \sum_{1 \leq i \leq n} \sigma_i(x)$$

This operator has the properties

$$\text{(VIII.0.6)} \qquad\qquad P^2 = P \ \text{ and } \ \sigma_i \circ P = P \circ \sigma_i = P \quad \text{for all } \sigma_i \in \mathcal{G}$$

We claim that $\dim[E : F] =: r \leq n$. Suppose towards a contradiction that $r > n$. If $\dim[E : F] = r$ is finite, choose any r basis vectors α_k for the extension $[E : F]$. In the case that $\dim[E : F] = \infty$, choose any $r > n$ and any r linearly independent vectors α_k from the extension $[E : F]$.

There would exist a nontrivial solution $x_k \in E$ with $1 \leq k \leq r$ for the system of equations

$$\text{(VIII.0.7)} \qquad\qquad \sum_{1 \leq k \leq r} x_k\, \sigma_i(\alpha_k) = 0 \quad \text{for all } \sigma_i \in \mathcal{G}$$

There exists an unknown $x_m \neq 0$, we may multiply all equations by x_m^{-1} and assume that $x_m = 1$. For all $\sigma_j \in \mathcal{G}$ we may multiply by this automorphism. By putting $\sigma_j \circ \sigma_i = \sigma_l$ and permuting the equations one gets

$$\sum_{1 \leq k \leq r} \sigma_j(x_k)\, \sigma_l(\alpha_k) = 0 \quad \text{for all } \sigma_l \in \mathcal{G}$$

and finally the averaged equations

$$\sum_{1 \leq k \leq r} (Px_k)\, \sigma_l(\alpha_k) = 0 \quad \text{for all } \sigma_l \in \mathcal{G}$$

There is at least one nonzero unknown $Px_m = P1 = 1$, and the trivial character with $\sigma_1(x) = x$ for all $x \in E$. Hence we have obtained a linear dependence relation for the α_k with nontrivial coefficients $Px_k \in F$ from the ground field. This contradicts the assumption that the α_k are r linearly independent vectors from the extension $[E : F]$. Hence the case $r > n$ is impossible. $\qquad\square$

Proposition 78. *Any n automorphisms define the functions $\sigma_i : E \setminus \{0\} \mapsto E \setminus \{0\}$ on the multiplicative group of E, which are also called* characters.
*   We claim that these are n linearly independent functions.*

Proof. We suppose towards a contradiction a linear dependence among the characters $\sigma_i(x)$ with $1 \leq i \leq m$ and make m minimal. Clearly $m \geq 2$ since the characters are nonzero.

There would exist a nontrivial solution $y_i \in E$ with $1 \leq i \leq m$ and $y_m \neq 0$ for the system of equations

$$(\text{VIII.0.8}) \qquad\qquad \sum_{1 \leq i \leq m} y_i\, \sigma_i(x) = 0 \quad \text{for all } x \in E$$

There exists an $a \in E$ such that $\sigma_1(a) \neq \sigma_m(a)$. We may replace $x \mapsto ax$ and produce from equation (VIII.0.8)

$$\sum_{2 \leq i \leq m} y_i \sigma_i(a)\, \sigma_i(x) = 0 \quad \text{for all } x \in E$$

Too, we may multiply equation (VIII.0.8) and subtract to get

$$\sum_{2 \leq i \leq m} y_i(\sigma_1(a) - \sigma_i(a))\, \sigma_i(x) = 0 \quad \text{for all } x \in E$$

Since $y_m(\sigma_1(a) - \sigma_m(a)) \neq 0$ this is a dependence relation between the $m-1$ characters $\sigma_i(x)$ with $2 \leq i \leq m$, contradicting the minimality of m. $\qquad\square$

We prove $|\mathcal{G}| \leq \dim[E : F]$. We know already that $\dim[E : F] = r$ is finite. We choose any r basis vectors α_k for the extension $[E : F]$.

Suppose towards a contradiction $n > r$. There would exist a nontrivial solution $y_i \in E$ with $1 \leq i \leq n$ for the system of equations

$$\sum_{1 \leq i \leq n} y_i\, \sigma_i(\alpha_k) = 0 \quad \text{for all } 1 \leq k \leq r$$

Hence

$$\sum_{1 \leq i \leq n} y_i \, \sigma_i(x) = 0 \quad \text{for all } x \in E$$

contradicting the linear independence of the characters. Hence $n \leq r$ as to be shown. $\square$

Corollary 40. *The matrix*

$$(\text{VIII.0.9}) \qquad \sigma_i(\alpha_k) \quad \text{with } 1 \leq i \leq n \text{ and } 1 \leq k \leq n$$

with the group of automorphisms σ_i of the extension E/F and the basis vectors α_k is nonsingular.

Theorem 30 (The universal splitting property I). *Let E be a field and $\mathcal{G}$ be a finite group of n automorphisms $\sigma_i : E \mapsto E$. Let $F = \mathrm{Fix}(E : G) \subseteq E$ be the fixed field of the group $\mathcal{G}$. Then the following holds*

(i) *For any $\alpha \in E$, there exists a polynomial $Q \in F[x]$ that splits completely in the extension field E, is irreducible, has simple zeros, and among them $Q(\alpha) = 0$.*

(ii) *Each element of $\alpha \in E$ is algebraic over F.*

(iii) *Each irreducible polynomial $P \in F[x]$ with one root $\alpha \in E$, has all its roots in E.*

Proof of item (i). Given is any $\alpha \in E$. From the images $\sigma_i(\alpha)$ for $1 \leq i \leq n$ we pick out those $r \leq n$ items which are different. We may say these are obtained for $1 \leq i \leq r$ and $\alpha = \alpha_1$. We map the polynomial

$$Q = \prod_{1 \leq i \leq r} (x - \sigma_i(\alpha))$$

by all automorphisms $\sigma_j \in \mathcal{G}$ to obtain

$$\sigma_j \, Q = \prod_{1 \leq i \leq r} (x - \sigma_j \circ \sigma_i(\alpha)) = \prod_{1 \leq l \leq r} (x - \sigma_l(\alpha)) = Q$$

Indeed for any fixed j, there are obtained r different images $\sigma_j \circ \sigma_i(\alpha)$ for $1 \leq i \leq r$. Since $r < \infty$, this is the complete set of images $\sigma_i(\alpha)$ for $1 \leq i \leq n$.

We conclude that all coefficients of polynomial $Q \in E[x]$ are invariant under all automorphism from the group $\mathcal{G}$. Hence $Q \in F[x]$. $\square$

Problem 166. *To finish the proof, check that the assumptions of theorem 30 imply that above constructed polynomial $Q \in F[x]$ is <u>irreducible</u> has simple zeros and splits in E. In other words, all roots of Q are algebraic over F.*

Solution. Clearly, the polynomial Q defined above has simple zeros and splits in E.

To check the irreducibility of Q, let $P \in F[x]$ be a divisor of Q with $\deg P \geq 1$. Hence there exists a zero: $P(\alpha_l) = 0$ among the $\alpha_l = \sigma_i(\alpha)$ with $1 \leq i \leq n$. Hence $1 \leq l \leq r$. By applying all automorphisms $\sigma_i \in \mathcal{G}$ one gets all the images $\sigma_i(\alpha_l) = \alpha_1 \ldots \alpha_r$, which are all the zeros of Q.

Since $\sigma_i P = P$ for all $\sigma_i \in \mathcal{G}$, these

$$P(\sigma_i(\alpha_l)) = \sigma_i(P(\alpha_l)) = \sigma_i(0) = 0$$

are all zeros of P. Hence $\deg P \geq r$ and $P = Q$, confirming that Q is irreducible. $\square$

Proof of item (ii). Since the element α from the proof of item (i) is chosen arbitrarily, each $\alpha \in E$ is the zero of a polynomial $Q \in F[x]$ and hence algebraic over F. $\square$

Proof of item (iii). Let the polynomial $P \in F[x]$ be irreducible and assume it has one root $\alpha \in E$ in the field extension. Too, we may assume that P is monic. Calculate the polynomial Q as in part (i) with this instance for α. Let $R = \gcd(P, Q)$. From the Euclidean algorithm, it is clear that $R \in F[x]$. Since α is a common root of P and Q, we get $R(\alpha) = 0$ and $\deg R \geq 1$. Since R is a divisor of the irreducible polynomial P and $\deg R \geq 1$ and R is monic, we conclude that $P = R = \gcd(P, Q)$. Hence P is a divisor of Q.

Since the irreducible polynomial P is a divisor of the irreducible polynomial Q and $\deg P \geq 1$ and P is monic, we conclude that $P = Q$. Hence all roots of P are in the extension field E, as to be shown. $\square$

Problem 167. *Of course polynomial Q and its degree r may depend on the given $\alpha \in E$.*

Prove that $r = \dim[F(\alpha) : F]$ and that r is divisor of the group order $|\mathcal{G}|$.

Solution. The constructed polynomial $Q \in F[x]$ is irreducible and has root $\alpha \in E$. Hence it is the minimal polynomial of α and the powers α^i with $0 \leq i < r$ span the single-root extension $F(\alpha)$. By the tower theorem

$$\dim[E : F] = \dim[E : F(\alpha)] \cdot \dim[F(\alpha) : F]$$

Hence $r = \dim[F(\alpha) : F]$ is a divisor of $|\mathcal{G}| = \dim[E : F]$.

Of course, $F = \mathrm{Fix}(E : G) \subseteq E$ is the fixed field of the group $\mathcal{G}$. $\square$

Another solution. The roots $\{\alpha_1, \ldots, \alpha_r\}$ of the constructed polynomial $Q \in F[x]$ are the orbit of the root $\alpha \in E$ under the group $\mathcal{G}$.

By the orbit-stabilizer theorem 159 holds the following:

For each $x = \alpha \in \{\alpha_1, \ldots, \alpha_n\} = X$ there is a natural bijection between its orbit $\mathcal{G} \cdot x = \{g \cdot x | g \in \mathcal{G}\} \subseteq X$, and the set of left cosets $\mathcal{G}/G_\alpha$ of its stabilizer subgroup G_α.

Hence the length of any group orbit is a divisor of the group order $n = |\mathcal{G}|$. Especially r is a divisor of n. $\qquad\square$

Problem 168. *Let E/F be a field extension as in theorem 30 about the universal splitting property I. Prove that there exists a <u>separable</u> polynomial $S \in F[x]$ such that S splits in E and E is the (smallest) <u>splitting field</u> for polynomial S.*

Solution. There exists a basis $\omega_1, \ldots \omega_n$ of the extension E/F. Hence $E = F(\omega_1, \ldots, \omega_n)$.

For each $1 \leq i \leq n$, we choose $\alpha := \omega_i$ and construct as in item (i) the irreducible polynomial Q_i with zero α and simple zeros. The polynomial

$$S = Q_1 \cdot Q_2 \cdots Q_n$$

is separable. It splits in E since each factor splits in E. Among its roots are all basis vectors $\omega_1, \ldots \omega_n$. Hence it cannot split in any smaller field extension. $\qquad\square$

Remark. Take the case of a field with characteristic zero. We may take for S a polynomial with simple zeros. One knows that

$$\deg S \leq \dim[E : F] \leq \deg S!$$

But for prime characteristic, the polynomial S may have multiple zeros. I see no possibility to give a nice upper bound for the degree of polynomial S.

Problem 169. *Prove that for any finite dimension extension E/F, the group order $|\mathrm{Aut}(E : F)|$ is a divisor of the dimension $\dim[E : F]$. Moreover*

$$|\mathrm{Aut}(E : F)| = \dim[E : F] \Leftrightarrow \mathrm{Fix}(E : \mathrm{Aut}(E : F)) = F$$

Proof. By the previous problem 169, there exists a <u>separable</u> polynomial $S \in F[x]$ such that E is the splitting field for polynomial S. By the fundamental Lemma 72 about the Galois group, the group order $|\mathrm{Aut}(E : F)|$ is finite.

From the definitions 50 and 46 one directly gets

$$L := \mathrm{Fix}(E : \mathrm{Aut}(E : F)) \supseteq F$$

By Artin's theorem 29 holds more: $|\mathrm{Aut}(E : F)| = \dim[E : L]$

Finally by the tower theorem

$$|\mathrm{Aut}(E : F)| \cdot \dim[L : F] = \dim[E : L] \cdot \dim[L : F] = \dim[E : F]$$

Hence the group order $|\mathrm{Aut}(E : F)|$ is a divisor of $\dim[E : F]$. Equality is equivalent to $L = F$ $\qquad\qquad\square$

Theorem 31 (Construction of a primitive element). *Let E be a field and $\mathcal{G}$ be a finite group of n automorphisms $\sigma_i : E \mapsto E$. Let $F = \mathrm{Fix}(E : G) \subseteq E$ be the fixed field of the group $\mathcal{G}$.*

Then there exists $\alpha \in E$ such that $E = F(\alpha)$.

Proof. By Artin's Theorem 29 holds $\dim[E : F] = |\mathcal{G}|$, especially the extension is finite dimensional. The proof has two distinguish two cases.

The field F is finite. If F contains q elements and $\dim[E : F] = n$, then E contains q^n elements. We have shown in proposition 75 that the multiplicative group $E \setminus \{0\}$ of a finite field is cyclic. Let α be a generator. In that case holds $E = F(\alpha)$.

The field F contains infinitely many elements. Since the vector space E is generated by any basis β_i for $1 \leq i \leq n$ it is enough to prove that the extension by adjunction of two elements $\alpha, \beta \neq F$ can be achieved by adjunction of one element γ. With the automorphisms σ_i from the group $\mathcal{G}$, we define the polynomial

$$f(x) = \prod_{1 \leq i < j \leq n} (\sigma_i(\alpha + \beta x) - \sigma_j(\alpha + \beta x))$$

It has the degree $\frac{n(n-1)}{2}$ and hence at most that many zeros. Since the field F is assumed to be infinite, there exists $t \in F$ such that $f(t) \neq 0$. Hence all n elements $\sigma_i(\alpha + \beta t)$ for $1 \leq i \leq n$ are different.

By proposition 30 item (i), for any γ_i, there exists a polynomial $Q_i \in F[x]$ that splits completely in the extension field E, is irreducible, has simple zeros, and among them $Q_i(\gamma_i) = 0$. We may assume that $\sigma_1 = \mathrm{id} \uparrow E$ and take $Q_i := \sigma_i(Q_1)$. Since $Q_1 \in F[x]$, we get even $Q_i = Q_1$ for all $1 \leq i \leq n$. Hence Q_1 has n different zeros and $\deg Q_1 \geq n$. Since Q_1 is irreducible, we get $\deg Q_1 = n$. Hence $\dim[F(\alpha + \beta t) : F] \geq n$.

On the other hand, $F(\alpha + \beta t) \subseteq E$ implies $\dim[F(\alpha + \beta t) : F] \leq n$. Hence $\dim[F(\alpha + \beta t) : F] = n$ and $F(\alpha + \beta t) = E$. Thus adjoining α and β may be replaced by adjoining $\gamma = \alpha + \beta t$. In the end it is enough to adjoin a single element to obtain E.

$\square$

Problem 170. *All n elements $\gamma_i := \sigma_i(\alpha + \beta t)$ for $1 \le i \le n$ are different. there are at least n homomorphisms $\sigma_i : E \mapsto E$. By proposition 78, these are n linearly independent functions.*

I claim that even the restrictions $\sigma_i : F(\alpha + \beta t) \mapsto E$. are n linearly independent functions.

The last two theorems 30 and 31 contain assumptions that are convenient to make a field extension E/F in a convenient sense "natural" or "without obstacles". Unhappily enough the relevant notion is difficult to pin down, but we have gather enough information to this goal.

Definition 51 (Galois field extension). A field extension E/F is called *Galois extension* if either one of the following assumptions holds

(a) The group $\mathcal{G} = \mathrm{Aut}(E : F))$ of automorphisms $\sigma_i : E \mapsto E$ is finite, and the subfield $F \subseteq E$ is the fixed field of the group $\mathcal{G}$. In other words $\mathrm{Fix}(E : \mathrm{Aut}(E : F)) = F$.

(b) There exists a polynomial $Q \in F[x]$ which splits completely in the extension field E, and E is the splitting field of Q. Moreover the polynomial Q is separable.

Corollary 41. *Each Galois extension is finite. Each Galois extension has a primitive element.*

Proof. By Artin's Theorem 29 holds $\dim[E : F] = |\mathcal{G}|$, especially the extension is finite dimensional.

To prove the second assertopk we use definition (a). The group $\mathcal{G} = \mathrm{Aut}(E : F))$ of automorphisms $\sigma_i : E \mapsto E$ is assumed to be finite, and the ground field $F \subseteq E$ equals the fixed field $\mathrm{Fix}(E : \mathrm{Aut}(E : F)) = F$.

Hence the extension E/F is finite dimensional and the construction of a primitive element for which $E = \mathrm{Fix}(E : \mathrm{Aut}(E : F))(\alpha)$ was done by theorem 31. $\square$

Corollary 42. *Let $E = F(\alpha_1, \ldots, \alpha_n)$ be an extension. Assume that the minimal polynomials Q_i for the α_i have only simple zeros. Then there exists a primitive element.*

Proof. We may use definition (b). The product $Q = \prod Q_i$ is a separable polynomial. Let $L \supseteq E$ be the splitting field of Q. The extension L/F is Galois. By Steinitz' theorem there exist only finitely many intermediate fields between F and L. A fortiori there exist only finitely many intermediate fields between F and E. By Steinitz' theorem there exists a primitive element such that $E = F(\beta)$ $\square$

Theorem 32. *The two defining assumptions for a Galois extension E/F given in definition 51 are equivalent.*

Proof. The implication (a) $\Rightarrow$ (b) has been shown in Problem 168 above. $\qquad\square$

Proof. The converse implication (b) $\Rightarrow$ (a) needs a different method and gets notoriously awkward. I include a proof by combining Emil Artin's classic with the newer textbook by Gerd Fischer.

We assume there exists a separable polynomial $S \in F[x]$ which splits completely in the extension field E, By definition 45 of a separable polynomial holds

$$(\text{VIII.0.10}) \qquad S(x) = c(x - \alpha_1) \cdots (x - \alpha_r)P_1(x) \cdots P_s(x)$$

where $r \geq 0, s \geq 0$ and $c, \alpha_1, \ldots, \alpha_r \in F$, and moreover, the irreducible polynomials $P_t(x) \in F[x]$ have degree at least two, and have no multiple roots.

We fix $n = \deg S$ and provide a proof successively for $r = n, n-1, n-2, \ldots, 1, 0$. In the case $r = n$, the splitting field is $E = F$ and the claim (a) holds obviously. Let now $r < n$, and assume the claim holds for any separable polynomial S with at least $r+1$ linear factors. We have to get claim for the case with S as in equation (VIII.0.10). Since $s \geq 1$, the factor P_1 exists. Let α_{r+1} be any root of P_1. In the appropriate extension $F(\alpha_{r+1})$, we may factor S further to obtain

$$S(x) = c(x - \alpha_1) \cdots (x - \alpha_r)(x - \alpha_{r+1})(x - \beta_1)(x - \beta_\mu)Q_1(x) \cdots Q_\nu(x)$$

with $\mu \geq 0, \nu \geq 0$ and the Q_i are irreducible and have no multiple zeros. We use now $F(\alpha_{r+1})$ as ground field. [24] By assumption, the given field E is the splitting field of S over F. Hence the field E is the splitting field of S over $F(\alpha_{r+1})$, too. Thus the induction assumption may be applied.

In order to check the claim,—whether $\text{Fix}(E : \text{Aut}(E : F)) \subseteq F$,—we take now any element $\theta \in \text{Fix}(E : \text{Aut}(E : F))$, and check whether $\theta \in F$. Thus $\theta \in E$ is fixed under all automorphisms in $\text{Aut}(E : F)$. Consequently θ is fixed is under all automorphisms in the smaller group

$$\text{Aut}(E : F(\alpha_{r+1})) \leq \text{Aut}(E : F)$$

By the induction assumption this implies that $\theta \in F(\alpha_{r+1})$.

Since $\theta \in F(\alpha_{r+1})$ and $P_1(\alpha_{r+1}) = 0$ holds

$$\theta = \sum_{0 \leq i < t} c_i \alpha_{r+1}^i$$

---

[24] I think this is permissible. We have fixed the degree n and hence the dimension of any field extension that possibly may occur is bounded by n.

358

with $t = \deg P_1$ and $c_i \in F$. Since P_1 is assumed to have only simple zeros holds

$$P_1(x) = \prod_{0 \le i < t} (x - \alpha_{r+1+i})$$

with all different α_{r+1+i} for $0 \le i < t$. Each one of the t transformations $\alpha_{r+1} \mapsto \alpha_{r+1+i}$ with $0 \le i < t$ may be extended to an automorphism $\sigma_i \in \mathrm{Aut}(F(\alpha_{r+1}) : F)$, for which $\sigma_i(\alpha_{r+1}) = \alpha_{r+1+i}$. This is a basic property that holds for any single root extension. I claim that one may even extend them to automorphisms $\sigma_i \in \mathrm{Aut}(E : F)$. Indeed such an extension is done by finitely many times adjoining a single root.

Secondly, we still have as assumed $\theta \in \mathrm{Fix}(E : \mathrm{Aut}(E : F))$, and hence $\theta \in \mathrm{Fix}(F(\alpha_{r+1}) : \mathrm{Aut}(E : F))$. Especially, θ is fixed under the automorphisms $\sigma_i \in \mathrm{Aut}(E : F)$ from above. Hence we see get $\sigma_i(\theta) = \theta$ for all $0 \le i < t$. This is the following system of equations

$$\theta = \sum_{0 \le i < t} c_i \alpha_{r+1+j}^i \quad \text{for } 0 \le j < t$$

Thus the polynomial

$$R(x) = \sum_{0 \le i < t} c_i x^i - \theta$$

has degree at most $t - 1$ but has t zeros since $R(\alpha_{r+1+j}) = 0$ for $0 \le j < t$. Hence $R = 0$ and $\theta = c_0 \in F$, as claimed.

We have checked the inclusion $\mathrm{Fix}(E : \mathrm{Aut}(E : F)) \subseteq F$. The reverse inclusion is obvious by Problem 165. Hence claim (a) is confirmed. $\square$

For field extensions E/F we have elaborated on the definitions of the two notations

splitting extension:　In proposition 71 are given three characterizations.

Galois extension:　In definition 51 are given two definitions. Their equivalence has been proved by theorem 32.

Are the two notions *splitting extension* and *Galois extension* really equivalent to each other?

Lemma 87. *For a field of characteristic zero, the notions of Galois extension is equivalent to the notion of splitting extension.*

For a field of prime characteristic, each splitting extension is a Galois extension, but the converse need not hold.

Proof. By definition 51, item(b), a Galois extension is the splitting extension of a *separable polynomial,*—hence indeed a splitting extension.

Assume the field F has characteristic zero. Assume conversely that the extension E/F is a splitting extension, say of polynomial P. This polynomial is a product of irreducible polynomials P_i. We may construct a product of irreducible polynomials Q_i which are divisors of the P_i such that the Q_i are relatively prime, and the products $\prod P_i$ and $\prod Q_i$ have the same set of zeros.

The extension E/F is a splitting extension of $\prod Q_i$, too. For characteristic zero, all the Q_i and their product $\prod Q_i$ are separable polynomials. Hence the extension E/F is the splitting extension of a *separable polynomial,*—and hence a Galois extension. $\square$

This begs for the case of prime characteristic a delicate question: Do there exist polynomials the splitting extension of which are not Galois extensions? Of course such polynomials are not separable.

Problem 171 (Subtle and misleading). *For a Galois extension necessarily the following conclusions hold. But are they sufficient to guarantee a Galois extension?*

(i) *For any $\alpha \in E$, there exists a polynomial $Q \in F[x]$ that splits completely in the extension field E, is irreducible, has simple zeros, and among them $Q(\alpha) = 0$.*

(ii) *Each element of $\alpha \in E$ is algebraic over F.*

(i) and (ii)

(i) and E/F finite dimensional

(iii) *Each irreducible polynomial $P \in F[x]$ with one root $\alpha \in E$, has all its roots in E.*

(iv) *There exists a primitive root.*

(iii) and (iv)

(iii) and (iv) *and the field has characteristic zero.*

Decide from the theorems above or by appropriate examples which ones of the above implicationsis strong enough to always guarantee that an extension is Galois.

Proof. **(i)** is not strong enough to guarantee a Galois extension. Counterexample: Ground field $F = \mathbf{Q}$ and extension E the field of all algebraic numbers.

(ii) is not strong enough to guarantee a Galois extension. Counterexample: Ground field $F = \mathbf{Q}$ and extension $E = F(\sqrt[3]{2})$.

(i) and (ii) is not strong enough to guarantee a Galois extension. Counterexample: Ground field $F = \mathbf{Q}$ and extension E the field of all algebraic numbers.

(i) and E/F finite dimensional are strong enough to guarantee a Galois extension.

(iii) is not strong enough to guarantee a Galois extension. Counterexample: See remark VIII.0.6. The extension is not Galois since no primitive element exists.

(iv) is not strong enough to guarantee a Galois extension. Counterexample: Ground field $F = \mathbf{Q}$ and extension $E = F(\sqrt[3]{2})$. This is a single element extension but not Galois.

(iii) and (iv) Assume item (iii) and item (iv) hold. Let $m \in F[x]$ be the minimal polynomial of the primitive root r. The minimal polynomial is known to be irreducible. Since $m(r) = 0$ and $r \in E$ item (iii) yields that the polynomial m splits in the extension E/F. There cannot exist any intermediate field $F \subset L \subset E$ such that m splits already in the smaller field L. Indeed $r \in L$ implies $L \supseteq F(r) = E$. Hence E is the splitting field of the minimal polynomial m. To get the conclusion that the extension E/F is Galois, we still need to confirm that m is separable. By definition 45, a polynomial $Q \in F[x]$ is called *separable* iff its irreducible factors have no multiple roots. So for the already irreducible polynomial m, we need to confirm that it has only simple zeros. That holds for fields of characteristic zero, but not for fields with prime characteristic. Item (iii) and (iv) are not strong enough to get a Galois extension.

(iii) and (iv) and the field has characteristic zero. These three assumptions together imply a Galois extension. The extension E/F is a Galois extension since it is the splitting extension of the *separable polynomial m* mentioned above..

$\square$

Problem 172 (Subtle and fine). *Assume the extension E/F to be finite dimensional, and assume that for any $\alpha \in E$, there exists a polynomial $Q \in F[x]$ that splits completely in the extension field E, is irreducible, has simple zeros, and among them $Q(\alpha) = 0$.*

Prove the extension to be Galois.

Proof. Let $E = F(\alpha_1, \ldots, \alpha_n)$ be the given extension. By assumption there exist the irreducible polynomials Q_i with simple zeros and $Q_i(\alpha_i) = 0$. For the product $S = \prod Q_i$, the extension E/F is the splitting extension. Too, the polynomial S is separable. Hence by definition 51 item (b), we get a Galois extension E/F. $\qquad\square$

Remark. Assume for the extension E/F to hold:

(iii) Each irreducible polynomial $P \in F[x]$ with one root $\alpha \in E$, has all its roots in E.

(iv) Assume there exists a primitive element α.

Especially the minimal polynomial P of the primitive element splits completely. Moreover since $\deg P = \dim[E : F]$ the polynomial cannot split in any intermediate field. Hence E/F is the splitting extension for the minimal polynomial P. For a field $F \supseteq \mathbf{Q}$ of characteristic zero, the irreducible polynomial P has simple zeros, hence according to definition 51 item (b), we get a Galois extension.

But what happens assuming just items (iii) and (iv) from above for a field of characteristic $p > 0$?

VIII.0.8   The Discriminant

Let $f \in F[x]$ be a polynomial and $E \supset F$ be a splitting extension. With the zeros $x_1 \ldots x_n$ holds the splitting

$$f(x) = c \prod_{1 \leq i \leq n} (x - x_i)$$

Definition 52 (Modified discriminant). The modified discriminant of any polynomial $f \in F[x]$ is defined as

$$(\text{VIII.0.11}) \qquad \Delta(f) = \prod_{1 \leq i < j \leq n} (x_i - x_j)^2$$

Lemma 88. *The modified discriminant is an element of the base field F. In terms of the formal derivative at the zeros holds*

$$(\text{VIII.0.12}) \qquad \Delta(f) = (-1)^{\frac{n(n-1)}{2}} c^{-n} \prod_{1 \leq i \leq n} f'(x_i) = (-1)^{\frac{n(n-1)}{2}} \prod_{1 \leq i \neq j \leq n} (x_i - x_j)$$

Proof. $\Delta(f)$ is a symmetric polynomial and hence by the main theorem about symmetric polynomials 26, it is a polynomial in the elementary symmetric functions of the x_i and by Viëta's formulas a polynomial in the coefficients of F, and especially a member of the base field $F \ni \Delta(f)$. The extension $[F(\delta) : F]$ is either one or two dimensional.

If one reorders the terms with $i > j$ to get those with $i < j$ a second time, there are $\frac{n(n-1)}{2}$ transpositions to be done. Hence

$$\prod_{1 \leq i > j \leq n} (x_i - x_j) = (-1)^{\frac{n(n-1)}{2}} \prod_{1 \leq i < j \leq n} (x_i - x_j)$$

$$(-1)^{\frac{n(n-1)}{2}} \prod_{1 \leq i \neq j \leq n} (x_i - x_j) = \prod_{1 \leq i < j \leq n} (x_i - x_j)^2 = \Delta(f)$$

The product rule holds for the formal derivative and implies

$$f'(x_i) = c \prod_{1 \leq j \leq n \text{ and } j \neq i} (x_i - x_j) \quad \text{for all } 1 \leq i \leq n$$

$$\frac{1}{c^n} \prod_{1 \leq i \leq n} f'(x_i) = \prod_{1 \leq i \neq j \leq n} (x_i - x_j) = (-1)^{\frac{n(n-1)}{2}} \Delta(f)$$

$\square$

Definition 53 (Discriminant root). I call the product

$$(\text{VIII.0.13}) \qquad\qquad \delta(f) = \prod_{1 \leq i < j \leq n} (x_i - x_j)$$

the *discriminant root*

Contrary to the discriminant, the root discriminant may or may not be a member of the base field F. But clearly $\delta(f)$ is a member of the field extension $F(\sqrt{\Delta(f)}) \subseteq E$, and hence of the splitting extension. This field extension may be either a one or two dimensional extension of the base field F. These two cases give a valuable information about the Galois group $\mathrm{Aut}(E : F)$.

I have never seen any good method to fix the sign of the discriminant root,—but this ambiguity does not raise any problem.

Theorem 33 (Discriminant theorem). *Assume that $f \in F[x]$ is irreducible in F, let E be the splitting field of f, and assume $\Delta(f) \neq 0$.*

(i) $\mathrm{Aut}(E : F(\delta)) \leq \mathcal{A}_n$

(ii) $G \le \mathcal{A}_n$ *iff* $\delta(f) \in F$ *iff* $\Delta(f)$ *is a square in* F

(iii) *If* $\Delta(f)$ *is not a square in* F, *then the group contains odd permutations and its order* $|G|$ *is even. Moreover*

$$\mathrm{Fix}(E : G \cap \mathcal{A}_n) = F(\delta)$$

(iv) *For the common case* $F = \mathbb{R}$. *If the real polynomial* f *has exactly one pair of conjugate complex roots, again the conclusions of item* (iii) *hold.*

Proof of item (i). By the assumption $\Delta(f) \ne 0$, the roots $x_1, \ldots x_n$ are all distinct and simple. Each $g \in G$ induces a permutation $\pi_g \in \mathcal{S}_n$ of the n different roots. For all permutations $\pi \in \mathcal{S}_n$ holds

$$\prod_{1 \le i < k \le n} (x_{\pi(i)} - x_{\pi(k)}) = \sigma(\pi) \prod_{1 \le i < k \le n} (x_i - x_k)$$

$$g(\delta) = \sigma(\pi)\delta$$

Any $g \in \mathrm{Aut}(E : F(\delta))$ fixes $\delta = g(\delta)$ while permuting the roots. Hence $\sigma(\pi_g) = 1$ and $g \in \mathcal{A}_n$. $\qquad\square$

Proof of item (ii).

$$\begin{aligned}
G \le \mathcal{A}_n &\Leftrightarrow \sigma(\pi_g) = 1 \text{ for all } g \in G \\
&\Leftrightarrow g(\delta) = \delta \text{ for all } g \in G \\
&\Leftrightarrow \delta \in \mathrm{Fix}(E : G) = F \\
&\Leftrightarrow \Delta(f) \text{ is a square in } F
\end{aligned}$$

By the definition of a Galois field extension 51, part (a) holds $\mathrm{Fix}(E : \mathrm{Aut}(E : F)) = F$.

    If $\Delta(f) = h^2$ is a square of $h \in F$, then $h = \pm\delta(f)$ and hence $\delta \in F$. $\qquad\square$

Proof of item (iii). Assume $\Delta(f)$ is not a square in F. Hence $\dim[F(\delta) : F] = 2$ and the tower theorem gives

$$|G| = \dim[E : F] = \dim[E : F(\delta)]\dim[F(\delta) : F] = 2\dim[E : F(\delta)]$$

which is even. The discriminant δ is mapped by the automorphism $g \in G$ to $g(\delta) = \sigma(\pi_g)\delta$ and hence fixed for $g \in G \cap \mathcal{A}_n$. Hence

$$F(\delta) \subseteq \mathrm{Fix}(E : G \cap \mathcal{A}_n) =: L$$

Let $H = G \cap \mathcal{A}_n$ and apply Artin's Theorem 29 to this group. Hence $\dim[E : L] = |H|$. By the tower theorem

$$|G| = \dim[E : F] = \dim[E : L]\dim[L : F] = |H|\dim[L : F] = 2|H|\dim[L : F(\delta)]$$

Since H is the kernel of the homomorphism $g \mapsto \sigma(\pi_g)$, we get from the isomorphism theorem that $G/H \cong \{\pm 1\}$ and hence $|G| = 2|H|$. Hence $\dim[L : F(\delta)] = 1$, hence $L = F(\delta)$, as claimed. $\qquad\square$

Proof of item (iv). Assume the real polynomial $f \in \mathbb{R}[x]$ has exactly one pair $x_1 \neq \overline{x_1} = x_2$ of conjugate complex roots. Hence

$$\delta(f) = (x_1 - \overline{x_1}) \prod_{3 \leq i \leq n} |x_1 - x_i|^2 \prod_{3 \leq i < k \leq n} (x_i - x_k)$$

Hence $\delta(f) = ir \neq 0$ with $r \in \mathbb{R}$, which is not a square. $\qquad\square$

Remark. In the case of characteristic zero, and additionally $F \subseteq \mathbb{R}$, the base field is ordered, and all perfect squares are positive.

We may of course nevertheless get $\Delta(f) < 0$,—since the squares in the definition of the modified discriminant are to be taken in the field *extension* E, and this field may contain complex numbers. Hence $\Delta(f) < 0$ implies that field extension $F(\sqrt{\Delta(f)})/F$ is two-dimensional and $\mathrm{Aut}(E : F)) \not< \mathcal{A}_n$.

But here the converse statement is not true. $\Delta(f) > 0$ does not exclude the case of a two dimensional extension. To exclude the two dimensional extension, we need to check that $\Delta(f)$ *is a perfect square in* E. This is a stronger assumption than simply positivity.

VIII.0.9 About the Characteristic $p > 0$

Definition 54 (The field K^{1/p^∞}). Let $\mathbf{Z}_p \subseteq K$ be any field with characteristic $p > 0$, for example the prime field $\mathbf{Z}_p$. The field K^{1/p^∞} is the union of the splitting fields of the polynomials $x^{p^m} - k$ for all $k \in K$ and nonnegative integers m.

Lemma 89. *A polynomial $P \in K[x]$ is irreducible in the field K of characteristic $p > 0$ if and only if $P(x) = Q(x^{p^m})$ where $m \geq 0$ and $Q \in K[x]$ is irreducible and has simple zeros.*

Proof. Assume the polynomial $P \in K[x]$ is irreducible. Let $Q := \gcd(P, P')$ be the greatest common divisor. Since $Q \in K[x]$ and $\deg Q \leq \deg P' < \deg P$ the irreducibility implies that $Q = c \in K$ is a constant. We distinguish the cases (referring to the splitting field)

polynomial P has a multiple root. At the multiple root α holds $P(\alpha) = P'(\alpha) = Q(\alpha) = 0$. Hence $c = 0$. I claim that even $P' = 0$. Otherwise $x - \alpha$ would be a divisor of Q, which is a contradiction. Let now

$$P(x) = \sum_{0 \le n \le \deg P} a_n x^n$$

$$P'(x) = \sum_{1 \le n \le \deg P} n a_n x^{n-1} = 0 \ \text{ hence}$$

$$n a_n = 0 \ \text{ for } 1 \le n \le \deg P \text{ hence either } a_n = 0 \text{ or } p \mid n$$

Hence there exists a polynomial Q such that $P(x) = Q(x^p)$. Moreover Q is irreducible and $\deg Q < \deg P$. We repeat the argument until Q has only simple zeros. Thus we obtain for some $m \ge 1$

$$P(x) = Q(x^{p^m}) \text{ where } Q \text{ is irreducible and has only simple zeros.}$$

polynomial P has only simple roots. We obtain the formula above with $m = 0$.

Conversely assume that $P(x) = Q(x^{p^m})$ where $m \ge 0$ and $Q \in K[x]$ is irreducible and has only simple zeros. All polynomials in the remaining proof are monic. In the splitting field of K one gets

$$P(x) = \prod_{1 \le j \le p^{-m} \deg P} (x - \beta_j)^{p^m}$$

I claim the polynomial P cannot be reducible. Indeed, we assume in $K[x]$ to exist a factoring $P = RS$ with R irreducible. Since R is irreducible in $K[x]$, we get $\gcd(R, S) = 1$. From the first part of the proof we know that $R(x) = T(x^{p^n})$ where $n \ge 0$ and $T \in K[x]$ is irreducible and has only simple zeros. In the splitting field of K one gets

$$T(x) = \prod_{1 \le j \le p^{-n} \deg T} (x - \beta_j)^{p^n}$$

We use now that in the splitting field T is a divisor of P. Hence the β_j reoccur. Moreover holds $n \le m$. Since $\gcd(R, S) = 1$ holds even $n = m$. Hence

$$\prod_{1 \le j \le p^{-m} \deg T} (x - \beta_j) \ \text{ is a divisor of } \ \prod_{1 \le j \le p^{-m} \deg P} (x - \beta_j)$$

in other words, $T(x)$ is a divisor of $Q(x)$. Since $Q \in K[x]$ is irreducible we conclude that either $T = 1$ or $T = Q$. Hence either $R = 1$ or $R = P$, confirming that P is irreducible. $\qquad\square$

366

Problem 173. *Convince yourself that any one of the polynomials $x^{p^m} - k$ with $k \in K$ and positive integer $m \geq 1$ has exactly one zero, but nevertheless is irreducible.*

Proof. Assume $\alpha^{p^m} = \beta^{p^m} = k$. Because of the characteristic $p > 0$ holds

$$(\alpha - \beta)^{p^m} = \alpha^{p^m} - \beta^{p^m} = 0, \quad \text{hence } \alpha = \beta$$

Let α be a zero in its splitting field. We get

$$x^{p^m} - k = x^{p^m} - \alpha^{p^m} = (x - \alpha)^{p^m}$$

Any nontrivial irreducible factor R would be with $n \leq m$ and a polynomial Q irreducible with simple zeros, and $r = p^{m-n}$:

$$R(x) = (x - \alpha)^{r \cdot p^n} = Q(x^{p^n})$$
$$(x^{p^n} - \alpha^{p^n})^r = Q(x^{p^n})$$
$$(t - \alpha^{p^n})^r = Q(t)$$

Because Q has simple zeros we get $r = 1$ and $n = m$. No nontrivial factor R is possible. $\square$

Problem 174. *Convince yourself that the definition 54 is a valid definition for a field.*

Solution. Obviously holds $\mathbf{Z}_p \subseteq K \subset K^{1/p^\infty}$ and hence $0, 1 \in K^{1/p^\infty}$.

Take any elements $\alpha, \beta \in K^{1/p^\infty}$. Hence there exist integers $m, n \geq 0$ such that $\alpha^{p^m} = k \in K$ and $\beta^{p^n} = r \in K$ We may assume $n \leq m$ and get $\beta^{p^m} = r^{p^{m-n}} = l \in K$, too. Hence (assuming in the second line $\beta \neq 0$) holds

$$(\alpha - \beta)^{p^m} = \alpha^{p^m} - \beta^{p^m} = k - l \in K \quad \text{and hence } \alpha - \beta \in K^{1/p^\infty}$$
$$(\alpha\beta^{-1})^{p^m} = \alpha^{p^m}\beta^{-p^m} = kl^{-1} \in K \quad \text{and hence } \alpha\beta^{-1} \in K^{1/p^\infty}$$

$\square$

Definition 55 (*p*-root complete field). I call a field F with characteristic $p > 0$ to be *p*-root complete if for all $\alpha \in F$ exists $\beta \in F$ such that $\beta^p = x\alpha$. In other words, the polynomial $x^p - \alpha \in F[x]$ always has a zero.

Problem 175. *Continuing the definition, convince yourself that for a p-root complete field holds $x^p - \alpha = (x - \beta)^p$.*

Problem 176. *Assume that polynomial $P \in K[x]$ is irreducible in the field K of characteristic $p > 0$, and the field is p-root complete. Convince yourself that there exists an irreducible polynomial R with simple zeros and $m \geq 0$ such that $P = R^{p^m}$.*

Proof. By lemma 89 there exists an irreducible polynomial $Q \in K[x]$ with simple zeros and $m \geq 0$ such that $P(x) = Q(x^{p^m})$. Let

$$Q(x) = \sum_{0 \leq i \leq \deg Q} q_i x^i$$

Since the field is p-root complete there exist $r_i \in K$ such that $r_i^{p^m} = q_i$ (using a self-evident induction). Hence

$$P(x) = \sum_{0 \leq i \leq \deg Q} r_i^{p^m} x^{i \cdot p^m} = \left(\sum_{0 \leq i \leq \deg Q} r_i x^i \right)^{p^m} = (R(x))^{p^m}$$

One still has to check that a double zero of R would imply a double zero of Q, as well, a factoring of R would imply a factoring of Q. Hence R is an irreducible polynomial with simple zeros. $\square$

Problem 177. *Prove that any finite field K of characteristic $p > 0$ is p-root complete.*

Proof. Assume that $|K| = p^m$ with $m \geq 1$. By proposition 75 the multiplicative group $\mathcal{G}$ of field K is cyclic. Take any element $\alpha \in K$. By Lagrange's Theorem we get $\alpha^{p^m - 1} = 1$ and hence $\alpha^{p^m} = \alpha$. We put

$$\beta := \alpha^{p^{m-1}}$$

and get $\beta^p = \alpha$ as required. $\square$

Lemma 90. *Assume that field F has characteristic $p > 0$ and is p-root complete. Assume for the extension E/F to hold:*

(iii) *Each irreducible polynomial $P \in F[x]$ with one root $\alpha \in E$, has all its roots in E.*

(iv) *Assume there exists a primitive element α.*

Then the extension E/F is Galois.

Proof. Especially the minimal polynomial P of the primitive element splits completely. Moreover since $\deg P = \dim[E : F]$ the polynomial cannot split in any intermediate field. Hence E/F is the splitting extension.

By problem 176 there exists an irreducible polynomial R with simple zeros and $m \geq 0$ such that $P = R^{p^m}$. According to definition 45, the polynomial P is separable. Hence according to definition 51 item (b), we get a Galois extension.

But which more funny examples may exist? $\square$

VIII.0.10   The Main Theorem of Galois Theory

Theorem 34. *Let E/F be a Galois extension. Let $\mathcal{L}$ be the set of intermediate fields $F \subseteq L \subseteq E$. Let $\mathcal{G} = \mathrm{Aut}(E : F)$ be the group of automorphisms with the ground field F fixed. The following items hold:*

(i) *There exists a bijection $\lambda : \mathcal{G} \mapsto \mathcal{L}$ which maps the subgroups $G \leq \mathcal{G}$ to their respective fixed fields: $\lambda(G) = \mathrm{Fix}(E : G)$.*

The inverse mapping $\lambda^{-1} : \mathcal{L} \mapsto \mathcal{G}$ maps the fixed fields L to their respective subgroups: $\lambda^{-1}(L) = \mathrm{Aut}(E : L)$.

(ii) *For each intermediate field $F \subseteq L \subseteq E$ holds the following*

    (a) *E/L is a Galois extension and*

$$(\mathrm{VIII.0.14}) \qquad \dim[E : L] = |\mathrm{Aut}(E : L)|$$
$$(\mathrm{VIII.0.15}) \qquad \dim[L : F] = \mathrm{index}(\mathrm{Aut}(E : F), \mathrm{Aut}(E : L))$$

    (b) *Each automorphism $\phi \in \mathrm{Aut}(E : F)$ maps any intermediate field L to an intermediate field $\phi(L)$ such that Hence*

$$\mathrm{Aut}(E : \phi(L)) = \phi \circ \mathrm{Aut}(E : L) \circ \phi^{-1}$$

    (c) *L/F is a Galois extension if and only if*

$$L = \phi(L) \quad \text{for all } \phi \in \mathrm{Aut}(E : F)$$

That in turn holds if and only if $\mathrm{Aut}(E : L)$ is a normal divisor of $\mathrm{Aut}(E : F)$.

    (d) *In the case that (c) holds positively, the restriction $\phi \uparrow_L$ produces the mapping*

$$\uparrow_L : \phi \in \mathrm{Aut}(E : F) \mapsto \phi \uparrow_L \in \mathrm{Aut}(L : F)$$

and the commutative diagram

$$
\begin{array}{ccc}
\mathrm{Aut}(E : F) & \xrightarrow{\;\equiv\;} & \mathrm{Aut}(E : F)/\mathrm{Aut}(E : L) \\
\Big\uparrow{\scriptstyle I} & & \Big\uparrow{\scriptstyle \simeq} \\
\mathrm{Aut}(E : F) & \xrightarrow{\;\uparrow_L\;} & \mathrm{Aut}(L : F)
\end{array}
$$

Proof. **(i)** Define the mapping $\lambda : \mathcal{G} \mapsto \mathcal{L}$ which maps the subgroups $G \leq \mathcal{G}$ to their respective fixed fields by $\lambda(G) = \mathrm{Fix}(E : G)$.

Define, too, the (reversing) mapping $\gamma : \mathcal{L} \mapsto \mathcal{G}$ which maps the fixed fields L to their respective subgroups by $\gamma(L) = \mathrm{Aut}(E : L)$.

We claim $G = \gamma(\lambda(G)) = \mathrm{Aut}(E : \mathrm{Fix}(E : G))$ for all subgroups $G \leq \mathcal{G}$. Especially, this equation confirms that λ is an injective mapping. Take any $G \leq \mathcal{G}$ and let $L := \lambda(G)$. Directly from the definitions one gets

$$\gamma(L) = \mathrm{Aut}(E : \mathrm{Fix}(E : G)) \geq G$$

Problem 178. *Convince yourself of the latter claim.*

Confirmation. By definition

$$\begin{aligned}
\sigma \in \gamma(L) &\Leftrightarrow \sigma(x) = x \ \text{ for all } x \in L \\
&\Leftrightarrow x \in \mathrm{Fix}(E : G) \Rightarrow \sigma(x) = x \\
&\Leftrightarrow \forall \sigma \in G \ \sigma(x) = x \Rightarrow \sigma(x) = x
\end{aligned}$$

For any element $\sigma \in G$ holds the last line. $\qquad\square$

By Artin's key theorem 29 holds for any finite subgroup $G \leq \mathrm{Aut}(E)$ and its fixed field $\mathrm{Fix}(E : G) = \lambda(G)$ the equality

$$\dim[E : \lambda(G)] = |G|$$

By problem 169, for any finite dimension extension E/L, the group order $|\gamma(L)| = |\mathrm{Aut}(E : L)|$ is a divisor of the dimension $\dim[E : L]$, and hence together

$$\dim[E : \lambda(G)] = |G| \leq |\gamma(L)| \leq \dim[E : L] = \dim[E : \lambda(G)]$$

enforcing equalities and hence $|G| = |\gamma(L)|$.
Finally $G = \gamma(L) = \gamma(\lambda(G))$.

We claim $\lambda(\gamma(L)) = \mathrm{Fix}(E : \mathrm{Aut}(E : L)) = L$ for all intermediate fields $L \in \mathcal{L}$. Especially, this equation confirms that λ is an surjective mapping. Take any intermediate field L and let $G := \gamma(L)$. Directly from the definitions one gets

$$\lambda(G) = \lambda(\gamma(L)) = \mathrm{Fix}(E : \mathrm{Aut}(E : L)) \supseteq L$$

Problem 179. *Convince yourself of the latter claim.*

Confirmation. By definition

$$x \in \lambda(G) \Leftrightarrow \sigma(x) = x \ \text{ for all } \sigma \in \mathrm{Aut}(E : L)$$
$$\Leftrightarrow (\sigma \in \mathrm{Aut}(E : L) \text{ and } x \in \lambda(G)) \Rightarrow \sigma(x) = x$$
$$\Leftrightarrow (\sigma \in \mathrm{Aut}(E : F) \text{ and } x \in \lambda(G) \text{ and } \forall\, x \in L\,\sigma(x) = x)$$
$$\Rightarrow \sigma(x) = x$$

For any element $x \in L$ holds the last line. $\qquad\qquad\square$

By Artin's key theorem 29 holds for any finite subgroup $G \le \mathrm{Aut}(E)$ and its fixed field $\mathrm{Fix}(E : G) = \lambda(G)$ the equality

$$\dim[E : \lambda(G)] = |G|$$

Problem 180. *Show that the extension E/L is Galois.*

The extension E/L is Galois. Hence $|G| = |\mathrm{Aut}(E : L)| = \dim[E : L]$. Together we get

$$|G| = \dim[E : \lambda(G)] \le \dim[E : L] = |G|$$

enforcing equalities and hence $\lambda(G) = L$. Finally get confirm $\lambda(\gamma(L)) = L$.

Together the above arguments confirm that λ and γ are bijections and inverse of each other.

Problem 181. *Using that the maps γ and λ are decreasing, show from the statements above that*

$$\gamma(\lambda(\gamma(L))) = L \ \text{ for all } L \in \mathcal{L}$$
$$\lambda(\gamma(\lambda(G))) = G \ \text{ for all } G \le \mathcal{G}$$

But this is still a too weak result.

(b) For each automorphism $\phi, \psi \in \mathrm{Aut}(E : F)$ and intermediate field L holds

$$\psi \in \mathrm{Aut}(E : \phi(L)) \Leftrightarrow \psi(\phi(L)) = \phi(L) \Leftrightarrow (\phi^{-1} \circ \psi \circ \phi)(L) = L$$
$$(\phi^{-1} \circ \psi \circ \phi) \in \mathrm{Aut}(E : L) \Leftrightarrow \psi \in \phi \circ \mathrm{Aut}(E : L) \circ \phi^{-1}$$

We may write the result as

$$\lambda(\phi(L)) = \phi \circ \lambda(L) \circ \phi^{-1}$$
$$\phi(\gamma(G)) = \gamma(\phi \circ G \circ \phi^{-1})$$

(c) Assume that $\mathrm{Aut}(E : L)$ is a normal divisor of $\mathrm{Aut}(E : F)$. This assumption holds if and only if

$$\Leftrightarrow \mathrm{Aut}(E : L) = \phi\,\mathrm{Aut}(E : L)\,\phi^{-1}$$
$$\Leftrightarrow \mathrm{Aut}(E : L) = \mathrm{Aut}(E : \phi(L))$$
$$\Leftrightarrow \gamma(L) = \gamma(\phi(L))$$
$$\Leftrightarrow L = \phi(L) \quad \text{for all } \phi \in \mathcal{G}$$

since by part (i) the mapping γ is injective.

Lemma 91. *Assume that the extension E/F is Galois, let L be an intermediate field $E \supseteq L \supseteq F$ for which the stability condition*

$$L = \phi(L) \quad \text{for all } \phi \in \mathrm{Aut}(E : F)$$

holds. Then the extension L/F is Galois.

Proof. To check that this stability implies that L/F is Galois, we use the definition 51 item (a). Let $F_1 = \mathrm{Fix}(L : \mathrm{Aut}(L : F))$.

Take a fixed element $x \in F_1$. Hence $\phi(x) = x$ for all $\phi \in \mathrm{Aut}(L : F)$. Each automorphisms $\phi \in \mathrm{Aut}(E : F)$ has, because of the stability assumption, a restriction $\phi \uparrow_L \in \mathrm{Aut}(L : F)$ for which holds

$$\phi(x) = \phi \uparrow_L (x) = x$$

Since E/F is assumed to be a Galois extension, we conclude that $x \in F$.

The last paragraph has checked that $F_1 = F$. By the definition 51 item (a), the extension L/F is Galois. $\qquad\square$

Problem 182. *Conversely, under the assumption that L/F is Galois holds. $L = \phi(L)$ for all $\phi \in \mathrm{Aut}(E : F)$. Prove this claim.* [25]

Solution. There exist by Artin's key theorem 29 $j = \dim[L : F]$ automorphisms $\phi \uparrow_L \in \mathrm{Aut}(L : F)$.

The extension E/L is known to be Galois. Hence by Artin's key theorem 29 there exist $k = \dim[E : L]$ automorphisms $\psi \in \mathrm{Aut}(E : L)$.

The compositions $\chi := \phi \circ \psi$ are automorphisms in $\mathrm{Aut}(E : F)$ and <u>are all different</u>. Hence the entire group of order $|\mathrm{Aut}(E : F)| = k \cdot j = \dim[E : F]$ is covered. For all of them holds $\chi(L) = L$ as claimed. $\qquad\square$

---

[25] This proof I really missed in all texts.

(d) Under the assumption of stability the restriction $\phi\uparrow_L$ produces the mapping

$$\uparrow_L\colon \phi \in \mathrm{Aut}(E:F) \mapsto \phi\uparrow_L \in \mathrm{Aut}(L:F)$$

which is a homomorphism. Its kernel is $\mathrm{Aut}(E:L)$. Let $\mathrm{Im}\uparrow_L$ be its image. The main theorem about group homomorphisms yields an isomorphism $\mathrm{Im}\uparrow_L \simeq \mathrm{Aut}(E:F)/\mathrm{Aut}(E:L)$.

We claim that the above homomorphism is surjective, in other words $\mathrm{Im}\uparrow_L = \mathrm{Aut}(L:F)$. This is confirmed by counting:

$$|\mathrm{Im}\uparrow_L\,| = \frac{|\mathrm{Aut}(E:F)|}{|\mathrm{Aut}(E:L)|} = \frac{\dim[E:F]}{\dim[E:L]} = \dim[L:F] = |\mathrm{Aut}(L:F)|$$

The diagram below is commutative.

$$
\begin{array}{ccc}
\mathrm{Aut}(E:F) & \xrightarrow{\;\equiv\;} & \mathrm{Aut}(E:F)/\mathrm{Aut}(E:L) \\[4pt]
I\uparrow & & \uparrow\simeq \\[4pt]
\mathrm{Aut}(E:F) & \xrightarrow{\;\uparrow_L\;} & \mathrm{Aut}(L:F)
\end{array}
$$

$\square$

Proposition 79. *Let $f \in K[x]$ be a separable polynomial over K. Let E/K be the splitting extension of polynomial f. Let F/K be any field extension, typically an extension introduced for convenience when finding the roots of polynomial f. Let $\mathrm{span}(E,F)$ be the composite of E and F.*

Then $\mathrm{span}(E,F)/F$ is the splitting extension of the polynomial interpreted as $f \in F[x]$, and hence the Galois extension $\mathrm{Gal}(f:F)$. Moreover

$$\mathrm{Aut}(\mathrm{span}(E,F):F) \simeq \mathrm{Aut}(E, E\cap F)$$

are isomorphic. Hence $\mathrm{Gal}(f:F)$ is a subgroup of $\mathrm{Gal}(f:K)$, and

$$\mathrm{Gal}(f:F) = \mathrm{Gal}(f:K) \Leftrightarrow K = E\cap F$$

Proof. In the field $\mathrm{span}(E,F)$ the polynomial interpreted as $f \in F[x]$ is splitting. Too, the splitting field needs to be larger than both E and F. Hence $\mathrm{span}(E,F)$ is the splitting field of $f \in F[x]$. Since this polynomial remains to be separable with the extension F/K of its ground field we see that,

$$\mathrm{Aut}(\mathrm{span}(E,F):F) = \mathrm{Gal}(f:F)$$

The point of the proposition is to remove the extension of the ground field from the Galois group. The restriction $\uparrow_E$ is an homomorphism

$$\uparrow_E \colon \phi \in \mathrm{Aut}(\mathrm{span}(E, F) : F) \mapsto \phi \uparrow_E \in \mathrm{Aut}(E, E \cap F)$$

The kernel of this homomorphism consists of the mappings such that $\phi(x) = x$ for all $x \in E$. Since $\phi(x) = x$ for all $x \in F$ is assumed, we see that the kernel consists only of the identity mapping. Hence the restriction mapping $\uparrow_E$ is injective.

Too, the restriction mapping is surjective. Here is the reasoning given by Artin: Any image $\psi \in \mathrm{Aut}(E, E \cap F) \leq \mathrm{Aut}(E, K)$ induces a permutation of the zeros of the separable polynomial $f \in K[x]$. This is a permutation of the zeros of polynomial $f \in F[x]$, which is remaining to be separable. By the Fundamental Lemma 72 about the Galois-group, the permutation is the restriction of some mapping $\phi \in \mathrm{Aut}(\mathrm{span}(E, F) : F)$. Hence $\psi = \phi \uparrow_E$.

The Fundamental Theorem about homomorphisms of groups now implies the isomorphism

$$\mathrm{Aut}(\mathrm{span}(E, F) : F) \simeq \mathrm{Aut}(E, E \cap F)$$

Clearly $\mathrm{Aut}(E, E \cap F) \leq \mathrm{Aut}(E, K)$ is a subgroup, and the latter larger group is by definition $\mathrm{Aut}(E, K) = \mathrm{Gal}(f : K)$ By the Fundamental Theorem 34 of Galois Theory the intermediate field are one-to-one to the groups of automorphisms. Hence $\mathrm{Aut}(E, E \cap F) = \mathrm{Aut}(E, K) \Leftrightarrow E \cap F = K$. $\qquad\square$

Lemma 92. *Let E/K be a Galois extension and F/K be any (convenient) field extension. Then $\mathrm{span}(E, F)/F$ is a Galois extension.*

Problem 183. *Is the following argument correct?*[26] *Too, the restriction mapping is surjective. Indeed we may decompose $F = \mathrm{span}(F_1, F_2)$ with $F_1 \subseteq E$ and $F_2 \cap E = \{0\}$. and get*

$$\mathrm{Aut}(\mathrm{span}(E, F) : F) = \mathrm{Aut}(\mathrm{span}(E, F_2) : \mathrm{span}(F_1, F_2)) \ \ and$$
$$\mathrm{Aut}(E, E \cap F) = \mathrm{Aut}(E, E \cap F_1)$$

Any image point $\psi \in \mathrm{Aut}(E, E \cap F_1)$ may be extended by setting $\psi(x) = x$ for all $x \in F_2$ to produce an extension in $\mathrm{Aut}(\mathrm{span}(E, F) : F)$.

---

[26]I did not find a mistake, but prefer to follow Artin's argument as done in the proof above.

VIII.1   Galois Theory of Cubic and Quadratic Equations

VIII.1.1   The Galois Group for an Irreducible Cubic

Proposition 80. *Let $P(x) = x^3 + a_2x^2 + a_1x + a_0$ be a monic irreducible cubic polynomial with rational coefficients. The group of automorphisms of the splitting field is either the symmetric group S_3 or the cyclic group $\mathbf{Z}_3$. The latter case is described by either one of the following equivalent assumptions:*

(a) *The splitting field F of polynomial P is a three dimensional extension of the rationals.*

(b) *Adjoining any one root of $P(x)$ yields its splitting field.*

(c) *The group of automorphisms of the splitting field for $P(x)$ is a subgroup of the alternating group A_3.*

(d) *The antisymmetric polynomial*

$$(x_1 - x_2)(x_1 - x_3)(x_2 - x_3)$$

from the roots x_1, x_2, x_3 of polynomial $P(x)$ is rational. Moreover, the roots are real.

Proof. Let the reduction of polynomial P be

$$Q(x) = P(x - \tfrac{a_2}{3}) = x^3 - 3px + 2q$$

All three items (a),(b) and (c) are equivalent to the corresponding items concerning Q. Too, the polynomial P is irreducible if and only if the polynomial Q is irreducible. Hence it is enough to prove the claims for polynomial Q. $\qquad\square$

We show (a) $\Rightarrow$ (b). Assume the splitting field F is a three dimensional extension of the rationals. The field extension E obtained by adjoining a single root x_1 has as dimension the degree of the irreducible polynomial with root x_1. Hence E is a three dimensional extension, too. From the inclusion $E \subseteq F$, we obtain $E = F$. $\qquad\square$

We show (b) $\Rightarrow$ (c). Assume that adjoining any one root x_1 of $P(x)$ yields its splitting field. Hence the splitting field F is a three dimensional extension of the rationals. Specifying the image $\sigma x_1 = x_i$ determined the automorphism σ uniquely. Hence the group of its automorphisms has order either three or one. In the latter case, the roots would be fixed by all automorphisms and hence would be rational. This contradicts the assumption that the polynomial is irreducible. Hence the group of its automorphisms of field F is the cyclic group of order three $\mathbf{Z}_3 = A_3$. $\qquad\square$

We show (c) $\Rightarrow$ (d). Assume that the group of automorphisms of the splitting field for $P(x)$ is a subgroup of the alternating group A_3. Hence the antisymmetric polynomial $(x_1 - x_2)(x_1 - x_3)(x_2 - x_3)$ from the roots is invariant under all automorphisms of the splitting field, and hence it is rational. $\qquad\square$

We show (d) $\Rightarrow$ (c). We define the complex number r by requiring

$$\sqrt{-3}r = \sqrt{-p^3 + q^2}$$

and make $r \geq 0$ in the case that r is real. We solve the equation $x^3 - 3px + 2q = 0$ with Cardano's formula and substitute

$$p^3 = q^2 + 3r^2$$

A mathematica computation yields

$$(x_1 - x_2)(x_1 - x_3)(x_2 - x_3) = -18r$$

Assume that this antisymmetric polynomial is rational. Hence r is rational and nonnegative. Moreover, we conclude that the three zeros are all real.

Finally, the antisymmetric polynomial $(x_1 - x_2)(x_1 - x_3)(x_2 - x_3)$ from the roots is invariant under all automorphisms of the splitting field since this claim holds for the rational number r. Hence the group of automorphisms of the splitting field for $Q(x)$ is a subgroup of the alternating group A_3. $\qquad\square$

We show that (d) *and* (c) $\Rightarrow$ (b). We proceed as in the last proof. Since the polynomial is assumed to be irreducible, it has no multiple zeros, and hence $r > 0$. A mathematica computation yields

$$\begin{bmatrix} 1 & x_1 & x_1^2 \\ 1 & x_2 & x_2^2 \\ 0 & x_3 & x_3^2 \end{bmatrix}^{-1} \cdot \begin{bmatrix} x_1 & x_2 & x_3 \\ x_2 & x_3 & x_1 \\ x_3 & x_1 & x_2 \end{bmatrix} = \begin{bmatrix} 0 & \frac{p^2}{r} & -\frac{p^2}{r} \\ 1 & -\frac{q+r}{2r} & \frac{q-r}{2r} \\ 0 & -\frac{p}{2r} & \frac{p}{2r} \end{bmatrix}$$

and hence holds

$$\begin{bmatrix} x_1 & x_2 & x_3 \\ x_2 & x_3 & x_1 \\ x_3 & x_1 & x_2 \end{bmatrix} = \begin{bmatrix} 1 & x_1 & x_1^2 \\ 1 & x_2 & x_2^2 \\ 0 & x_3 & x_3^2 \end{bmatrix} \begin{bmatrix} 0 & \frac{p^2}{r} & -\frac{p^2}{r} \\ 1 & -\frac{q+r}{2r} & \frac{q-r}{2r} \\ 0 & -\frac{p}{2r} & \frac{p}{2r} \end{bmatrix}$$

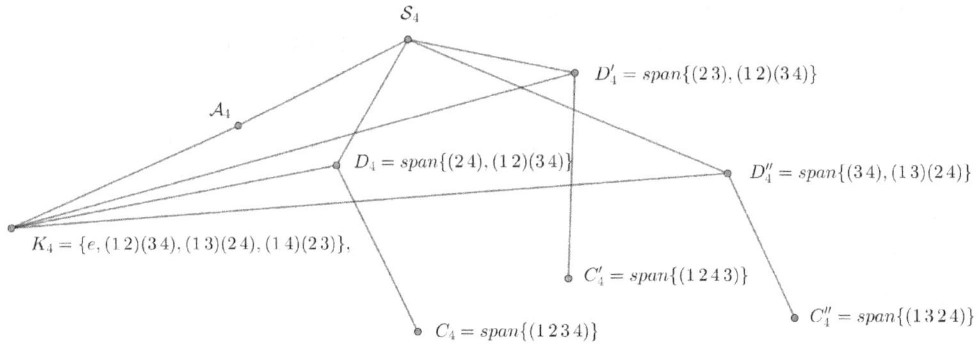

Figure 5: Transitive subgroups of $\mathcal{S}_4$.

The first row of the right-hand side lies in the field extension E obtained by adjoining the single zero x_1. From the second identity, we see that all three roots x_1, x_2, x_3 lie in this extension, In other words, E is the splitting extension F, which hence has dimension 3 as claimed. $\qquad\square$

VIII.1.2   The Galois Group for an Irreducible Quartic

At this point, I make the assumption that the polynomial $h(x)$ is irreducible. It may be taken about any base field F of characteristic different from 2 and 3. The irreducibility implies $r \neq 0$.

For characteristic different from 2 holds $h'(x) = q + 2px + 4x^3 \neq 0$. Hence the roots are simple. About the Galois group $\mathrm{Gal}(h, F)$ is known that it acts transitively on the roots x_1, x_2, x_3, x_4. By the Bahn lemma 159, also called orbit-stabilizer theorem, the order $|G|$ is divisible by 4. [27]

The group $\mathrm{Gal}(h, F) \leq \mathcal{S}_4$ is a subgroup of the permutations of the roots x_1, x_2, x_3, x_4. In the figure on page 376 we have gathered the possible subgroups that satisfy the above requirements. As a subgroup of the alternating group $\mathcal{A}_4$, we get only one Klein four-group $K_4 = \{e, (1\,2)(3\,4), (1\,3)(2\,4), (1\,4)(2\,3)\}$. Since $K_4 \lhd \mathcal{A}_4$ is a normal divisor of the alternating group, there are no conjugate ones.

The dihedral group $D_4 = span\{(2\,4), (1\,2)(3\,4)\}$ is not a subgroup of the alternating group. It has eight elements, which are given by the rotations and reflections that map a square with the vertices $1, 2, 3, 4$ to itself. The two conjugate groups D_4' and D_4'' are obtained by different enumeration of the vertices of the square, in the

---

[27] Theorem 16.6.6 in Michael Artin's book [2]

orders $1, 2, 4, 3$ and $1, 3, 2, 4$, respectively. The rotation groups $C_4 \lhd D_4$, $C_4' \lhd D_4'$, and $C_4'' \lhd D_4''$ are the corresponding rotations of the square. The Klein four-group K_4 is a subgroup of all three dihedral groups. Moreover

$$K_4 = D_4 \cap D_4' \cap D_4''$$

All these subgroups of the symmetric group occur as Galois group of an irreducible quartic polynomial.

We want now to determine the Galois group as exactly as possible. Each automorphism $\sigma \in \mathrm{Gal}(h, F)$ induces an automorphism $\Phi(\sigma) \in \mathrm{Gal}(R, F)$ for the Descartes resolvent IX.1.6 $R(v) = -q^2 + (p^2 - 4r)v + 2pv^2 + v^3$. has the zeros

$$[v_1, v_2, v_3] = [x_1 x_2 + x_3 x_4 - p, \; x_1 x_3 + x_2 x_4 - p, \; x_1 x_4 + x_2 x_3 - p]$$

Look at the zeros of the resolvent under the action of the Galois group $\mathrm{Gal}(R, F)$. $v_1 = x_1 x_2 + x_3 x_4 - p$ is fixed under all elements of the dihedral group $D_4'' = span\{(3\,4), (1\,3)(2\,4)\}$ since it is fixed under the generating elements. One checks that v_1 is not fixed under any further permutations of the roots x_1, x_2, x_3, x_4. Indeed holds $Stab(\mathcal{S}_4, v_1) = D_4''$. As easily is checked

$$Stab(\mathrm{Gal}(h, F), v_1) = D_4'' \cap \mathrm{Gal}(h, F)$$
$$Stab(\mathrm{Gal}(h, F), v_2) = D_4 \cap \mathrm{Gal}(h, F)$$
$$Stab(\mathrm{Gal}(h, F), v_3) = D_4' \cap \mathrm{Gal}(h, F)$$

and we state

Lemma 93. *The order of a stabilizer* $Stab(\mathrm{Gal}(h, F), v_i)$ *is not divisible by* 3.

For $i = 1, 2, 3$, the orbits $\mathrm{Gal}(R, F) \cdot v_i$ each contain at most 3 elements. I distinguish the following cases

(a) At least one of the orbits $\mathrm{Gal}(R, F) \cdot v_i$ contains at most 2 elements.

(b) The three orbits $\mathrm{Gal}(R, F) \cdot v_i$ each contain 3 elements.

(a) Assume that one orbit $\mathrm{Gal}(R, F) \cdot v_i$ contains at most 2 elements. If the orbit $\mathrm{Gal}(R, F) \cdot v_1 = \{v_1, v_2\}$ contains two elements, the orbit $\mathrm{Gal}(R, F) \cdot v_3 = \{v_3\}$ contains only one element. In all subcases, there exists an orbit $\mathrm{Gal}(R, F) \cdot v_i = \{v_i\}$ with only one element. Assume v_1 is fixed under the entire Galois group $\mathrm{Gal}(R, F)$. Now $F = \mathrm{Fix}[E : \mathrm{Gal}(h, F)]$ implies $v_i \in F$. That means that the

Descartes resolvent is reducible,— in the base field F. The Bahn Lemma 159 implies that

$$|\mathcal{S}tab(v_i)|\,|\mathrm{Gal}(R, F) \cdot v_i| = |\mathrm{Gal}(R, F)|$$
$$|\mathcal{S}tab(v_i)| = |\mathrm{Gal}(R, F)|$$
$$\mathcal{S}tab(v_i) = \mathrm{Gal}(R, F)$$

By the above lemma 93, the order $|\mathcal{S}tab(v_i)|$ is not divisible by 3. From the list of possible Galois groups, we are left with the possibilities that $\mathcal{S}tab(v_1) = \mathrm{Gal}(R, F)$ is either C_4'', D_4'' or K_4.

(b) The three orbits $\mathrm{Gal}(R, F)\cdot v_i$ each contain 3 elements. Hence $|\mathrm{Gal}(R, F)\cdot v_1| = 3$. The action of the Galois group is transitive on the orbit $[v_1, v_2, v_3]$. Hence the resolvent is irreducible. Moreover, the Bahn Lemma tells that

$$|\mathcal{S}tab(v_i)|\,|\mathrm{Gal}(R, F) \cdot v_i| = |\mathrm{Gal}(R, F)|$$
$$|\mathcal{S}tab(v_i)| = \frac{|\mathrm{Gal}(R, F)|}{3}$$

Hence the group order $|G|$ is divisible by three. That implies that either $\mathrm{Gal}(R, F) = \mathcal{A}_4$ or $\mathrm{Gal}(R, F) = \mathcal{S}_4$.

We may now combine the above information with the discriminant theorem 33 and obtain.

Proposition 81. *Let $f \in F[x]$ be a polynomial of degree 4 over a base field of characteristic different from 2. Assume that the polynomial is irreducible. Then we get the following four cases about the Galois group $\mathrm{Gal}(f, F)$.*

case (a) *Assume the discriminant is not a perfect square in the base field F, and the resolvent is irreducible. In that case holds $\mathrm{Gal}(f, F) = \mathcal{S}_4$. The Galois group is the entire symmetric group.*

case (b) *Assume the discriminant is not a perfect square in the base field F, and the resolvent is reducible. In that case holds $\mathrm{Gal}(f, F) = K_4$. The Galois group is the Klein four-group.*

case (c) *Assume the discriminant is a perfect square in the base field F, and the resolvent is irreducible. In that case holds $\mathrm{Gal}(f, F) = \mathcal{A}_4$. The Galois group is the alternating group.*

case (d) *Assume the discriminant is a perfect square in the base field F, and the resolvent is reducible. In that case holds either $\mathrm{Gal}(f, F) = D_4$ or $\mathrm{Gal}(f, F) = C_4$. The Galois group is either the dihedral group of order 8 or cyclic group of order 4.*

The polynomial need not be put into the depressed form since the transformation to achieve this is a simple shift of the zeros and does not alter the Galois group. For the actual application of this theorem, we go back to the general monic polynomial

$$f(x) = a + bx + cx^2 + dx^3 + x^4$$

For the mathematica computations see the file quartic.nb. The replacement $x \to x - d/4$ leads to the depressed quartic

$$h(x) = r + qx + px^2 + x^4$$
$$= a - \frac{bd}{4} + \frac{cd^2}{16} - \frac{3d^4}{256} + \left(b - \frac{cd}{2} + \frac{d^3}{8}\right) x + \left(c - \frac{3d^2}{8}\right) x^2 + x^4$$

and the resolvent

$$R(v) = -q^2 + (p^2 - 4r)v + 2pv^2 + v^3$$
$$= -\left(b - \frac{cd}{2} + \frac{d^3}{8}\right)^2 + \left(-4a + c^2 + bd - cd^2 + \frac{3d^4}{16}\right) v + 2\left(c - \frac{3d^2}{8}\right) v^2 + v^3$$

The discriminant is

$$\Delta(f) = 256a^3 - 27b^4 + 144ab^2c - 128a^2c^2 - 4b^2c^3 + 16ac^4 - 192a^2bd + 18b^3cd$$
$$-80abc^2d - 6ab^2d^2 + 144a^2cd^2 + b^2c^2d^2 - 4ac^3d^2 - 4b^3d^3 + 18abcd^3 - 27a^2d^4$$

Problem 184. *We can the symmetric group S_4 also visualize in three dimensions by the rotations and reflections of a tetrahedron, again mapping the vertices among themselves. Explain the geometrical meaning of the different possible Galois groups.*

VIII.2   Galois Theory in the Constructible Field

Theorem 35 (The Galois group of numbers in the constructible field extension $K + iK$). *Equivalent are*

(a) *The complex number $z \in \mathbb{C}$ lies in the constructible field extension $K + iK$.*

(b) *There exists a tower of two-dimensional extensions*

$$(\text{VIII.2.1}) \qquad \mathbf{Q} = \mathbf{F}_0 \subset \mathbf{F}_1 \subset \mathbf{F}_2 \subset \cdots \subset \mathbf{F}_n \subset \mathbb{C}$$

such that $z \in \mathbf{F}_n$.

(c) *The Galois group* $\mathrm{Gal}(f, \mathbf{Q})$ *of the minimal polynomial* $f \in \mathbf{Q}[z]$ *has the order* 2^r, *a power of* 2.

Corollary 43 (The Galois group of numbers in the constructible field). *Equivalent are*

(a) *The real number* $k \in \mathbb{R}$ *lies in the constructible field* K.

(b) *There exists a tower of two-dimensional extensions*

$$(\text{VIII.2.2}) \qquad \mathbf{Q} = \mathbf{F}_0 \subset \mathbf{F}_1 \subset \mathbf{F}_2 \subset \cdots \subset \mathbf{F}_n \subset \mathbb{R}$$

such that $k \in \mathbf{F}_n$.

(c) *The Galois group* $\mathrm{Gal}(f, \mathbf{Q})$ *of the minimal polynomial* $f \in \mathbf{Q}[k]$ *has the order* 2^r, *a power of* 2.

(a) $\Rightarrow$ (c). Let $z \in K + iK$. From the definition one gets a tower VIII.3.1 of two-dimensional extensions. Let $f \in \mathbf{Q}[z]$ be the minimal polynomial of the number k. Let E is the splitting field over the rational field of the minimal polynomial f. Clearly holds $\mathbf{Q}(z) \subseteq E$, but one may not expect equality to hold. At least we get

$$\deg f = \dim[\mathbf{Q}(z) : \mathbf{Q}] \leq \dim[E : \mathbf{Q}]$$

From $\dim[\mathbf{F}_{i+1} : \mathbf{F}_i] = 2$ we get, as explain in lemma 47 above, for all $0 \leq i \leq n - 1$ existence of $D_i \in \mathbf{F}_i$ such that $\mathbf{F}_{i+1} = \mathbf{F}_i(\sqrt{D_i})$.

Lemma 94. *The extension* $\mathrm{span}(E, F_n)/\mathbf{Q}$ *is a Galois extension.*

Proof. The above extension is generated as $\mathbf{Q}(\sqrt{D_0}, \ldots \sqrt{D_{n-1}}, z) =: \mathbf{Q}(\alpha_0, \ldots, \alpha_n)$. Let $Q_i \in \mathbf{Q}[x]$ be the minimal polynomials of α_i, for all $0 \leq i \leq n$. These are irreducible polynomials with simple zeros such that $Q_i(\alpha_i) = 0$ for $0 \leq i \leq n$. These polynomials split completely in the extension field $\mathrm{span}(E, F_n)$.

We use the idea from the problem 172 "Subtle and fine]". For the product $S = \prod Q_i$, the extension $\mathrm{span}(E, F_n)/\mathbf{Q}$ is the splitting extension. Too, the polynomial S is separable. Hence by definition 51 item (b), we get a Galois extension $\mathrm{span}(E, F_n)/\mathbf{Q}$. $\qquad \square$

Given any isomorphism $\text{span}(E, \mathbf{F}_n) \mapsto \mathbf{F}' \subseteq \mathbb{C}$. One gets a corresponding tower of two-dimensional extensions

$$\mathbf{Q} = \mathbf{F}_0 \subset \sigma(\mathbf{F}_1) \subset \sigma(\mathbf{F}_2) \subset \cdots \subset \sigma(\mathbf{F}_n) = \mathbf{F}'$$

For the two-dimensional extension, the midnight formula to solve a quadratic equation is used once more. We apply σ to the formula $\mathbf{F}_{i+1} = \mathbf{F}_i(\sqrt{D_i})$ and get $\sigma(\mathbf{F}_{i+1}) = \sigma(\mathbf{F}_i)(\sqrt{\sigma(D_i)})$.

Note that $D_i' := \sigma(D_i)$ need not all to be positive, indeed they may be complex. Inductively we prove that $D_i' \in K + iK$ holds for $0 \le i \le n - 1$. The induction start is clear since $D_0' = D_0 \in \mathbf{Q}$.

Here is the induction step $"i \to i + 1"$:

We use the formula for the complex square root

$$(\text{VIII.2.3}) \qquad \sqrt{x + iy} = \sqrt{\frac{\sqrt{x^2 + y^2} + x}{2}} + i \,\text{sign}\,(y) \sqrt{\frac{\sqrt{x^2 + y^2} - x}{2}}$$

and obtain $\sqrt{D_i'} = E_i + iF_i$ with $E_i, F_i \in K$ and hence $\sqrt{D_i'} = E_i + iF_i \in K + iK$. One gets $\sigma(\mathbf{F}_{i+1})\sigma(\mathbf{F}_i)(\sqrt{\sigma(D_i)}) \subseteq K + iK$. Hence $D_{i+1}' = \sigma(D_{i+1}) \in \sigma(\mathbf{F}_{i+1}) \subset K + iK$.

From the last step is obtained $\sigma(\mathbf{F}_n) \subseteq K + iK$.

By item (ii) from the three characterizations of the splitting field 71 holds $\sigma(E) \subseteq E$. Hence

$$\sigma \restriction E \in \text{Gal}(f, \mathbf{Q})$$
$$(\text{VIII.2.4}) \qquad \sigma(E \cap \mathbf{F}_n) \subset E \cap (K + iK)$$

have been shown for any isomorphism $\sigma : \text{span}(E, \mathbf{F}_n) \mapsto \mathbf{F}' \subseteq \mathbb{C}$. and especially for any automorphism $\sigma \in \text{Aut}[\text{span}(E, \mathbf{F}_n), \mathbf{Q})$.

In the beginning was shown that the extension $\text{span}(E, F_n)/\mathbf{Q}$ is a Galois extension with the splitting polynomial $S = \prod Q_i$. The factor $Q_n \in \mathbf{Q}[z]$ is the minimal polynomial of the constructible number k. Since the minimal polynomial Q_n is irreducible, we may invoke the proposition 70 about splitting extension, item (c) to obtain.

> For any algebraic conjugate z_i of the constructible number k, there exists
> an automorphism $\sigma \in \text{Aut}[\text{span}(E, \mathbf{F}_n), \mathbf{Q})$ such that $\sigma(k) = z_i$.

From the above result (VIII.2.4) we conclude $z_i \in E \cap (K + iK)$ holds for all the conjugates. Since the conjugates span the splitting field E, we get $E \subset K + iK$.

Take any basis β_k of the splitting extension $E/\mathbf{Q}$. Since $\beta_k \in K + iK$ each base element is contained in some field $\mathbf{F}_n$ which is obtained from the basic field $\mathbf{Q}$ by a finite number of two-dimensional extensions. The union of these finitely many extension towers produces one tower of two-dimensional extensions, and the top field $\mathbf{F}_N$ of this union tower contains all base elements β_k. Hence $E \subseteq \mathbf{F}_N$. We know that $\dim[\mathbf{F}_N : \mathbf{Q}]$ is 2^N, a power of two. The tower theorem implies that the dimension $\dim[E : \mathbf{Q}]$ is a divisor of 2^N, and hence again a power of 2.

The Galois group $\mathrm{Gal}(f, \mathbf{Q}) = \mathrm{Aut}[E : \mathbf{Q}]$ of the minimal polynomial $f \in \mathbf{Q}[z^*]$ has the order $\dim[E : \mathbf{Q}]$, which is a power of 2. $\square$

For the proof of the converse, some propositions from group theory are needed.

Lemma 95. *Assume that G is a group of order $|G| = p^r$ with p prime. Then the center Z is a nontrivial subgroup. Especially there exists a cyclic subgroup $Z_p \lhd G$ which is of order p and a normal divisor of group G.*

Proof. The class equation of a group adds the orders of the disjoint equivalence classes. The elements of the center are the one-element equivalence classes. The equivalence class of any element $g \in G$ is the set

$$C_g = \{hgh^{-1} : h \in G\}$$

The number of elements is

$$|C_g| = \frac{|G|}{|g'|} \quad \text{with} \quad g' = \{h \in G : hg = gh \in G\}$$

Hence $|C_g|$ is a divisor of $|G|$, and under the present assumptions a prime power. From the partition into disjoint classes is obtained the class equation

$$G = \bigcup_{z \in Z} \{z\} \cup \bigcup_{g \in G \backslash Z} C_g$$

$$|G| = |Z| + \sum \{\frac{|G|}{|g'|} \ : \ \text{classes } C_g \text{ with } |C_g| \geq 2\}$$

Under the present assumptions the order $|G|$ and all summand of the right-hand sum are divisible by prime p. Hence $|Z|$ is divisible by p. Under the present assumptions holds $|Z| = p^l$ with $1 \leq l \leq r$. Hence all elements of Z have a prime power as order. There exists an element $z_p \in Z$ of the order p, which generates the subgroup $Z_p < Z$ of order p. Since the elements of Z_p commute with all group elements, this subgroup Z_p is a normal divisor. $\square$

Proposition 82. *Assume that G is a group of order $|G| = p^r$ with p prime. Then there exists a chain of subgroups*

$$\{e\} = G_0 \lhd G_1 \lhd G_2 \lhd \cdots \lhd G_{r-1} \lhd G_r = G$$

such that the quotient groups of successive terms all have order p. All $G_t \lhd G$ are normal divisors of the original group.

Proof. The series is constructed inductively starting at the lower end. The group G_1 is obtained by the previous lemma. Assume that the series has been constructed up to and including some G_t with $1 \le t < r$. We apply the lemma to the quotient group $\tilde{G} := G/G_t$. There exists normal divisor $\tilde{C} \lhd \tilde{G}$ of order p. Let $\tilde{C}$ be generated by the remainder class $c_p G_t = G_t c_p$. Hence $\tilde{C}$ consists of the remainder classes $(c_p)^j G_t = G_t (c_p)^j$ for $0 \le j < p$. We get these p disjoint remainder classes since $c_p^p \in G_t$ is the lowest positive power in G_t. We use the canonical homomorphism

$$\phi : G \mapsto \tilde{G} \quad \text{which maps } g \in G \text{ to } \phi(g) = gG_t = G_t g, \text{ the remainder class}$$

The preimage is

$$\phi^{-1}(\tilde{C}) = \{h \in G : hG_t = (c_p)^j G_t \text{ with } 0 \le j < p\}$$
$$= \{h \in G : \text{ there exists } 0 \le j < p \text{ such that } h(c_p)^{-j} \in G_t\}$$
$$= \{(c_p)^j g_t : \text{ with } 0 \le j < p \text{ and } g_t \in G_t\}$$

- This is a normal divisor of G since for all $s \in G$ holds

$$h(c_p)^j \in G_t \Rightarrow s\,h(c_p)^j s^{-1} \in sG_t s^{-1} \Rightarrow shs^{-1}(c_p)^j \in G_t$$
$$h \in \phi^{-1}(\tilde{C}) \Rightarrow shs^{-1} \in \phi^{-1}(\tilde{C})$$

- Moreover $G_t \lhd \phi^{-1}(\tilde{C})$ as the reader may check.

- $\phi^{-1}(\tilde{C})$ has $p \cdot |G_t| = p^{t+1}$ elements since the third description given above is unique.

The reader knows already that we put $G_{t+1} := \phi^{-1}(\tilde{C})$ to complete the induction step. $\square$

(c) $\Rightarrow$ (a). Let $z \in \mathbb{C}$ be any number, and assume that the Galois group $\mathrm{Gal}(f, \mathbf{Q})$ of the minimal polynomial $f \in \mathbf{Q}[z]$ has the order 2^r. From the above proposition about groups with order a prime power, we obtain the chain of subgroups

$$\{e\} = G_0 \lhd G_1 \lhd G_2 \lhd \cdots \lhd G_{r-1} \lhd G_r = \mathrm{Gal}(f, \mathbf{Q})$$

where $G_t \lhd \mathrm{Gal}(f, \mathbf{Q})$ is a normal divisor of order 2^t. By the Main Theorem 34 of Galois Theory, we get the corresponding intermediate fields between the splitting field E of the minimal polynomial f and the rational numbers $\mathbf{Q}$.

$$E_t = \lambda(G_{r-t}) = \mathrm{Fix}(E : G_{r-t})$$
$$G_t = \lambda^{-1}(E_{r-t}) = \mathrm{Aut}(E : E_{r-t})$$

They provide the tower of fields, now ordered with increasing sizes

$$\mathbf{Q} = E_0 \subset E_1 \subset E_2 \subset \cdots \subset E_{r-1} \subset E_r = E$$

Each step is a field extension of dimension 2. Hence all elements are members of the complex extension of the constructible field. Especially $E \subset K + iK$ and $z \in K + iK$.

Under the additional assumption that $z \in \mathbb{R}$ holds $z \in K$. In other words, the number z is in the constructible field. In that case, we may cut down the above chain to a chain of subfields of the constructible field. Let l be the highest index for which holds $E_l \subset \mathbb{R}$.

The splitting field E contains all algebraic conjugates z_i of z, and these need not all be real. Hence even under the additional assumption that $z \in \mathbb{R}$ need not hold $l = r$. Instead we have the more complicated picture

$$\mathbf{Q} = E_0 \subset E_1 \subset \cdots \subset E_l = \mathbb{R} \cap E_{l+1} \subseteq \mathbb{R} \cap E_{l+2} \subseteq \cdots \subseteq \mathbb{R} \cap E_r = \mathbb{R} \cap E \subset K$$

I claim: the extensions $(\mathbb{R} \cap E_{l+2})$, $(\mathbb{R} \cap E_{l+1})$ and the following ones may be either one or two-dimensional. Why?

We use item (c) of the Main Theorem of Galois Theory. The extensions $E_t/\mathbf{Q}$ is a Galois extension iff $\mathrm{Aut}(E : E_t)$ is a normal divisor of $\mathrm{Aut}(E : \mathbf{Q})$. Indeed, we know the second assumption to be true.

Moreover, the extensions $E_t/\mathbf{Q}$ is a Galois extension iff $E_t = \phi(E_t)$ holds for all $\phi \in \mathrm{Aut}(E : \mathbf{Q})$ We may put $\phi = \tau$, the complex conjugation. Under the additional assumption $z \in \mathbb{R}$, its minimal polynomial f has real coefficients. Hence the splitting field E contains all conjugate complex pairs among the algebraic conjugates of z. Hence $\tau \in \mathrm{Aut}(E : \mathbf{Q}) = \mathrm{Gal}(f, \mathbf{Q})$. The complex conjugation is a member of the Galois group. Via the Main Theorem we conclude that $E_t = \tau(E_t)$. We see that complex conjugation acts in E_t, with the set of fixed elements being $\mathbb{R} \cap E_t$.

Problem 185. *Explain why the dimension* $\dim[E_t : \mathbb{R} \cap E_t] \leq 2$ *is either one or two.*

Proof. Solution Assume that $a + ib \in E_i$ and $b \neq 0$, then $a - ib \in F_i$ and $ib \in E_i$. A generating set of E_i consists of ib and otherwise only real vectors. from $\mathbb{R} \cap E_i$. $a + ib$ is solution of a quadratic equation with coefficients in $\mathbb{R} \cap E_i$. Hence $E_i = (\mathbb{R} \cap E_i)(a + ib)$ is a two-dimensional extension. $\qquad\square$

We have now confirmed $\dim[E_t : \mathbb{R} \cap E_t] = 2$ and $\dim[E_t : E_{t-1}] = 2$ for all $l < t \leq r$. Hence we deduct that

$$\dim[\mathbb{R} \cap E_{t+1} : \mathbb{R} \cap E_t] = 2 \quad \text{for all } l < t \leq r$$

going through the "ladder" of two-dimensional extensions.

$$
\begin{array}{ccccccc}
\mathbb{R} \cap E_{l+1} & \subseteq & \mathbb{R} \cap E_{l+2} & \subseteq & \cdots & \subseteq & \mathbb{R} \cap E_r = \mathbb{R} \cap E \subset K \\
\cap & & \cap & & & & \cap \\
E_{l+1} & \subset & E_{l+2} & \subset & \cdots & \subset & E_r = E \subset K + iK
\end{array}
$$

We may delete the one-dimensional step $E_l = \mathbb{R} \cap E_{l+1}$ to obtain a chain of $r - 1$ two-dimensional extensions up to the field $\mathbb{R} \cap E$, and thus we obtain any real element in E. $\qquad\square$

Problem 186. *Let $f \in \mathbf{Q}[z]$ be an irreducible polynomial of degree 4. Prove that the roots of f are in the constructible field extension $K + iK$ if and only if the Descartes resolvent of f is reducible.*

Solution. The shortest solution is based on proposition 81 together with the theorem 35 about the Galois group of numbers in the constructible field extension $K + iK$.

The Descartes resolvent is reducible iff the Galois group is either $\mathrm{Gal}(f, \mathbf{Q}) = \mathcal{K}_4, D_4$ or $\mathrm{Gal}(f, \mathbf{Q}) = C_4$. These groups have orders $4, 8$ and 4, respectively, which are powers of 2. By theorem 35 the latter claim holds iff the roots are in the constructible field extension $K + iK$.

The Descartes resolvent is irreducible iff the Galois group is either $\mathrm{Gal}(f, \mathbf{Q}) = \mathcal{S}_4$ or $\mathrm{Gal}(f, \mathbf{Q}) = \mathcal{A}_4$. These groups have orders 24 and 12, respectively, which are not powers of 2. By theorem 35 the latter claim implies that the roots are not in the constructible field extension $K + iK$. $\qquad\square$

Problem 187. *Let $f \in \mathbf{Q}[z]$ be an irreducible polynomial of degree 4. Convince yourself more directly, without the use of the Galois group that the Descartes resolvent of f being reducible implies that the roots of polynomials f are in the constructible field extension $K + iK$.*

Answer. The Descartes resolvent is a polynomial of degree 3 with rational coefficients. Such a reducible polynomial has a rational root. Factoring the resolvent yields as second factor a polynomial of degree two with rational coefficients. By the midnight formula, its zeros lie in the field extension $K + iK$. Hence all three roots v_1, v_2, v_3 of the resolvent lie in $K + iK$. The zeros of the polynomial f are now obtained by Descartes formula VII.3.4, which involves only square roots. Hence the zeros of polynomial f are in the real case in the constructible field K. In the complex case, one may use the formula VIII.2.3, and gets that the roots are the field extension $K + iK$. $\qquad\square$

Problem 188. *Let $p(x) = x^4 - x^3 - 5x^2 + 1$. Given the information from above, the following steps allow one confirm in a rather pedestrian way that the roots are not in the constructible field, in spite of the fact the single roots extensions have dimension 4.*

(a) *Show that $p(x)$ has four distinct real roots $\alpha, \beta, \gamma, \delta$.*

(b) *Show that neither of them can be rational.*

(c) *Show that no factoring $p(x) = (x^2 + ax + 1)(x^2 + bx + 1)$ or $p(x) = (x^2 + ax - 1)(x^2 + bx - 1)$ with integer a, b exists.*

(d) *Show that the polynomial $p \in \mathbf{Q}[x]$ is irreducible.*

(e) *What is the dimension $[\mathbf{Q}(\alpha) : \mathbf{Q}]$.*

(f) *How can one confirm that the order of the Galois group $\mathrm{Gal}(p, \mathbf{Q})$ is not a power of 2.*

(g) *Calculate the Descartes resolvent of p and check that it is reducible.*

(h) *Why are the roots not constructible?*

Explanations. For this problem, I have even restrained from the (further) use of mathematica. That is a bid too little help by mathematica. But on the other hand, mathematica cannot solve this problem *completely*, as the beginner may tend to think.

By the rule of Descartes 2, the polynomial $p(x) = x^4 - x^3 - 5x^2 + 1$, has either 2 or 0 positive roots. Since $p(1) = -4 < 0$, we have indeed 2 positive simple roots. Since $p(-x) = x^4 + x^3 - 5x^2 + 1$, we get either 2 or zero negative roots. Since $p(-1) = -2 < 0$, we have indeed 2 negative simple roots.

By the proposition 31 about the rational roots of a polynomial form $\mathbf{Z}[z]$, the only possible rational roots of p are 1 and -1. Hence there are no rational roots. By Gauss' lemma to confirm that the polynomial p is irreducible, one has only to show that no factoring $p(x) = (x^2 + ax + 1)(x^2 + bx + 1)$ or $p(x) = (x^2 + ax - 1)(x^2 + bx - 1)$ with integer a, b does exist. Since p has no linear term, we get $b = -a$.

$$(x^2 + ax + 1)(x^2 - ax + 1) = (x^2 + 1)^1 - a^2 x^2$$

has no term with x^3. Same situation for $(x^2 + ax - 1)(x^2 - ax - 1)$. Hence the polynomial p is irreducible.

The minimal polynomial for any single root is p,—indeed there are no proper factors that could be the minimal polynomial. Hence $\dim[\mathbf{Q}(\alpha) : \mathbf{Q}] = 4$.

To answer part (f), I use proposition 81. As already clear from the two previous problems, one needs to determine whether the Descartes resolvent is irreducible. I have given above the formula for general polynomial $a + bx + cx^2 + dx^3 + x^4$

$$R(v) = -\left(b - \frac{cd}{2} + \frac{d^3}{8}\right)^2 + \left(-4a + c^2 + bd - cd^2 + \frac{3d^4}{16}\right)v + 2\left(c - \frac{3d^2}{8}\right)v^2 + v^3$$

Plug in $a = 1, b = 0, c = -5, d = -1$ and obtain

$$R(v) = -(441/64) + (419v)/16 - (43v^2)/4 + v^3$$
$$64R(w/4) = -441 + 419w - 43w^2 + w^3$$

One needs to confirm this is irreducible. To this end it is enough to check that is does not have any rational root. By the proposition 31 about the rational roots of a polynomial form $\mathbf{Z}[z]$, the only possible rational roots are $\pm$ the divisors of 441. There are no negative roots. One factors $441 = 3^2 \cdot 7^2$ and get divisors $1, 3, 7, 9, 21, 49\ldots$ and calculates for $-441 + 419w - 43w^2 + w^3$

$$-64, 456, 728, 576, -1344, 34496, \ldots$$

then I may stop the computations since the polynomial of degree three cannot have any further sign change for larger values. The conclusion from this numerical computation is: the resolvent has no rational root. For a polynomial of degree *two or three*, this is enough to confirm its irreducibility. Now proposition 81 comes into play. It yields that the Galois group $\mathrm{Gal}(p, \mathbf{Q})$ is either the symmetric group or the alternating group, acting on the four roots. In both cases, the group order is divisible by three, and not a power of 2.

By theorem 35, the roots are not in the constructible field K. $\square$

VIII.3   Galois Theory for Two-Marked Ruler

Theorem 36 (Nicomedes, Cardano, Bombelli, Burnside, Hartshorne). *Here are equivalent statements*

(a) *there exists a tower of extensions of dimension at most four*

(VIII.3.1) $\mathbf{Q} \subseteq F = \mathbf{F}_0 \subset \mathbf{F}_1 \subset \mathbf{F}_2 \subset \cdots \subset \mathbf{F}_n \subset \mathbb{R}$

such that $x \in \mathbf{F}_n$.

(b) *In the Euclidean coordinate field F^2, the quantity x is constructible by means of compass and a two-marked straightedge,—with the restriction that the marks of the straightedge are only allowed to be put onto straight lines, but not onto circles.*

(d) *The Galois group $\mathrm{Gal}(f, F)$ of the minimal polynomial $f \in F[x]$ for the quantity x has the order $2^a 3^b$ with natural numbers $a, b \geq 0$.*

Lemma 96. *Assume that any of the statements* (a), (b), *or* (c) *from theorem 22 is true for the real quantity x,—starting from the real base field $F \subseteq \mathbb{R}$. Then the dimension $\dim[F(x) : F]$ of the one-element field extension $F(x)/F$ equals $2^a 3^b$ for some natural $a, b \geq 0$.*

Proof. All extensions in the tower (VIII.3.1) have dimensions $2^a 3^b$ for some natural $a, b \geq 0$. Hence $\dim[\mathbf{F}_l : F]$ equals $2^a 3^b$ for some natural $a, b \geq 0$. The same statement holds by the tower theorem 23 for the intermediate field $F \subseteq F(x) \subseteq \mathbf{F}_l$. $\qquad\square$

Proving (a) $\Rightarrow$ (d). We use the idea from the problem 172 "Subtle and fine". For the extensions of degree 4 in the tower (VIII.3.1), one needs to put in an intermediate field of degree 2, if such a field exists. After that change, the above extension is generated as a series of one-element extensions $\mathbf{F}_{i+1} = \mathbf{F}_i(\alpha_i)$ for $0 \leq i \leq n-1$, which are all of degree $2, 3$ or 4. In the end holds $x \in \mathbf{F}_n = F(\alpha_0, \ldots, \alpha_{n-1})$.

Let $Q_i \in F[x]$ be the minimal polynomials of α_i, for all $0 \leq i \leq n-1$. Let $Q_n \in F[x]$ be the minimal polynomials of $x =: \alpha_n$, all over the base field F. These are irreducible polynomials with simple zeros such that $Q_i(\alpha_i) = 0$ for $0 \leq i \leq n$. Let E be the splitting field of x. These polynomials split completely in the combined extension field $\mathrm{span}(E, \mathbf{F}_n)$.

For the product $S = \prod Q_i$, the extension $\mathrm{span}(E, \mathbf{F}_n)$ is the splitting extension. Too, the polynomial S is separable. Hence by definition 51 item (b), we get a Galois extension $\mathrm{span}(E, \mathbf{F}_n)/F$.

Given is any isomorphism

$$\sigma : \mathrm{span}(E, \mathbf{F}_1, \ldots \mathbf{F}_n) \mapsto \mathbf{F}' \subseteq \mathbb{C}$$

with fixed field F. One gets a corresponding tower of $2, 3$ or 4-dimensional extensions

$$\mathbf{Q} \subseteq F = \mathbf{F}_0 \subset \sigma(\mathbf{F}_1) \subset \sigma(\mathbf{F}_2) \subset \cdots \subset \sigma(\mathbf{F}_n) = \mathbf{F}'$$

By item (ii) from the three characterizations of the splitting field 71 holds $\sigma(E) \subseteq E$. Hence

$$\sigma \restriction E \in \mathrm{Gal}(f, F)$$

holds for any isomorphism $\sigma : \mathrm{span}(E, \mathbf{F}_n) \mapsto \mathbf{F}' \subseteq \mathbb{C}$. and especially for any automorphism $\sigma \in \mathrm{Aut}[\mathrm{span}(E, \mathbf{F}_n), F)$.

In the beginning was shown that the extension $\mathrm{span}(E, F_n)/F$ is a Galois extension with the splitting polynomial $S = \prod Q_i$. The factor $Q_n \in F[x]$ is the minimal polynomial of the marked-straightedge constructible number x. Since the minimal polynomial Q_n is irreducible, we may invoke the proposition 70, item (c) about the existence and uniqueness properties of the splitting field to obtain

> For any Galois conjugate z_i of the constructible number x over the base field F, there exists an automorphism $\sigma \in \mathrm{Aut}[\mathrm{span}(E, \mathbf{F}_n), F)$ such that $\sigma(x) = z_i$.

Note that many of the z_i may be complex. That can happen even for the case of classically constructible extensions considered in the previous section. It shall always happen for the angle trisections, and extraction of cube roots. I need the following definition to clarify the possible range of the above isomorphism.

Definition 56. Let *restricted marked straightedge* denote a straightedge with two marks, which are only to be put at a given point or onto a given line, but not onto a given circle.

Let $\mathcal{M}K : F$ denote the field of quantities that are constructible with compass and restricted marked straightedge, in the Euclidean coordinate field F^2, from any finite number of given quantities from the base field F.

Let $(\mathcal{M}K : F)(\omega, i)$ denote the field extension obtained by adjoining the third root of unity ω and the forth root of unity i.

Any isomorphism $\sigma \in \mathrm{Aut}[\mathrm{span}(E, \mathbf{F}_n), F)$ does map into the field $(\mathcal{M}K : F)(\omega, i)$. Since the conjugates span the splitting field E, we get $E \subset \mathrm{span}(E, \mathbf{F}_n), F)$. Hence

$$\sigma \upharpoonright E \in \mathrm{Gal}(f, F)$$

(VIII.3.2) $$\sigma(E \cap \mathbf{F}_n) \subset E \cap (\mathcal{M}K : F)(\omega, i)$$

Take any basis β_k of the splitting extension E/F. Each base element is contained in some field $\sigma(\mathbf{F}_n)$ which is obtained from the basic field $F(i, \omega)$ by a finite number of at most four-dimensional extensions. The union of these finitely many extension towers produces one tower of two-dimensional extensions, and the top field $\mathbf{F}_N$ of this union tower contains all base elements β_k. Hence $E \subseteq \mathbf{F}_N$. We know that $\dim[\mathbf{F}_N : \mathbf{Q}]$ is $2^a 3^b$ for some $a, b \geq 0$. The tower theorem implies that the dimension $\dim[E : F]$ is $2^a 3^b$ for some $a, b \geq 0$. $\qquad\square$

(d) $\Rightarrow$ (a) *and* (b). For the quantity x is assumed that the Galois group $\mathrm{Gal}(f, F)$ of its minimal polynomial $f \in F[x]$ has the order $2^a 3^b$ for some natural numbers $a, b \geq 0$. We have to use Burnside's Theorem (37) which is proved is the next section. A group the order of which is the product of two prime powers, is solvable. Thus the assumption (d) implies that the Galois group $\mathrm{Gal}(f, F)$ is solvable. By proposition (83), there exists a subnormal series

$$\{e\} = G_0 \lhd G_1 \lhd \cdots \lhd G_r = \mathrm{Gal}(f, F)$$

where all factor groups G_{i+1}/G_i have either order 2 or 3. Hence they are not only Abelian but cyclic, too. By the Main Theorem 34 of Galois Theory, we get the corresponding intermediate fields between the splitting field E of the minimal polynomial f and the base field F.

$$E_t = \lambda(G_{r-t}) = \mathrm{Fix}(E : G_{r-t})$$
$$G_t = \lambda^{-1}(E_{r-t}) = \mathrm{Aut}(E : E_{r-t})$$

They provide the tower of fields, now ordered with increasing sizes

$$F = E_0 \subset E_1 \subset E_2 \subset \cdots \subset E_{r-1} \subset E_r = E$$

Each step is a field extension of dimension either 2 or 3. Since the Galois group acts on the zeros of polynomial f, and $f(x) = 0$ is assumed from the beginning, we get $x \in E$. The severe flaw in the ointment is that the splitting field E is only in rare cases contained in the real field!

Here is my strategy to a stepwise real extension up to a field containing the given quantity x. One begins with the tower

$$F \subset F(i) \subset F(i, \omega) \subset E_1(i, \omega) \subset E_2(i, \omega) \subset \cdots \subset E_{r-1}(i, \omega) \subset E(i, \omega)$$

One adjoins the imaginary unit i in order to be lateron able to take the real- and imaginary parts. The third root of unit ω is adjoined because it appear the computer-Cardano formula to solve a third order equation.

Next is constructed a possible much longer tower.

case(A) $\dim[E_{i+1} : E_i] = 2$.

There exists $\alpha_i \in E_i$ such that $E_{i+1} = E_i(\sqrt{\alpha_i})$. But note that $\alpha_i = x + iy$ is in general complex. Since $x, y \in E_i(i)$, we may use the formulas for the complex square root

$$(\text{VIII.2.3}) \qquad \sqrt{x + iy} = \sqrt{\frac{\sqrt{x^2 + y^2} + x}{2}} + i\,\mathrm{sign}\,(y)\sqrt{\frac{\sqrt{x^2 + y^2} - x}{2}}$$

With three quadratic extensions by real square roots, one gets a (possibly larger) extensions $E'_{i+1} \supseteq E_{i+1}$ such that $\Re\sqrt{\alpha_i}, \Im\sqrt{\alpha_i} \in E'_{i+1} \cap \mathbb{R}$.

case(B) $\dim[E_{i+1} : E_i] = 3$:

We have again a one-element extension $\mathbf{F}_i(\alpha)$ since there cannot be any intermediate fields. The minimal polynomial $p \in \mathbf{F}_i[x]$ of α has degree 3. We may assume the reduced form

(VII.2.1) $$p(x) = x^3 + b\,x + c = 0$$

The discriminant is

(VII.2.5) $$D = \frac{c^2}{4} + \frac{b^3}{27}$$

Since the minimal polynomial is irreducible, the case $D = 0$ with a multiple zero cannot occur. But the quantities $b, c,$ and D are now in general complex. The first step are up to three quadratic extensions to produce a field containing both $\Re\sqrt{D}$ and $\Im\sqrt{D}$. Here we need the imaginary unit i to be able to separate real- and imaginary parts. In place of

$$\alpha = \sqrt[3]{-\frac{c}{2} + \sqrt{D}} - \frac{b}{3\sqrt[3]{-\frac{c}{2} + \sqrt{D}}}$$

is next adjoined the third root $\sqrt[3]{-\frac{c}{2} + \sqrt{D}}$. From the computer-ready formula VII.2.16, we see that the resulting field E'_{i+1} the polynomial p splits. Here we use that the third unit root ω has been adjoined in the beginning.

We obtained a longer tower with the following property. Each two-dimensional extension is produced by a real square root, Each three-dimensional extension is a Galois extension, achieved by adjoing a complex third root $\sqrt[3]{x + iy}$. We use the formulas

$$x + iy = \sqrt{x^2 + y^2}\,(\cos\theta + i\sin\theta)$$

$$\sqrt[3]{x + iy} = \sqrt{\sqrt[3]{x^2 + y^2}}\left(\cos\tfrac{\theta}{3} + i\sin\tfrac{\theta}{3}\right)$$

and get $\cos\theta$ and $\sin\theta$ by real quadratic extensions. Next we use only real third roots and angle trisection.

Hence one may substitute each three dimensional extensions by a sequence of extensions by real square roots, real cube roots, or by angle trisections, as in part (b), item (i) (ii) and (iii) of proposition 65. $\qquad\Box$

VIII.4   Burnside's Theorem

Definition 57 (Subnormal series, normal series, solvable group). Let G be a group and denote the identity element by e. A *subnormal series* of G is a finite sequence of subgroups $G_0 = \{e\}, G_1, \ldots, G_k = G$ such that $G_{i-1} \trianglelefteq G_i$ for $1 \leq i \leq k$. Usually one summarized the subnormal series in the following form:

$$\{e\} =: G_0 \trianglelefteq G_1 \trianglelefteq \cdots \trianglelefteq G_k := G$$

In this case, the quotient groups G_i/G_{i-1} are well-defined and are called the *factors* of the subnormal series. If in addition holds $G_i \trianglelefteq G$ for all i, this series is called a *normal series*. The length of such a series is the number of strict inclusions (equivalently, the number of non-trivial factors).

Definition 58 (Solvable group). The group G is called *solvable* if and only if it has a subnormal series whose factors are all Abelian.

Lemma 97. *Assume that $N \triangleleft G$ is a proper normal divisor and that the quotient group G/N is finite and Abelian. Then there exists a further normal divisor of H of G such that*

$$N \triangleleft H \triangleleft G$$

and $|H/N|$ is a prime number.

Proof. Take any $a \neq e$ in the quotient group G/N and choose $b = a^q$ such that the order of b is a prime p. Let H' be the cyclic group generated by b. Let $\kappa : G \mapsto G/N$ be the canonical homomorphism and put $H := \kappa^{-1}(H')$. We get the following diagram, where the down arrows indicate the homomorphism κ.

$$
\begin{array}{ccccc}
N & \triangleleft & H & \triangleleft & G \\
\downarrow & & \downarrow & & \downarrow \\
\{\bar{e}\} & \triangleleft & H' & \triangleleft & G/N
\end{array}
$$

The last line is correct since G/N is Abelian. Clearly holds $H' = \kappa(H)$. By means of the canonical homomorphism one checks that there occur normal divisors in the first line, too. Hence $|H/N| = |H'| = p$ is a prime. $\qquad\square$

Proposition 83. *A group G is finite and solvable if and only if there exists a subnormal series where all factor groups have prime order.*

Proof. Assume the group G is finite and solvable. Assume that the quotient group G_i/G_{i-1} for some i does not have prime order. By inserting H between G_{i-1} and G_i as explained in the lemma, one obtains again a subnormal series with Abelian quotients, and at least one factor group of prime order. This procedure can be repeated until all factor groups have prime order. The converse is even more obvious since any group of prime order is Abelian. □

Lemma 98. *Each subgroup of a solvable group is solvable.*

Indication of the reason. One iterates the commutator groups $\mathrm{Kom}^k(G)$ for the subgroup $H < G$. The iterated commutator groups $\mathrm{Kom}^k(H)$ go down to the unit group after finitely many steps. One obtains the finite series of Abelian quotient groups $\mathrm{Kom}^{i-1}(H)/\mathrm{Kom}^i(H)$ for $1 \leq i \leq J$ in the subnormal series

$$\{e\} = \mathrm{Kom}^J(H) \triangleleft \mathrm{Kom}^{J-1}(H) \triangleleft \cdots \triangleleft \mathrm{Kom}(H) \triangleleft H$$

□

Proposition 84. *Assume the group G has normal divisor N. The group G is solvable if and only if both groups N and G/N are solvable.*

Proof. Assume the group G is solvable and has the normal divisor N. Let $\kappa : G \mapsto G/N$ be the canonical homomorphism. We use that

$$\mathrm{Kom}^k G/N = \mathrm{Kom}^k \kappa(G) = \kappa(\mathrm{Kom}^k G)$$

for all $k \geq 0$. Since G is solvable, there exists $K \in \mathbf{N}$ such that $\mathrm{Kom}^K G = \{e\}$. The relation above implies $\mathrm{Kom}^K G/N = \{e\}$, and hence G/N is solvable. The subgroup N is solvable by the previous lemma 98.

Conversely, assume that the normal divisor N and the quotient group G/N are solvable. Hence there exist natural numbers J and K such that

$$\mathrm{Kom}^J(N) = \{e\} \quad \text{and} \quad \mathrm{Kom}^K G/N = \{\bar{e}\} \simeq N$$

The latter relation implies

$$\kappa(\mathrm{Kom}^K G) = \mathrm{Kom}^K G/N = \{\bar{e}\}$$
$$\mathrm{Kom}^K G \subseteq N$$
$$\mathrm{Kom}^{K+J} G = \{e\}$$

as to be shown. □

Corollary 44. *If the group G has a subnormal series for which all quotient groups are solvable, then the group is solvable.*

Theorem 37 (Burnside's Theorem). *A group the order of which is the product of two prime powers, is solvable.*

Let $|G| = p^a q^b$ with primes p, q and any integers $a, b \geq 0$. The case $p = q$ or $a = 0$ or $b = 0$ was confirmed by proposition 82 above. We may assume $p \neq q$, $a \geq 1$ and $b \geq 1$.

Proceeding by contradiction, we assume there exists a minimal counterexample. That is a group G of minimal order $p^a q^b$ which is not solvable. This group has to be simple, because of proposition 84. Moreover holds $a \geq 1$, $b \geq 1$, and $|G| \geq 6$.

We shall use the character table of the group. To this end, let h be the number of conjugacy classes in G and $\chi_1, \ldots, \chi_h$ be the list of inequivalent irreducible characters of G over $\mathbb{C}$. Here $\chi_1 \equiv 1$ denotes the unit representation. The character table produces the orthonormal matrix

$$\text{(VIII.4.1)} \qquad \frac{\sqrt{|C_\alpha|}}{\sqrt{|G|}} \chi_\beta(g_\alpha)$$

with $1 \leq \alpha \leq h$ enumerating the classes and $1 \leq \beta \leq h$ the representations

Here $g_\alpha \in C_\alpha$ is any group element from the conjugacy class C_α. Because $\rho(g)^{|G|} = \rho(g^{|G|}) = \mathcal{I}d_n$ holds for any n-dimensional representation ρ, we see that the character values $\chi(g)$ are sums of $|G|$-th roots of unity, and hence the characters are algebraic integers. More astonishing is the following fact.

Claim. Let q be any prime. For each $g \neq e$ there exists an irreducible group representation ρ different from the unit representation, such that $\chi(g) \neq 0$ and the dimension $n = \chi(e)$ is not divisible by q.

Proof. The orthogonality of the columns gives

$$\sum_{2 \leq \beta \leq h} \frac{\chi_\beta(e)}{q} \chi_\beta(g) = \frac{-1}{q} + \frac{1}{q} \sum_{1 \leq \beta \leq h} \chi_\beta(e) \chi_\beta(g) = \frac{-1}{q} \quad \text{for all } g \neq e$$

The $\chi_\beta(g)$ are sums of roots of unity and hence algebraic integers. Since $\chi_\beta(e)$ is the dimension of the representation, it is a positive integer.

If all terms $\frac{\chi_\beta(e)}{q}$ for $2 \leq \beta \leq h$ and $\chi_\beta(g) \neq 0$ would be integers, the left-hand side would be an algebraic integer. But $-1/q$ is not an algebraic integer. Hence there exists a term $\chi_\beta(g) \neq 0$ and $2 \leq \beta \leq h$ which is not an algebraic integer. $\square$

Here are a few remarks about representations of finite groups. The set $\mathbb{C}^G$ of complex valued functions on the group gets an algebra with its natural vector space structure and the convolution $\star$ as multiplication. This is the natural extension of the group multiplication to $\mathbb{C}^G$. I mention the nice formula

$$(f \star g)(t) := \sum_{s \in G} f(s)g(s^{-1}t) = \sum_{s \in G} f(ts^{-1})g(s)$$

The center of the group *algebra* consists of the functions which are constant on the conjugacy classes of the group. It has the dimension h and is spanned by the quantities

$$z_\alpha := \sum_{g \in C_\alpha} g \quad \text{for } 1 \le \alpha \le h$$

For any member k in the class C_γ let $a_{\alpha,\beta,\gamma}$ be the number of possibilities to write k as a product $k = fg$ with $f \in C_\alpha$ and $g \in C_\beta$. This count is constant over all $k \in C_\gamma$ since $sks^{-1} = sfs^{-1}sgs^{-1}$ gives the corresponding product for the equivalent member sks^{-1}. For the numbers $a_{\alpha,\beta,\gamma}$ holds

$$C_\alpha C_\beta = \sum_\gamma a_{\alpha,\beta,\gamma}\, C_\gamma$$

(VIII.4.2)
$$z_\alpha \star z_\beta = \sum_\gamma a_{\alpha,\beta,\gamma}\, z_\gamma$$

as equations for *multisets* and for the corresponding sums.

Claim. Take any irreducible representation ρ, with character χ and dimension n. We claim that $|C_g|\chi(g)/n$ is an algebraic integer for all $g \in G$.

Proof. We may assume that ρ is not the unit representation, otherwise the claim is true anyway. Too, we may assume that $\chi(g) \ne 0$ and $g \ne e$, otherwise the claim is true anyway. The representation mapping ρ is naturally extended to a mapping from the group algebra $\mathbb{C}^G$ to the linear mappings $\mathbb{C}^n \mapsto \mathbb{C}^n$. One simply defines

$$\rho\left(\sum_{g \in G} f(g)g\right) := \sum_{g \in G} f(g)\rho(g)$$

I keep the notation ρ for the extension. This is an homomorphism of algebras since

$$\rho(f) \circ \rho(g) = \rho(f \star g) \quad \text{for all } f, g \in \mathbb{C}^G$$

Use the notation $\rho = \rho_\alpha$ for some index $2 \le \alpha \le h$ and choose g such that $\chi(g) \ne 0$.

Let $k := \sum_{h \simeq g} \rho(h)$ and hence $k = \rho(z_\alpha)$. Now $\chi(g) \neq 0$ implies $\operatorname{trace} k = |C_g|\chi(g) \neq 0$ and $k \neq 0$. It is easy to check that

$$\sum_{h \simeq g} \rho(s)\rho(h)\rho(s^{-1}) = \sum_{h \simeq g} \rho(h)$$

$$\rho(s) \sum_{h \simeq g} \rho(h) = \sum_{h \simeq g} \rho(h)\rho(s)$$

holds for all $s \in G$. Since the representation ρ is irreducible, the (easy part of the) Lemma of Schur implies there exists λ such that

$$\rho(z_\alpha) = \sum_{h \sim g} \rho(h) = \lambda\,\mathcal{I}d_n$$

Take traces to obtain

(VIII.4.3) $$|C_g|\chi_\alpha(g) = n \cdot \lambda$$

On the other hand the relation (VIII.4.2) tells

$$\rho(z_\alpha) \circ \rho(z_\beta) = \sum_\gamma a_{\alpha,\beta,\gamma}\, \rho(z_\gamma)$$

$$\lambda\rho(z_\beta) = \sum_\gamma a_{\alpha,\beta,\gamma}\, \rho(z_\gamma)$$

Since $k = \rho(z_\alpha) \neq 0$ the latter equations with $1 \leq \beta \leq h$ are an eigenvalue problem for the matrix $a_{\alpha,\cdot,\cdot}$ with eigenvector $\rho(z_\cdot)$ and eigenvalue λ. We get the eigenvalue λ, which is an eigenvalue of an integer valued matrix, and hence an algebraic integer. Because of equation (VIII.4.3), we see that $|C_g|\chi(g)/n$ is an algebraic integer. $\qquad\square$

Claim. Assume that a minimal counterexample exists. There exists a group element $g \neq e$ and there exists a irreducible representation different from the unit representation, of dimension n such that $0 \neq \chi(g)/n$ is an algebraic integer.

Proof. To get rid of the factor $|C_g|$, we need to invoke the Sylow Theorem. The group G of order $p^a q^b$ has a Sylow subgroup S of order p^a. Only very rarely, this happens to be a normal divisor. But by proposition 82 above, this subgroup has a nontrivial center. This center contains an element $g \neq e$ of order p. The group element g commutes with all elements of S. The normalizer $N_g = \{h \in G : hg = gh\}$ contains the subgroup S. Hence $|N_g| = p^a q^c$ with $0 \leq c \leq b$. One puts $d = b - c \geq 0$. The size of the conjugacy class is

$$|C_g| = \frac{|G|}{|N_g|} = q^{b-c} = q^d$$

Since $g \neq e$ the claim (VIII.4) shows existence of an irreducible group representation ρ different from the unit representation, such that $\chi(g) \neq 0$ and q does not divide the dimension $n = \chi(e)$.

By claim (VIII.4), $|C_g|\chi(g)/n$ is an algebraic integer. We have now confirmed that $q^d\chi(g)/n = |C_g|\chi(g)/n = \lambda$ is an algebraic integer.

Since the prime q does not divide n, we see q^d and n are relatively prime. The extended Euclidean algorithm yields (rational) integers $x, y \in \mathbf{Z}$ such that $xq^d + yn = 1$. Multiply by $\frac{\chi(g)}{n}$ to get

$$\frac{\chi(g)}{n} = x\frac{q^d\chi(g)}{n} + y\chi(g)$$

The characters of any representation are algebraic integers. Indeed, they are sums of eigenvalues, which is turns are $|G|$-th roots of unity and hence all algebraic integers. We see that the right-hand side of the above equation is an integer combination of algebraic integers, and hence an algebraic integer. $\qquad\square$

Proposition 85. *Given is any irreducible representation ρ of a finite group G. Let $n = \rho(e)$ be its dimension. If for any g the fraction $0 \neq \frac{\chi(g)}{n}$ is an algebraic integer, then $\rho(g)$ is a multiple of the identity $\mathcal{I}d_n$, and necessarily $\rho(g) = \frac{\chi(g)}{n}\mathcal{I}d_n$.*

Proof. By assumption $\alpha = \frac{\chi(g)}{n}$ is an algebraic integer. Let m be its monic minimal polynomial. Let $\lambda_1, \ldots, \lambda_n$ be the eigenvalues of $\rho(g)$, not necessarily distinct. As already above, we use that the λ_i are $|G|$-th roots of unity and hence algebraic integers, and $|\lambda_i| = 1$. The trace property and the triangle inequality yield

(VIII.4.4) $$\alpha = \frac{\lambda_1 + \cdots + \lambda_n}{n}$$

(VIII.4.5) $$|\alpha| \leq \frac{|\lambda_1| + \cdots + |\lambda_n|}{n} = 1$$

Let $L/\mathbf{Q}$ be a Galois extension in which the polynomial m and the minimal polynomials of all λ_i for $1 \leq i \leq n$ split. Since the polynomial m is irreducible, for each algebraic conjugate α_j, there exists an automorphism $\sigma_j \in \mathrm{Aut}[L, \mathbf{Q}]$ such that $\alpha_j = \sigma_j(\alpha)$. The existence of such an conveniently extended automorphism is guaranteed by the splitting proposition 70.

398

We apply the automorphism σ_j on both sides of the trace equation (VIII.4.4). The images $\sigma_j(\lambda_i)$ are again $|G|$-th roots of unity and hence algebraic integers with absolute value 1. Hence corresponding to estimate (VIII.4.5) are obtained

$$(\text{VIII.4.6}) \qquad \sigma_j(\alpha) = \frac{\sigma_j(\lambda_1) + \cdots + \sigma_j(\lambda_n)}{n}$$

$$(\text{VIII.4.7}) \qquad |\sigma_j(\alpha)| \leq \frac{|\sigma_j(\lambda_1)| + \cdots + |\sigma_j(\lambda_n)|}{n} = 1$$

We multiply the latter estimates for $1 \leq j \leq \deg m$ and obtain for the norm of α

$$(\text{VIII.4.8}) \qquad |N(\alpha)| = \prod_{1 \leq j \leq \deg m} |\sigma_j(\alpha)| \leq 1$$

By assumption α is an algebraic integer. By theorem 25 the norm of an algebraic integer,—over the field $\mathbf{Q}$,—is an integer. Moreover holds $\alpha \neq 0$ by assumption. Hence we conclude that $N(\alpha) = \pm 1$ and $|N(\alpha)| = 1$. Hence estimate (VIII.4.8) is indeed an equality. Now estimate (VIII.4.7) implies that equalities $|\sigma_j(\alpha)| = 1$ for $1 \leq j \leq \deg m$.

Hence all the estimates (VIII.4.7) are indeed equalities. Especially with $j = 1$ we get the equality sign in the triangle inequality:

$$|\lambda_1 + \cdots + \lambda_n| = |\lambda_1| + \cdots + |\lambda_n|$$

It is well known that the equality holds for the triangle inequalities among complex numbers if and only if all summands are nonnegative multiples of the same quantity. Since we know additionally that $|\lambda_i| = 1$ for all $1 \leq i \leq n$, we conclude that all the eigenvalues λ_i are equal. Hence the representation value $\rho(g)$ is a nonzero multiple of the identity $\mathcal{I}d_n$, as claimed. $\qquad \square$

End of the proof of the Burnside Theorem. Proceeding by contradiction, we assume there exists a minimal counterexample. That is a group G of minimal order $p^a q^b$ which is not solvable. This group has to be simple, because of proposition 84. Moreover holds $a \geq 1$ and $b \geq 1$.

Take the group element $g \neq e$ and the irreducible representation ρ different from the unit representation as in claim VIII.4.

Let $N = \ker(\rho)$. Since the group G is simple and ρ is not the unit representation holds $N = \{e\}$. The Main Theorem about group homomorphisms tells that $\text{Im}(\rho) \simeq G$. For any group element g with $\rho(g) = \mu \mathcal{I}d_n$ holds $\rho(s)\rho(g) = \rho(g)\rho(s)$ for all $d \in G$. In other words, $\rho(g)$ lies in the center of $\text{Im}(\rho)$.

- Since $\mathrm{Im}(\rho) \simeq G$ is a simple group, we conclude that $\rho(g) \neq \mu \mathcal{I}d_n$ for all $g \neq e$.

- By claim VIII.4, there exists a group element $g \neq e$ and there exists an irreducible representation different from the unit representation, of dimension n such that $0 \neq \chi(g)/n$ is an algebraic integer.

  By proposition 85, the latter claim implies that $\rho(g)$ is a multiple of the identity $\mathcal{I}d_n$, and necessarily $\rho(g) = \frac{\chi(g)}{n}\mathcal{I}d_n$.

We have a contradiction. There does not exist a minimal counterexample to Burnside's Theorem,—the proof is complete. $\qquad\square$

VIII.5   Two-Marked Ruler and Equations of Fifth Order

Theorem 38 (Baragar). *There exist fifth order polynomials with integer coefficients, the zeros of which can be constructed with compass and two-marked straightedge but not with radicals. Especially, the roots can be neither obtained by solid construction, as specified by definition 36.*

Here are the rules for constructions with compass and two-marked straightedge. The compass and straightedge are allowed to be used in the classical way. But additionally, there are two marked points P and Q on the straightedge with distance $|PQ| = 1$, and the straightedge may be used in the following way. One puts the two marks onto already constructed circles or straight lines and let the straightedge pass through an already constructed point, to get the newly constructed points P and Q at the marks on the straightedge.

For the explanation of such constructions, it is useful to introduce a special curve, called the *conchoid*. For a given point O and guiding line l, not passing through point O, the points of the conchoid are obtained by the second marked point Q on a two-marked straightedge, which passes through the point O and has the first marked point P lying on the guiding line. To obtain an equation for the conchoid, we take the origin of the Cartesian coordinate system to be the point O, and the guiding line to be the line with equation $x = a$. In polar coordinates (r, ϕ), the guiding line has the equation $r = a/\cos\phi$. The second marked point Q has the distance $|OQ| = \pm 1 + a/\cos\phi$ from the origin. The equation for the conchoid in polar coordinates is hence

$$r = \pm 1 + \frac{a}{\cos\phi}$$

To obtain the equation in Cartesian coordinates (x, y), we use the formulas $x = r \cos \phi$, $y = r \sin \phi$. From $r = \pm 1 + ar/x$ is obtained $r(x - a) = \pm 1$, $r^2(x - a)^2 = 1$, and finally

(VIII.5.1)
$$(x - a)^2(x^2 + y^2) = 1$$

as the equation of the conchoid in Cartesian coordinates.

To obtain examples addressed in the above theorem 38 of Baragar. one constructs a circle with center Z with Cartesian coordinates (b, c) and radius s. The Cartesian equation of this circle is

(VIII.5.2)
$$(x - b)^2 + (y - a)^2 = s^2$$

We shall take integer values for the numbers b, c, s. By means of the two-marked straightedge, points on the circle are constructed in the following way. One puts one mark of straightedge onto the guiding line $x = a$, and other mark onto the circle, and let the straightedge pass through the origin O. The Cartesian coordinates of the points constructible in this way are the solutions (x, y) of the system with the two equations (VIII.5.1) and (VIII.5.2). From the second equation is obtained $x^2 + y^2 = 2bx + 2cy + s^2 - b^2 - c^2$. We plug this expression into the first equation (VIII.5.1) and obtain an equation that can be solved for y

$$(x - a)^2(2bx + 2cy + s^2 - b^2 - c^2) = 1$$

$$y =$$

$$\frac{a^2(b^2 + c^2 - s^2) + x\left(-2a^2b - 2ab^2 - 2ac^2 + 2as^2\right) + x^2\left(4ab + b^2 + c^2 - s^2 + 1\right) - 2bx^3}{2c(a - x)^2}$$

The latter expression for y may be plugged into the equation (VIII.5.2) for the circle. Thus is obtained an equation of sixth order with the only unknown x.

To reduce to an equation of only degree five, Baragar suggests to choose the radius of the circle such that it passes through the point with coordinates $(a + 1, 0)$. Since this point lies on the conchoid, too, we get one solution $x = a + 1$ and the factor $(x - a - 1)$ for the equation of sixth order. Baragar's restriction leads to the equation

$$(a + 1 - b)^2 + c^2 = s^2$$

and hence $b^2 + c^2 - s^2 = b^2 - (b - a - 1)^2 = (1 + a)(-1 - a + 2b)$. This value for s is constructible for all rational a, b, c. One may now obtain an equation of fifth degree for the x-coordinate of the remaining intersection points of the conchoid and the circle, which are the points obtained by our construction with two-marked straightedge. I do not want to bore the reader with the entire general computation, but only give two examples.

- Baragar's choice in the article [4] is $a = 2, b = c = 1$ and hence $b^2 + c^2 - s^2 = (1+a)(-1-a+2b) = -3$ and $s = \sqrt{5}$. One obtains

$$(x-2)^2(2x + 2y + 3) = 1$$

$$y = -\frac{(x-3)(x^2-2)}{(x-2)^2}$$

One plugs into the equation $(x-1)^2 + (y-1)^2 = 5$ of the circle and obtains

$$\frac{2(x-3)\left(x^5 - 4x^4 + 2x^3 + 4x^2 + 2x - 6\right)}{(x-2)^4} = 0$$

Thus is obtained an equation of sixth order with the only unknown x. Besides the obviously constructed solution $x = 3$, we have five real or complex solution of the factor $x^5 - 4x^4 + 2x^3 + 4x^2 + 2x - 6$.

Lemma 99. *The polynomial $f(x) = x^5 - 4x^4 + 2x^3 + 4x^2 + 2x - 6$ is irreducible. It has three real and two conjugate complex solutions.*

Reason. The irreducibility can be checked with mathematica. Too, we may use the Eisenstein criterium 37. The polynomial $f(x)$ is monic. All its coefficients except for the leading one are divisible by the prime $p = 2$. The constant coefficient is not divisible by the square p^2. Hence the Eisenstein criterium yields that the polynomial is irreducible.

Descartes rule of signs shows that there are at most three positive and no negative real solutions. Geometrically, one sees that there exist three real solutions. Alternatively, one checks that $f(1) = -1 < 0$, $f(1.4) > 0, f(2) = -2 < 0$, $f(3) = 9 > 0$. Hence the intermediate value theorem tells that there are at least three positive solution. Taken together, we conclude that there are exactly three real solution of $f(x) = 0$. The existence of exactly one pair of conjugate complex solutions follows now from the Fundamental Theorem of Algebra. $\square$

My second choice is $a = 2, b = 0, c = 1$ and hence $b^2 + c^2 - s^2 = (1+a)(-1 - a + 2b) = -9$ and $s = \sqrt{10}$. One obtains

$$(x-2)^2(2y + 9) = 1$$

$$y = -\frac{2(x-3)(2x-3)}{(x-2)^2}$$

One plugs into the equation $x^2 + (y-1)^2 = 10$ of the circle and obtains

$$\frac{(x-3)\left(x^5 - 5x^4 + 24x^3 - 100x^2 + 180x - 108\right)}{(x-2)^4} = 0$$

Thus is obtained an equation of sixth order with the only unknown x. Besides the obviously constructed solution $x = 3$, we have five real or complex solution of the factor $g(x) = x^5 - 5x^4 + 24x^3 - 100x^2 + 180x - 108$.

Lemma 100. *The polynomial $g(x)$ is irreducible. It has three real and two conjugate complex solutions.*

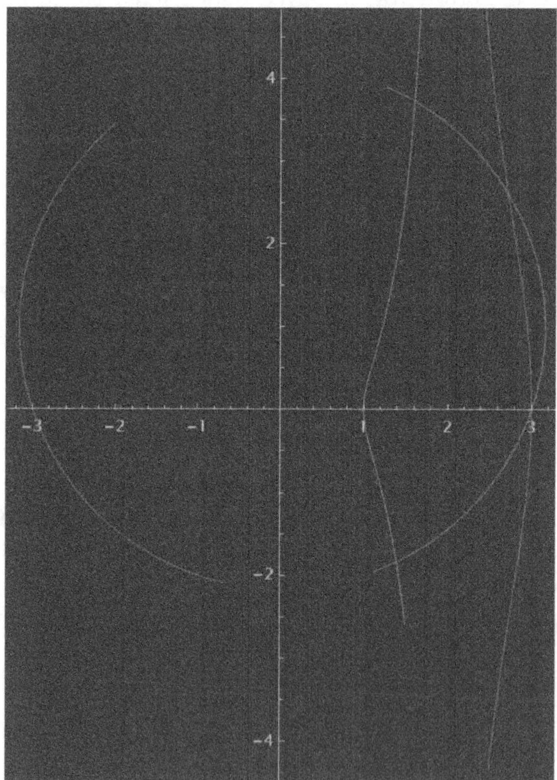

Figure 6: A two-marked ruler construction yields points not given by radicals

The proof of this lemma is easy done with mathematica and is left to the reader. The intersecting conchoid and circle are shown in the figure on page 402.

End of the proof of theorem 38. Since the polynomial is irreducible, the Galois group $G \subseteq \mathcal{S}_5$ of the polynomial $f(x)$ or $g(x)$ over the rational field acts transitively on the set X of the five complex roots. By problem 160, $5 = |X|$ is a divisor of the group order $|G|$. Since 5 is a prime number Cauchy's theorem 76 tells the Galois group contains an element of order 5, Since there exists exactly one pair of conjugate complex solutions, the complex conjugation is a transposition in the Galois group.

One easily checks that only subgroup of the symmetric group $\mathcal{S}_5$ which contains both a transposition and an element of order 5 is the entire symmetric group $\mathcal{S}_5$. We see that the Galois group of the polynomial $f(x)$ is the entire symmetric group $\mathcal{S}_5$. We know that this group is not solvable. By Abel's Main Theorem, the roots of the polynomial $f(x)$ are not solvable by radicals. Especially, the roots can be neither obtained by solid construction, as specified by definition 36. $\square$

Theorem 39 (Abel's Main Theorem). *Let F be a field of characteristic zero. A polynomial $f \in F[x]$ can be solved by radicals if and only if the Galois group $\mathrm{Gal}(f, F)$ of the polynomial f over the field F is a solvable group, as defined in definition 58.*

Let F be a field of prime characteristic zero. If a polynomial $f \in F[x]$ can be solved by radicals then the Galois group $\mathrm{Gal}(f, F)$ of the polynomial f over the field F is a solvable group, Almost conversely, If the Galois group $\mathrm{Gal}(f, F)$ is a solvable group, the polynomial $f \in F[x]$ can be solved by radicals together with modified radicals.

Modified radials are solutions of the equation $x^p - x = a$. This theorem is stated without its proof. The case of characteristic zero is proofed by Gerd Fischer [8]. The main steps of the general case are contained in Artin [1].

VIII.6   Get More Polynomials not Solvable By Radicals

Abel's Main Theorem is the salient step to get polynomials the roots of which cannot be solved by radicals. Since the Galois group is a subgroup of a symmetric group, and acts on the roots of the polynomial, the following proposition is useful to pin down actual examples for polynomial the roots of which cannot be solved with radicals.

Proposition 86 (Galois' old lemma). *Let p be a prime, $G < \mathcal{S}_p$ be a transitive subgroup of the symmetric group. If the group G is solvable, then one may renumber the digits $\{0, 1, \ldots, p-1\}$ such that the permutations in G are all affine transformations $i \mapsto a \cdot i + b \mod p$.*

Lemma 101. *Any normal divisor of a transitive group acting on a set X of prime order is either the one-element group, or it is transitive, too.*

Proof of the lemma. Let G be a transitive group acting on the set X. Because of transitivity the group orbit of any digit is the entire set X. The Bahn lemma implies that $|X|$ is a divisor of the group order $|G|$. Hence $p = |X|$ is a divisor of the group order $|G|$. By Cauchy's theorem, a cyclic group G_1 of order p is a subgroup of G. We may now renumber the set X in a manner such that the permutation $\sigma : i \mapsto i+1$ mod p generates G_1. Take the group orbit $Y \subseteq X$ for the subgroup H which contain the digit 0. Hence $Y = H(0)$. For any $s \in G$ holds $sHs^{-1} = H$ because H is assumed to be a normal subgroup. The set

$$sY = sH(0) = Hs(0)$$

is a group orbit, too, and has the same cardinality as Y. Hence X is partitioned into group orbits of H which have all the same cardinality. Hence $|Y|$ is a divisor of the prime $|X| = p$. Hence holds either $|Y| = 1$, in which case H is the one-element group, or $Y = X$, in which case H is transitive. $\qquad\square$

Lemma 102. *Let N be normal divisor of G and C be the commutator group of N. Then C a normal divisor of G.*

Proof. Take any $s \in G$ and $n, m \in N$.

$$s\,nmn^{-1}m^{-1}\,s^{-1} = n_s m_s n_s^{-1} m_s^{-1}$$

holds with $n_s = sns^{-1}, m_s = sms^{-1}$. Hence for any element $c = nmn^{-1}m^{-1} \in C$ holds $scs^{-1} \in C$. The commutator group of N is generated by these elements. Hence for all elements x in the commutator, the entire equivalence class sxs^{-1} with arbitrary $s \in G$ is contained in the commutator C. Hence C is a normal divisor of G. $\qquad\square$

Corollary 45. *For any solvable G, the iterated commutator groups produce not only a subnormal, but even a normal series. In other words, all subgroups in the series are normal divisors of the original group G.*

Proof. One begins with the group G. One iterates the commutator groups. Recall that the commutator group $\mathrm{Kom}(G)$ of any group G is the subgroup generated by the elements $ghg^{-1}h^{-1}$ with $g, h \in G$. For $k = 0, 1, 2, \ldots$ one defines $\mathrm{Kom}^0(G) = G$ and $\mathrm{Kom}^{k+1} = \mathrm{Kom}(\mathrm{Kom}^k(G))$. The commutator group is a normal divisor. It is the smallest normal divisor for which the quotient group is commutative. Since the group G is assumed to be solvable, one arrives with the iteration at smaller and smaller groups, and reaches the one element group $\{e\}$ after finitely many steps. By construction, this is only a subnormal series. The Lemma 102 implies inductively that any intermediate subgroup is not only a subgroup of the next larger subgroup, but even a normal subgroup of the original group G. $\qquad\square$

Proof of Galois' old lemma. Let the group G act on the digits $X = \{0, 1, \ldots, p-1\}$. We shall calculate with these digits modulo the prime p. We may built the *normal* series in the following form:

$$\{e\} =: G_0 \trianglelefteq G_2 \trianglelefteq \cdots \trianglelefteq G_r := G$$

by iterating the commutator group, as explained in the corollary 45 above. Moreover the group G_2 obtained in the second last step is a commutative group. By the lemma 101 G_2 is transitive. Hence the group order of any digit is the entire set X. The Bahn lemma implies that $|X|$ is a divisor of the group order $|G_2|$. Hence the prime $p = |X|$ divides the group order $|G_2|$. By Cauchy's theorem, a cyclic group G_1 of order p is a subgroup of G_2. We insert the group G_1 into the above obtained normal series. Because of the commutativity of G_2, the new series is normal, too. We may now renumber the set X in a manner such that the permutation $\sigma : i \mapsto i+1$ mod p generates G_1.

Remark (Not needed). The factor group G_2/G_1 has an order which is a divisor of $|G_r|/p$ and hence of $p!/p$ because G_r is a subgroup of the symmetric group $\mathcal{S}_p$. The order of this symmetric group is not divisible by p^2, but only by primes smaller than p. Hence the order of G_2/G_1 is not divisible by p.

Take any $\tau \in G_2$. Since G_2 is commutative holds

$$\tau\sigma(i) = \sigma\tau(i) \ \text{ for all } i \in X$$
$$\tau(i+1) = \tau(i) + 1 \quad \bmod p$$

with the addition to be understood modulo p. Inductively one gets

$$\tau(i+j) = \tau(i) + j \quad \bmod p \ \text{ for all } i, j \in X$$

We may simply restrict to $i = 0$, and see that each permutation in G_2 is an affine transformation, indeed with $a = \tau(0), b = 1$.

Take any $\tau \in G_r$. Since G_2 is a normal divisor of G_r there exists γ such that

$$\tau\sigma\tau^{-1} = \gamma \in G_2$$
$$\tau\sigma(i) = \gamma\tau(i) \ \text{ for all } i \in X$$

Since $\gamma \in G_2$, we have checked that $\gamma(j) = a + j$ holds for all $j \in X$, and we obtain with $j = \sigma(i) = i+1$

$$\tau(i+1) = \tau(i) + a \quad \bmod p \ \text{ for all } i \in X$$

with the addition to be understood modulo p. Inductively one gets

$$\tau(i+j) = \tau(i) + j \cdot a \quad \mod p \ \text{ for all } i,j \in X$$

We may simply restrict to $i = 0$, and see that the permutation in G_r is an affine transformation, indeed now with $b := a, a := \tau(0)$. $\qquad\square$

Remark. The induction is not needed in the main proof, but hidden in the general Corollary about solvable groups. I took the pain to work out these details because I found the explanations in the otherwise excellent book [1] by Artin not sufficient to lead to a proof. A remark on the internet tells that the result goes back to Galois himself, and was one of the motivations for his invention of finite fields.

Theorem 40. *For an irreducible separable polynomial of prime degree which is solvable by radicals, the splitting field is generated by any pair of distinct roots.*

Proof. Take two roots $\alpha_0 \neq \alpha_1$ of the solvable polynomial P. Let F be the base field and E be the splitting extension of P. Since the polynomial is separable, it is a Galois extension. By Abel's Main Theorem, the Galois group $\mathrm{Aut}(E, F)$ is solvable.

Since the polynomial is irreducible, the Galois group $\mathrm{Aut}(E, F)$ acts transitively on the set of the p roots. By assumption these roots are simple.

By Galois' old lemma proposition 86, all group elements are affine transformation modulo p. Especially, they have either no or one fixed point, or are equal to the identity.

The group $\mathrm{Aut}(E, F(\alpha_0, \alpha_1))$ is a subgroup, and all elements of this subgroup have at least *two* fixed points. Hence the subgroup is $\mathrm{Aut}(E, F(\alpha_0, \alpha_1)) = \{e\}$ the one element group, and hence $E = F(\alpha_0, \alpha_1)$. The splitting field is generated by any two roots, as has been claimed. $\qquad\square$

Corollary 46. *A real polynomial of prime degree, which is irreducible, has at least two real roots, and is solvable by radicals, has only real roots.*

Proof. Let $\alpha_0 \neq \alpha_1$ be two distinct real roots. If there would exist complex roots, the complex conjugation τ would generate an element of the Galois group different from the identity with two fixed points $\alpha_0 \neq \alpha_1$. Such a permutation cannot be an affine mapping modulo p. Hence there exist no complex roots. $\qquad\square$

Corollary 47. *A real polynomial of prime degree, which is irreducible, has at least two real roots, and has complex roots, too, is not solvable.*

VIII.7   About the General Polynomial Equation

Proposition 87 (Viëta's formulas). *Let K be any field and $f \in K[x]$ any monic polynomial of degree* $\deg f = n \geq 1$

$$f = \sum_{0 \leq r < n} (-1)^{n-r} a_{n-r}\, x^r + x^n$$

Let x_r for $1 \leq r \leq n$ be the zeros in any extension $K(x_1, \ldots, x_n)$. The coefficients and zeros are related by Viëta's formulas

$$\text{(VIII.7.1)} \qquad a_r = \sum_{1 \leq i_1 < i_2 < \cdots < i_r \leq n} x_{i_1} \cdots x_{i_r} \quad \text{for } 1 \leq r \leq n.$$

The r-th symmetric polynomial in the zeros of the polynomial f equals $(-1)^{n-r}$ times the coefficient of the power x^{n-r}.

Concerning a formula giving the zeros of a polynomial, the (too ambiguous) goal is to cover the case of arbitrary coefficients. We have to formalize the idea of a <u>general</u> polynomial $f \in K[x]$. Let $\mathcal{R}at(K; S_1, \ldots, S_n)$ be the field of rational functions with the variables S_r for $1 \leq r \leq n$ over the ground field K. To get the general polynomial, we use the polynomial ring $\mathcal{R}at(K; S_1, \ldots, S_n)[X]$. In this ring we define

$$\text{(VIII.7.2)} \qquad f := \sum_{0 \leq r < n} (-1)^{n-r} S_{n-r} X^r + X^n$$

to be the *polynomial with general coefficients.*
In the splitting field $E := \mathcal{R}at(K; S_1, \ldots, S_n)(x_1, \ldots, x_n)$ [28] one gets the zeros x_i and hence the factoring

$$f = \prod_{1 \leq i \leq n} (X - x_i)$$

Viëta's formulas now yield

$$\text{(VIII.7.3)} \qquad S_r = \sum_{1 \leq i_1 < i_2 < \cdots < i_r \leq n} x_{i_1} \cdots x_{i_r} \quad \text{for } 1 \leq r \leq n$$

Note that in this subsection I use capital letters for any indeterminate, but lower case letters for elements of any field.

---

[28] I insist to write, a bid awkwardly $E = \mathcal{R}at(K; S_1, \ldots, S_n)(x_1, \ldots, x_n)$. Most authors write simply $K(x_1, \ldots, x_n)$, I find this confusing.

Theorem 41 (The Galois group of the general polynomial). *The polynomial with general coefficients $f \in \mathcal{R}at(K; S_1, \ldots, S_n)[X]$, given by formula (VIII.7.2) has the Galois group*

$$\text{(VIII.7.4)} \qquad \text{Gal}(f : \mathcal{R}at(K; S_1, \ldots, S_n)) = \text{Aut}(E : \mathcal{R}at(K; S_1, \ldots, S_n)) \simeq \mathcal{S}_n$$

isomorphic to the symmetric group for n elements.

Developing the ideas. As a clever tool is defined a second polynomial $g \in \mathcal{R}at(K; X_1, \ldots, X_n)[X]$ to be

$$\text{(VIII.7.5)} \qquad g := \prod_{1 \leq i \leq n} (X - X_i)$$

Here the zeros X_i appear as new indeterminate. The polynomial g is called the polynomial with <u>general zeros</u>. Let

$$\text{(VIII.0.1)} \qquad \mathfrak{s}_r = \sum_{1 \leq i_1 < i_2 < \cdots < i_r \leq n} X_{i_1} \cdots X_{i_r} \ \text{ for } 0 \leq r \leq n$$

be the elementary symmetric polynomials in n variables. One puts $\mathfrak{s}_0 = 1$. The $\mathfrak{s}_r$ are neither indeterminate nor only elements of some field. Instead they act as expression <u>to be substituted</u>, thus introducing the general zeros X_i into any respective expression $h \in \mathcal{R}at(K; S_1, \ldots, S_n)$ or $h \in K[S_1, \ldots, S_n]$.

Let $L \subseteq \mathcal{R}at_{sym}(K; X_1, \ldots, X_n)$ be the field of symmetric polynomials, with the indeterminates X_i, and consisting of the expressions $h(\mathfrak{s}_1, \ldots, \mathfrak{s}_n)$ with arbitrary $h \in \mathcal{R}at(K; S_1, \ldots, S_n)$.
We denote this field by $L := \mathcal{R}at(K; \mathfrak{s}_1, \ldots, \mathfrak{s}_n)$.

Proposition 88. *The polynomial with <u>general zeros</u> by formula (VIII.7.5) is*

$$\text{(VIII.7.6)} \qquad g = \sum_{0 \leq r \leq n} (-1)^{n-r} \mathfrak{s}_{n-r} X^r$$

and hence indeed a polynomial from the ring $\mathcal{R}at(K; \mathfrak{s}_1, \ldots, \mathfrak{s}_n)[X]$.

Proposition 89. *The extension $\mathcal{R}at(K; X_1, \ldots, X_n)/\mathcal{R}at(K; \mathfrak{s}_1, \ldots, \mathfrak{s}_n)$ is Galois with*

$$\dim[\mathcal{R}at(K; X_1, \ldots, X_n) : \mathcal{R}at(K; \mathfrak{s}_1, \ldots, \mathfrak{s}_n)] = n!$$

and its group of automorphisms is

$$\text{(VIII.7.7)} \qquad \text{Gal}(g : \mathcal{R}at(K; \mathfrak{s}_1, \ldots, \mathfrak{s}_n))$$
$$= \text{Aut}(\mathcal{R}at(K; X_1, \ldots, X_n) : \mathcal{R}at(K; \mathfrak{s}_1, \ldots, \mathfrak{s}_n)) \simeq \mathcal{S}_n$$

Proof. The symmetric group $\mathcal{S}_n$ acts on the list of zeros $X_1, \ldots, X_n$ by permutation, and similarly on the elements of the field $\mathcal{R}at(K; X_1, \ldots, X_n)$.

Let $S := \mathcal{R}at_{sym}(K; X_1, \ldots, X_n)$ be the field of symmetric rational function with the variables X_i for $1 \leq i \leq n$ over the ground field K. As a formula we may state

$$(\text{VIII.7.8}) \qquad \mathrm{Fix}(\mathcal{R}at(K; X_1, \ldots, X_n) : \mathcal{S}_n) = \mathcal{R}at_{sym}(K; X_1, \ldots, X_n)$$

Problem 189. *Convince yourself that from the action of the symmetric group $\mathcal{S}_n$ acting on the elements of the field $\mathcal{R}at(K; X_1, \ldots, X_n)$ follows*

$$(\text{VIII.7.9}) \qquad \mathrm{Aut}(\mathcal{R}at(K; X_1, \ldots, X_n) : S) \simeq \mathcal{S}_n$$

Because of (VIII.7.9) and the fact (VIII.7.8), via Artin's theorem 29 we get

$$|\mathcal{S}_n| = \dim[\mathcal{R}at(K; X_1, \ldots, X_n) : S]$$

On the other hand, from the elementary symmetric function we have constructed the (possibly smaller) subfield $L := \mathcal{R}at(K; \mathfrak{s}_1, \ldots, \mathfrak{s}_n) \subseteq S$.

By proposition 88 the polynomial with *general zeros* is really in the (smaller) ring $g \in L[X]$. Via the formula (VIII.7.5), we see that g splits in the field extension $\mathcal{R}at(K; X_1, \ldots, X_n)/L$, but no smaller extension of L. Moreover it has simple zeros and hence g is separable. From the definition 51 item(b) one concludes that the extension $\mathcal{R}at(K; X_1, \ldots, X_n)/L$ is Galois. Hence from item (a)

$$|\dim[\mathcal{R}at(K; X_1, \ldots, X_n) : L]| = |\mathrm{Aut}(\mathcal{R}at(K; X_1, \ldots, X_n) : L)|$$

Any automorphism $\mathrm{Aut}(\mathcal{R}at(K; X_1, \ldots, X_n) : L)$ is uniquely determined by its permutating action on the roots X_i. Hence the order of the group is at most

$$|\mathrm{Aut}(\mathcal{R}at(K; X_1, \ldots, X_n) : L)| \leq |\mathcal{S}_n| = n!$$

We get the sharp bounds

$$\dim[\mathcal{R}at(K; X_1, \ldots, X_n) : L] = |\mathrm{Aut}(\mathcal{R}at(K; X_1, \ldots, X_n) : L)|$$
$$\leq |\mathcal{S}_n| = \dim[\mathcal{R}at(K; X_1, \ldots, X_n) : S] \leq \dim[\mathcal{R}at(K; X_1, \ldots, X_n) : L]$$

enforcing equalities and hence $L = S$. $\qquad\qquad\qquad\qquad\qquad\square$

Corollary 48. *For each symmetric rational function $R \in \mathcal{R}at_{sym}(K; X_1, \ldots, X_n) = S$ there exists $h \in \mathcal{R}at(K; S_1, \ldots, S_n)$ such that $R = h(\mathfrak{s}_1, \ldots, \mathfrak{s}_n)$ In short*

$$\mathcal{R}at_{sym}(K; X_1, \ldots, X_n) = \mathcal{R}at(K; \mathfrak{s}_1, \ldots, \mathfrak{s}_n)$$

Especially, for each symmetric polynomial in $P \in K_{sym}[X_1, \ldots, X_n]$ exists $h \in K[S_1, \ldots, S_n]$. such that $P = h(\mathfrak{s}_1, \ldots, \mathfrak{s}_n)$.

In this manner, we have obtained a second proof for the Main Theorem 26 about symmetric polynomials.

The original goal to prove theorem 41 has been lost of sight. To go on, we need to relate the polynomial f with general coefficients to the polynomial g with general zeros. $\qquad\square$

Finish the proof of Theorem 41. The evaluation homomorphisms which map by $\phi(S_i) = \mathfrak{s}_i$ and by $\Psi(X_i) = x_i$ are related in the diagram

$$E = \mathcal{R}at(K; S_1, \ldots, S_n)(x_1, \ldots, x_n) \xleftarrow{\;\Psi\;} \mathcal{R}at(K; X_1, \ldots, X_n)$$

$$\uparrow \supset \qquad\qquad\qquad\qquad\qquad\qquad \supset \uparrow$$

$$\mathcal{R}at(K; S_1, \ldots, S_n) \xrightarrow{\;\phi\;} \mathcal{R}at(K; \mathfrak{s}_1, \ldots, \mathfrak{s}_n)$$

I claim that $\phi(f) = g$ and $\Psi(g) = f$. The reader should check.

To confirm $\Psi \circ \phi = $ idendity, it is enough to check $\Psi \circ \phi(S_r) = S_r$ for $1 \leq r \leq n$.

$$\Psi \circ \phi(S_r) = \Psi(\mathfrak{s}_r) = \Psi\left(\sum_{1 \leq i_1 < i_2 < \cdots < i_r \leq n} X_{i_1} \cdots X_{i_r}\right)$$

$$= \sum_{1 \leq i_1 < i_2 < \cdots < i_r \leq n} x_{i_1} \cdots x_{i_r} = S_r$$

In last step we have used Viëta's formulas (VIII.7.3).

By the Main Theorem about symmetric polynomials 26, generalized from the ring case to quotient fields, one may check that ϕ is surjective. We have reproved in Corollary 48 above that

$$\Im\phi = \mathcal{R}at(K; \mathfrak{s}_1, \ldots, \mathfrak{s}_n) = \mathcal{R}at_{sym}(K; X_1, \ldots, X_n)$$

From the above result $\Psi \circ \phi = $ idendity we see that ϕ is injective, and hence bijective. Moreover the above result $\Psi \circ \phi = $ idendity and ϕ bijective together imply that Ψ is injective, and finally Ψ is confirmed to be a prolongation of ϕ^{-1}.

Problem 190. *To prove that*

$$(\text{VIII.7.10}) \qquad \dim[\mathcal{R}at(K; S_1, \ldots, S_n)(x_1, \ldots, x_n) : \mathcal{R}at(K; S_1, \ldots, S_n)] \leq n!$$

we define the intermediate fields

$$E_i = \mathcal{R}at(K; S_1, \ldots, S_n)(x_1, \ldots, x_i) \quad \text{for } 0 \leq i \leq n$$

Solution. The extension E_1/E_0 adjoins the zero x_1 of polynomial f, hence one gets $\dim[E_1 : E_0] \leq \deg f = n$. The next extension E_2/E_1 adjoins the zero x_2 of divided polynomial $f/(X - x_1)$, hence $\dim[E_2 : E_1] \leq n - 1$. In this way we get $\dim[E_i : E_{i-1}] \leq n - i$ for $0 \leq i < n$. From the tower theorem one concludes the estimate (VIII.7.10). $\qquad\square$

Lemma 103. *I claim additionally that Ψ is surjective.*

Proof. From VIII.7.7 one concludes

$$\dim[\mathcal{R}at(K; X_1, \ldots, X_n) : \mathcal{R}at(K; \mathfrak{s}_1, \ldots, \mathfrak{s}_n)] = n!$$
$$\dim[\Psi\mathcal{R}at(K; X_1, \ldots, X_n) : \phi^{-1}\mathcal{R}at(K; \mathfrak{s}_1, \ldots, \mathfrak{s}_n)] = n!$$
$$\dim[\Psi\mathcal{R}at(K; X_1, \ldots, X_n) : \mathcal{R}at(K; S_1, \ldots, S_n)] = n!$$

Since $\Psi\mathcal{R}at(K; X_1, \ldots, X_n) \subseteq \mathcal{R}at(K; S_1, \ldots, S_n)(x_1, \ldots, x_n)$
from equation (VIII.7.10) we get the sharp (squeezing) estimate

$$n! = \dim[\Psi\mathcal{R}at(K; X_1, \ldots, X_n) : \mathcal{R}at(K; S_1, \ldots, S_n)]$$
$$\leq \dim[\mathcal{R}at(K; S_1, \ldots, S_n)(x_1, \ldots, x_n) : \mathcal{R}at(K; S_1, \ldots, S_n)] \leq n! \quad \text{hence}$$
$$\Psi\mathcal{R}at(K; X_1, \ldots, X_n) = \mathcal{R}at(K; S_1, \ldots, S_n)(x_1, \ldots, x_n)$$

and finally Ψ is surjective. $\qquad\square$

In the end one gets from (VIII.7.7) the corresponding Galois group for polynomial f:

$$\mathrm{Gal}(f : \mathcal{R}at(K, S_1, \ldots, S_n))$$
$$= \mathrm{Aut}(\mathcal{R}at(K; S_1, \ldots, S_n)(x_1, \ldots, x_n) : \mathcal{R}at(K; S_1, \ldots, S_n))$$
$$= \mathrm{Aut}(\Psi\mathcal{R}at(K; X_1, \ldots, X_n) : \phi^{-1}\mathcal{R}at(K; \mathfrak{s}_1, \ldots, \mathfrak{s}_n))$$
$$= \mathrm{Aut}(\mathcal{R}at(K; X_1, \ldots, X_n) : \mathcal{R}at(K; \mathfrak{s}_1, \ldots, \mathfrak{s}_n))$$
$$= \mathrm{Gal}(g : \mathcal{R}at(K; \mathfrak{s}_1, \ldots, \mathfrak{s}_n)) \simeq \mathcal{S}_n$$

$$\square$$

Corollary 49. *The function $\mathfrak{s}_r$ for $1 \leq r \leq n$ are algebraically independent over field K.*

Proof. Suppose the assertion to be false. There would exist a polynomial $P \in K[S_1, \ldots, S_n]$ such that $P \neq 0$ and $P(\mathfrak{s}_1, \ldots, \mathfrak{s}_n) = 0$, in other words an algebraic relation between $\mathfrak{s}_1, \ldots, \mathfrak{s}_n$.

By the evaluation ϕ one gets

$$\phi P = P(\mathfrak{s}_1, \ldots, \mathfrak{s}_n) = 0$$

Since $\phi 0 = 0$ and ϕ is injective, we conclude that $P = 0$, contrary to the assumption. Hence there does not exist any algebraic relation between the elementary symmetric functions $\mathfrak{s}_1, \ldots, \mathfrak{s}_n$. $\qquad\square$

Corollary 50. *For degrees $n \geq 5$ and a field with characteristic zero, the general polynomial equation has no solutions with radical. Neither holds general solvability under the assumption that the discriminant is a perfect square.*

Indication of reason. For degrees $n \geq 5$, neither the symmetric group nor the alternating group are solvable. Hence the result follows from Abel's Main Theorem 39. $\qquad\square$

VIII.8   The Fundamental Theorem of Algebra

Main Theorem 3. *Every nonconstant polynomial with complex coefficients has at least one complex zero.*

Corollary 51. *Every nonconstant polynomial $P(z) \in \mathbb{C}[z]$ with complex coefficients of degree n has a factorization*

$$(\text{VIII.8.1}) \qquad\qquad P(z) = c \prod_{1 \leq i \leq n} (z - z_i)$$

with complex c and z_i. Hence the polynomial P exactly n zeros, counting their multiplicities.

Proof of the corollary. Let $P(z)$ be a polynomial with complex coefficients of degree $n \geq 1$. The assertion is proved by an induction on the degree n. The assertion is true for $n = 1$. As an induction assumption, suppose the corollary holds for all polynomials of degree less than n. By the main theorem 3 the polynomial $P(z) \in \mathbb{C}[z]$ has a complex zero z_1. We divide $P(z)$ by the polynomial $z - z_1$, with remainder. Since $P(z_1) = 0$, the remainder term turns out to be zero. Hence one obtains the factorization

$$P(z) = Q(z)(z - z_1)$$

where $Q(z)$ is a complex polynomial of degree $n - 1$. By the induction assumption the polynomial Q has a factorization

$$(\text{VIII.8.2}) \qquad\qquad Q(z) = c \prod_{2 \leq i \leq n} (z - z_i)$$

with complex c and z_i. Hence the assertion (VIII.8.1) follows via $P(z) = Q(z)(z - z_1)$. $\square$

Gauss has provided several proofs of the main theorem of algebra. Today, the most popular proofs are the two ones which are based on complex analysis. A very short, almost cryptic proof exploits Liouville's theorem that every bounded entire complex function is constant. A second, and I find more appealing proof uses the logarithmic residue theorem.

There exist further proofs, that do not use complex analysis, but instead rely on Galois theory. Below is given such a proof, going back to Lagrange. One uses four basic lemmas, the first one of which uses some fundamental real analysis.

Lemma 104. *Every polynomial with real coefficients of odd degree has at least one real zero.*

Proof. We may assume the polynomial to be monic and put

$$P(x) = \sum_{0 \le k \le n-1} a_k x^k + x^n$$

For real $|x| \ge 1 + \sum_{0 \le k \le n-1} |a_k| =: M$ holds

$$|P(x)| \ge |x|^n - \sum_{0 \le k \le n-1} |a_k||x|^k \ge |x|^n - \sum_{0 \le k \le n-1} |a_k||x|^{n-1}$$

$$= \left(|x| - \sum_{0 \le k \le n-1} |a_k| \right) |x|^{n-1} \ge 1$$

hence the polynomial P has no zeros with $|x| \ge M$. For $x \ge M$ we even get

$$P(x) \ge |x|^n - \sum_{0 \le k \le n-1} |a_k||x|^k \ge 1$$

For $x \le -M$ we get

$$P(x) \le -|x|^n + \sum_{0 \le k \le n-1} |a_k||x|^k \le -1$$

since for odd n holds $x^n = -|x|^n$. Because of $P(-M) < 0$ and $P(M) > 0$ and $P(x)$ being continuous, the intermediate value theorem implies that there exists at least one real zero of P with $|x_1| < M$. $\square$

Lemma 105. *Every quadratic polynomial with complex coefficients has at least one complex zero.*

Proof. Let $P(z) = az^2 + bz + c$ with complex a, b, c and $a \neq 0$. The zeros of P are given by the midnight formula. The complex square root $\sqrt{b^2 - 4ac}$ exists since it can be expressed by real square root as shown by the formula VIII.2.3 from the section on the square root of complex numbers. $\quad\square$

Lemma 106. *Any polynomial $R(z) \in \mathbb{C}[z]$, which takes for all real z only real values, actually has real coefficients.*

Proof. It is important that this remark does not use the main theorem of algebra. Hence I give the proof by induction on the degree $n = \deg R$. Assume that $R(z) \in \mathbb{C}[z]$ and $R(z) \in \mathbb{R}$ for all $z \in \mathbb{R}$.

Induction start for $n = 0$: The polynomial is a constant $R = a$ and the assumption of the lemma implies $a \in \mathbb{R}$. Hence $R \in \mathbb{R}[z]$.

Induction step for $n - 1 \to n$: Given is the polynomial

$$R(z) = \sum_{0 \leq k \leq n} a_k z^k$$

with $a_k \neq 0$ and degree $\deg R = n$. The derivative

$$R'(z) = \sum_{1 \leq k \leq n} k a_k z^{k-1}$$

is a polynomial with degree $\deg R' < n$. Because of the definition of the derivative the assumption of the lemma implies $R'(z) \in \mathbb{C}[z]$ and $R'(z) \in \mathbb{R}$ for all $z \in \mathbb{R}$. Hence the induction assumption implies that all coefficients of R' are real. For all $k \geq 1$ we get from $k a_k \in \mathbb{R}$ that $a_k \in \mathbb{R}$ holds.

Too, the assumption of the lemma implies $R(0) = a_0 \in \mathbb{R}$.

$\quad\square$

Lemma 107. *Assume that we know that every nonconstant polynomial $Q(z) \in \mathbb{R}[z]$ with real coefficients has at least one real or complex zero. Then every nonconstant polynomial $P(z) \in \mathbb{C}[z]$ with complex coefficients has at least one complex zero.*

Proof. Take any nonconstant polynomial $P(z) \in \mathbb{C}[z]$ with complex coefficients. Let

(VIII.8.3) $$Q(z) = P(z)P(\overline{z})$$

where $\overline{z}$ is the conjugate complex of z. From the rules to calculate with the conjugate complex, one checks that $P(\overline{z}) = \overline{P(z)}$ and hence $Q(z)$ is real for all $z \in \mathbb{C}$. Note that $Q(z) \notin \mathbb{C}[z]$, this is not a polynomial with complex coefficients. But the restriction $R(z) := Q(z) \upharpoonright \mathbb{R}$ defines a polynomial. At first sight, we see only that its coefficients are complex and we get only $R(z) \in \mathbb{C}[z]$. But the above lemma 106 implies that even $R(z) \in \mathbb{R}[z]$ holds.

Moreover we may check that $\deg R = 2 \deg P$. Since P is assumed to be nonconstant, we get $\deg R = 2 \deg P \geq 2$ and the polynomial R is nonconstant, too. Now the assumption of the lemma yields that R has at least one complex zero z_1. The definition (VIII.8.3) implies that either $P(z_1) = 0$ or $\overline{P(z_1)} = P(\overline{z_1}) = 0$. In both cases we obtain a complex zero of the polynomial P, as has been claimed. $\square$

Lagrange's proof of the main theorem of algebra. Given is a nonconstant polynomial $P(z) \in \mathbb{R}[z]$ of degree $n = \deg P \geq 1$. Note that because of lemma 107, we may restrict to real polynomials. We may assume the polynomial to be irreducible. Since the real field has characteristic zero, this implies that the zeros of P are simple. We may factor the degree as

$$n = 2^r m$$

with $r \geq 0$ and $m \geq 1$ odd. The proof is done by induction on r.

Induction start for $r = 0$:   In this case the degree n is odd. The claim has been proved by lemma 104.

Induction step up to r with $r \geq 1$:   We use the existence of roots in the splitting extension. At first the real field $\mathbb{R}$ is extended to the complex field $\mathbb{C}$, by splitting the polynomial $1 + x^2$. In other words, we define here the complex field as the quotient of the real field by the principal ideal $(1 + x^2)$ in $\mathbb{R}[x]$. Since the polynomial $1 + x^2$ is irreducible over the reals, we have a maximal ideal. Hence the quotient ring $\mathbb{R}/(1 + x^2)$ is a field.

For the convenience of the reader, still another reminder. The elements of this quotient ring are $a + bX$ with $a, b \in \mathbb{R}$ and the indeterminant X now satisfying $X^2 = -1$. So the indeterminant takes the role of the imaginary unit i.

The main next step is to split the polynomial $P(z)$. By abstract algebra we have already constructed a splitting field

$$K := \mathbb{C}[z_1, z_2, \ldots, z_n] \supseteq \mathbb{C}$$

416

This is a field extension of $\mathbb{C}$ the dimension of which is at most the factorial $n!$. Our goal is to prove that actually holds $K = \mathbb{C}$,—but we need to start with the abstract extension K. At least *as elements of K* we have the simple zeros $z_1, z_2, \ldots, z_n$.

We use the parameter $\lambda \in \mathbb{R}$. Lagrange now defines

$$(\text{VIII.8.4}) \qquad y_{ij} := x_i + x_j + \lambda x_i x_j \quad \text{for } 1 \leq i < j \leq n$$

$$(\text{VIII.8.5}) \qquad L_{ij}(y) := \prod_{1 \leq i < j \leq n} (y - y_{ij}) \in K[y]$$

The degree of the latter polynomial is

$$(\text{VIII.8.6}) \qquad \deg L = \frac{n(n-1)}{2} = 2^{r-1} m(n-1)$$

Since $r \geq 1$ the numbers $n - 1$ and m are odd and hence $m(n-1)$ is odd. So we may use the induction assumption for the polynomial L. But wait, for this stroke of genius to really work, we need at first to check that indeed holds $L \in \mathbb{C}[y]$.

I claim that the coefficients of the polynomial L are invariants under all permutation of the simple zeros $z_1, z_2, \ldots, z_n$. It is left to reader to check this claim. Any automorphism $\sigma \in \mathrm{Aut}[K : \mathbb{C}]$ and even of the field extension $K/\mathbb{R}$ induces such a permutation of the simple zeros $z_1, z_2, \ldots, z_n$. Hence the coefficients of the polynomial L are in the fixed field

$$\mathrm{Fix}(K : \mathrm{Aut}[K : \mathbb{C}] = \mathbb{C}$$

Here only the inclusion " $\supseteq$ " is elementary. The really needed inclusion " $\subseteq$ " holds only for the splitting extension. One needs theorem 32 telling that two defining assumptions for a Galois extension are equivalent.

We may now apply the induction assumption for the polynomial L. There exists a *complex* zero $y_{ij} \in \mathbb{C}$. This statement holds for arbitrary parameter $\lambda \in \mathbb{R}$. Since there exist infinitely many real numbers, but only finitely many pairs i, j we get even two reals numbers $\lambda \neq \mu$ such that for the *same pair i, j* holds both

$$x_i + x_j + \lambda x_i x_j \in \mathbb{C} \quad \text{and} \quad x_i + x_j + \mu x_i x_j \in \mathbb{C}$$

We conclude that both $b := -(x_i + x_j) \in \mathbb{C}$ and $c := x_i x_j \in \mathbb{C}$. Put $a := 1$. The polynomial $az^2 + bz + z \in \mathbb{C}[z]$ has the complex zeros x_i and x_j. Hence $x_i \in \mathbb{C}$ and $x_j \in \mathbb{C}$. We have shown that two of the abstract roots $x_i \in K$ of the polynomial P are indeed complex numbers.

$\square$

Part IX

The Lunes of Hippocrates and Transcendental Numbers

IX.1   The Lunes of Hippocrates

IX.1.1   The Historic Lunes

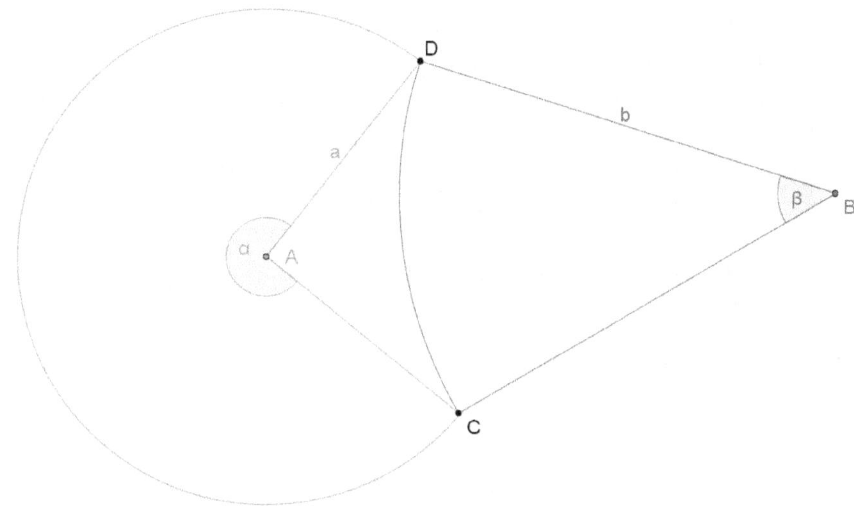

Figure 7:   Just a lune.

Definition 59 (Lune). A *lune* consists of two circular arcs having a common chord and lying on the same side of this chord. The interior of the lune is the the crescent shaped area formed by the difference of the interiors of the corresponding circles. It is bounded by the lune's two arcs.

Hippocrates of Chios (ca. 430 B.C.) posed the problem:

(**Hippocrates Problem**). *Find the lunes which are constructible and squarable with straightedge and compass.*

He gave three examples of constructible lunes. They are obtained beginning with the following two assumptions:

(a) The two circular sectors corresponding to the lune's arcs have the same area.

(b) The central angles of the two circular arcs are commensurable.

Hippocrates of Chios is credited with discovering three such lunes; two more were discovered in the 18th century. In the 20th century Tschebatorev and Dorodnov (1947) proved these fives are the only ones.

Lemma 108. *Assumption* (a) *implies the lune is squarable.*

Proof. Let CD be the common chord of the two arcs of the lune. Let $\mathcal{S}_A$ be the circular sector corresponding to the lune's longer arc and have center A, and $\mathcal{S}_B$ be the circular sector corresponding to the lune's shorter arc and have center B. Both sectors are delimited by the radiuses from their respective center to the endpoints of common chord CD. The sector $\mathcal{S}_B$ lies on the other side of the lune's shorter arc and does not intersect its interior area. We add to the lune the latter sector and then subtract the sector $\mathcal{S}_A$, and obtain the kite $\square ACBD$. Since by assumption (a) the two sectors have the same area, the lune has the same area as the kite and hence is squarable. $\qquad\square$

Lemma 109. *Assumption* (a) *and* (b) *together imply*

$$(\text{IX.1.1}) \qquad\qquad \frac{\alpha}{\beta} = \frac{n}{m} = \frac{b^2}{a^2}$$

with integers $n, m \geq 1$. *Indeed* $n < m$ *becomes possible only if one or both or angles* α *and* β *are more than* $360°$.

Proof. We now use assumption (b). Let the angle ϵ be the (greatest) common measure of the central angles α and β of the lune's longer and shorter arcs around centers A and B. Hence $\alpha = n\epsilon$ and $\beta = m\epsilon$ with integers $n > m \geq 1$. Subdivision of the sector $\mathcal{S}_A$ yields n congruent sectors with center A, and similarly, subdivision of the sector $\mathcal{S}_B$ yields m congruent sectors with center B. All sectors we have obtained have the same central angle ϵ. Hence they are similar.

By assumption (a), the two circular sectors $\mathcal{S}_A$ and $\mathcal{S}_A$ have equal area, hence the n small sectors of radius a around center A with central angle ϵ have together the same area as m similar sectors with central angle ϵ and radius b around center B. Since the areas of similar figures are proportional to the square of their linear dimension, we conclude $na^2 = mb^2$, and finally get equation (IX.1.1). $\qquad\square$

Definition 60 (Circular segment). A circular *segment* is bounded by an arc and a chord. A segment of central angle ϵ is obtained from the circular sector with the same angle and arc by subtraction or addition of the triangle with vertices at the endpoints of its circular arc and the center of the circle. For a short arc $0 < \epsilon < 180°$, the triangle is subtracted, for a long arc $180° < \epsilon < 360°$, the triangle is added to the sector.

Lemma 110. *Two circular segments with the congruent central angles and which are both bounded by the long, or both by the short arc are similar. There areas have the same ratio as the squares of their radius.*

Lemma 111. *The areas of similar figures are proportional to the square of their linear dimension.*

Proposition 90. *Assumption* (a) *and* (b) *together imply the lune has the same area as a polygon with vertices on its arcs.*

Proof. After subdivision of the arcs α and β, we obtain not only similar circular sectors, but also *similar segments* with the central angle ϵ. The endpoints of these segments divide the arc α of the lune into n arcs with central angle ϵ, and the inner arc β of the lune m arcs with central angle ϵ.

The n circular segments around center A have together the same area as the m similar circular segments around center B. This is clear from equation (IX.1.1) and Lemma 110.

We add to the lune the m segments on the other side of the lune's inner arc β, and subtract the n segments with vertices on the lune's external arc α. We obtain a polygon with vertices on the arcs of the lune which has the same area as the lune. $\square$

IX.1.2   The Lune Equation

Proposition 91. *Any lune satisfying assumption* (a) *and* (b) *satisfies the lune equation*

$$\text{(IX.1.2)} \qquad\qquad \frac{\sin(n\epsilon/2)}{\sin(m\epsilon/2)} = \pm\sqrt{\frac{n}{m}}$$

with integers $n, m \geq 1$.

Proof. The common chord of the two arcs of the lune has the length

$$|CD| = 2a\sin(\alpha/2) = 2b\sin(\beta/2)$$

Since $\alpha = n\epsilon$ and $\beta = m\epsilon$ and $na^2 = mb^2$, the equation is easy to confirm. $\square$

IX.1.3   The Constructible Lunes

For which values of the integers n and m can we <u>construct</u> squarable lunes? Let $\alpha = n\epsilon$ and $\beta = m\epsilon$ be the angles of the lune arcs. Hippocrates has found a squarable

422

and constructible lune for the following three cases:

$$\text{(a)} \quad n = 2, m = 1$$
$$\text{(b)} \quad n = 3, m = 1$$
$$\text{(c)} \quad n = 3, m = 2$$

The case (a) is easiest to guess: one puts $\alpha = 180°, \beta = 90°$. The longer arc of the lune is the circum-circle of an isosceles right triangle $\triangle CED$ and hence its center A is the midpoint of the hypothenuse CD. The shorter arc has its center B in the forth vertex of square $\square CEDB$. Hence we know that $b/a = \sqrt{2}$ as required.

Euler has found around 1771 two further squarable and constructible lunes for the following three cases:

$$\text{(e)} \quad n = 5, m = 1$$
$$\text{(f)} \quad n = 5, m = 3$$

A thorough account of these cases in given in Rothe [21].

IX.1.4   Lunes with Solvable Equations

Viëta has considered a lune leading to a third order equation with

$$\text{(g)} \quad n = 4, m = 1$$

Hence $x = 2\cos(\epsilon/2)$ satisfies the cubic equation $x^3 - 2x - 2 = 0$. This cubic has one real root, which one may obtain by Cardano's formula in its historic shape.

$$2\cos(\epsilon/2) = \sqrt[3]{-\frac{c}{2} + \sqrt{\frac{c^2}{4} + \frac{b^3}{27}}} + \sqrt[3]{-\frac{c}{2} - \sqrt{\frac{c^2}{4} + \frac{b^3}{27}}} = \sqrt[3]{1 + \sqrt{\frac{19}{27}}} + \sqrt[3]{1 - \sqrt{\frac{19}{27}}}$$

A construction with marked ruler and compass is possible by means of Nicomedes' method to construct third real roots. This method is explained in Hartshorne [10] p. 263, proposition 30.3, or Rothe [21] p.210, Construction VII.35.5.

In this section is mainly considered the case

$$\text{(h)} \quad n = 9, m = 1$$

which the Russian mathematician Postnikov has shown to be remarkable, too. See the article [16] which is translated from Postnikov's 1963 Russian book on Galois theory. In this case the lune equation is

$$\text{(IX.1.3)} \qquad \frac{\sin(9\epsilon/2)}{\sin(\epsilon/2)} = 3\sigma$$

with $\sigma = \pm 1$.

Proposition 92 (Postnikov). *For $\sigma = 1$ the equation (IX.1.3) has only one real solution. One gets a lune which can be obtained exactly from an equation of forth order. For $\sigma = -1$ the equation (IX.1.3) has four complex solutions. One gets no lunes, in spite of the fact that one has obtained an equation of forth order which is solved exactly for using only square roots.*

Remark. The above lune becomes constructible if one allows the trisection of a constructible angle. One uses the method of Nicomedes for trisection of an angle with marked straightedge and compass. See for example Hartshorne [10] p. 262, proposition 30.2 or Rothe [21] p.203, VII.35.2 Trisection by Nicomedes.

Numerics and the figure. We use the Tschebyscheff polynomial of second kind

$$U_n(\cos t) = \frac{\sin(n+1)t}{\sin t}$$

for integers $n = 0, 1, 2, \ldots$. Here the parameters are $n = 8$ and $t = \epsilon/2$ and one gets the equation

$$U_8(\cos \epsilon/2) = 3\sigma$$

Use the explicit formulas [29]

$$U_8(z) = 1 - 40z^2 + 240z^4 - 448z^6 + 256z^8$$
$$h(x) := 1 - 10x + 15x^2 - 7x^3 + x^4$$

Since U_8 is an even polynomial one can reduce to an equation of degree 4. One puts $z = 4x^2$. From $z = \cos \epsilon/2$ follows $x = 2\cos \epsilon + 2$, so this is a useful variable for numeric computation, too. For the case $\sigma = +1$, one needs the zeros of equation $h(x) - 3 = 0$. The list of coefficients $[-4, -10, 15, -7, 1]$ has three sign changes. The rule of Descartes tells that one has either three or one positive zeros, counting their multiplicities. Graphing on the interval $x \in [0, 5]$ confirms that there is only one positive zero. Since $x = 2\cos \epsilon + 2 \geq 0$ holds for all real angles ϵ, the negative zero of equation $h(x) - 3 = 0$ does not yield any lune. The positive root yields the numerical value $x \approx 3.74719$ and hence $\epsilon = \arccos(-1 + x/2) \approx 29.1209°$.

For the geometric construction one needs the triangle $\triangle ABD$ with a (long) base side AB and the interior angles $\alpha = \epsilon/2$ and $\beta = 180° - 9\epsilon/2$, and the reflection of point D across the line AB. This construction and the check of the lune are depicted in the figure on page 424. Graphing on the interval $x \in [0, 5]$ confirms that the polynomial $h(x) + 3$ has no real zeros, so one does not get any lune in the case $\sigma = -1$. $\qquad\square$

---

[29]The monomials are given in increasing order, following the system of mathematica.

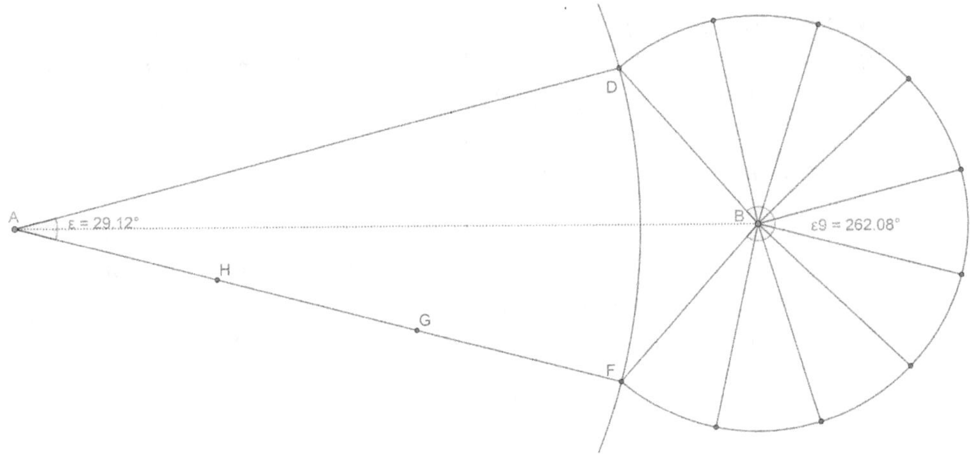

Figure 8:   Postnikov's $9:1$ lune

IX.1.5   Exact Solution for Postnikov's Equation

The equations $h(x) - 3 = 0$ and $h(x) + 3 = 0$ may be solved by radicals. Here is explained the case $h(x) + 3 = 0$, which turns out to be much simpler. From the above, we only know that all four solutions will be complex. The preliminary step is to get rid of the x^3 term in the polynomial. One uses the variable $y = (x + 7)/4$ and gets in the new variable the depressed quartic

$$256 \left(h((x+7)/4) - 3\sigma \right) = 333 - 768\sigma - 24x - 54x^2 + x^4 =: k(x)$$

Go on with Descartes' method and use a factoring

$$(IX.1.4) \qquad r + qx + px^2 + x^4 = (b + ax + x^2)(c - ax + x^2)$$

in the example with

$$333 - 768\sigma - 24x - 54x^2 + x^4 = r + qx + px^2$$

The new coefficients have to satisfy the equations

$$(IX.1.5) \qquad r = b \cdot c, \; q = -a(c - b), \; p = -a^2 + b + c$$

Elimination of b and c yields

$$R(a^2) = 0 \quad \text{with the Descartes-resolvent}$$
$$(IX.1.6) \qquad R(v) = -q^2 + (p^2 - 4r)v + 2pv^2 + v^3$$

One gets now a cubic equation for a^2, which is at least some progress.

To give an easier account, I now take up Postnikov's example with $\sigma = -1$. Thus one puts $[r, q, p] := [1101, -24, -54]$ and the resolvent becomes even reducible:

$$R[v] = -576 - 1488v - 108v^2 + v^3 = (12 + v)(-48 - 120v + v^2)$$

Go on with Descartes' method and the choice $a := 2\sqrt{-3}$. One may alternatively directly check with mathematica that the option

Extension -¿ Sqrt[-3]

makes the depressed quartic $1101 - 24x - 54x^2 + x^4$, and hence the original quartic $h(v) + 3$ reducible:

```
 In[193]:= Factor[h[v] + 3, Extension -> Sqrt[-3]]


Out[193]= 1/4   (-2 I-2 Sqrt[3] + (7 I+Sqrt[3])v - 2 I v^2)
                (2 I-2 Sqrt[3] + (-7 I+Sqrt[3])v + 2 I v^2)
```

So Descartes was able to explain this little wonder. Back to the general procedure, the next step is to calculate b and c from the conditions (IX.1.5). The first and third equation together give a quadratic equation with the solutions b and c. The second one determines in which order b and c are to be assigned to the solutions of the quadratic equation. One obtains from taking a square root and solving a quadratic equation two solutions of conditions (IX.1.5): the triplets (a, b, c) and $(-a, c, b)$. No matter which sign is chosen for a, the correct solution is given by

$$\{b, c\} = \{\tfrac{1}{2}(a^2 + p - \sqrt{a^4 + 2a^2p + p^2 - 4r}), \tfrac{1}{2}(a^2 + p + \sqrt{a^4 + 2a^2p + p^2 - 4r}]\}$$
$$\text{with } b \text{ and } c \text{ in the order such that } q = a(c - b)$$

The four roots of the depressed quartic are finally obtained by finding the roots for the two quadratic factors in Ansatz (IX.1.4). Thus one obtains the list

$$(IX.1.7) \quad [x_i] = \left[\frac{-a - \sqrt{a^2 - 4b}}{2}, \frac{-a + \sqrt{a^2 - 4b}}{2}, \frac{a - \sqrt{a^2 - 4c}}{2}, \frac{a + \sqrt{a^2 - 4c}}{2}\right]$$

The substitution $(a, b, c) \to (-a, c, b)$ exchanges the first and third, as well as the second and forth element of this list.

We now go back to the example and do this calculation for the case $\sigma = +1$. We have already chosen $a = 2\sqrt{-3}$ and have $[r, q, p] := [1101, -24, -54]$. Calculation, best with help of mathematica yield

$$\{c, b\} = -33 \pm 2\sqrt{-3}$$

426

We need now really to check the condition for q. One gets $a(c-b) = 2\sqrt{-3}(4\sqrt{-3}) = -24 = q$ as required. Hence c get the plus sign in front of the root. One plugs into equation (IX.1.7) and gets

$$x_k = -\,i\sqrt{3} - \sqrt{30 + 2i\sqrt{3}},\ -i\sqrt{3} + \sqrt{30 + 2i\sqrt{3}},$$

$$i\sqrt{3} - \sqrt{30 - 2i\sqrt{3}},\ i\sqrt{3} + \sqrt{30 - 2i\sqrt{3}}$$

$$y_k = \frac{x_k + 7}{4} = \frac{1}{4}\left(7 - i\sqrt{3} - \sqrt{30 + 2i\sqrt{3}}\right),\ \frac{1}{4}\left(7 - i\sqrt{3} + \sqrt{30 + 2i\sqrt{3}}\right),$$

$$\frac{1}{4}\left(7 + i\sqrt{3} - \sqrt{30 - 2i\sqrt{3}}\right),\ \frac{1}{4}\left(7 + i\sqrt{3} + \sqrt{30 - 2i\sqrt{3}}\right)$$

The latter y_k are the four solutions of $h(y_k) = -3$. This can be checked numerically and symbolically.

IX.2   Transcendental Numbers and the Lunes

IX.2.1   About Transcendental Numbers

A number α is called *algebraic* if there exists a nonzero integer polynomial $p \in \mathbf{Z}[x]$ such that $p(\alpha) = 0$. In that case α is called a *root of the polynomial p.*

The set of all real or complex algebraic numbers is denoted by $\mathbf{A}$. As stated in Theorem 8, the set of algebraic numbers is both a <u>countable</u> and a <u>field</u>. Obviously, all rational numbers are algebraic, but the converse is not true. We know that $\sqrt{2}$ and $i = \sqrt{-1}$ are two important examples of irrational but algebraic numbers.

The real or complex numbers, which are not algebraic are called *transcendental.*

Definition 61. The numbers $\alpha_1, \alpha_2, \ldots \alpha_n$ are called $\mathbf{Z}$-linearly independent if the dependence relation

$$k_1\alpha_1 + k_2\alpha_2 + \cdots + k_n\alpha_n = 0$$

with integers $k_1, k_2, \ldots k_n$ holds only for all $k_1 = k_2 = \cdots = k_n = 0$.

The numbers $\gamma_1, \gamma_2, \ldots \gamma_n$ are called linearly independent over the algebraic numbers if the dependence relation

$$\beta_1\gamma_1 + \beta_2\gamma_2 + \cdots + \beta_n\gamma_n = 0$$

with algebraic numbers $\beta_1, \beta_2, \ldots \beta_n$ holds only for all $\beta_1 = \beta_2 = \cdots = \beta_n = 0$.

Obviously, independence over the integers is equivalent to linear independence of the rationals $\mathbf{Q}$.

Theorem 42 (Lindemann-Weierstrass Theorem). *Let the algebraic numbers* $\alpha_1, \alpha_2, \ldots \alpha_n$ *be distinct. Then the exponentials* $e^{\alpha_1}, e^{\alpha_2}, \ldots e^{\alpha_n}$ *are linearly independent over the algebraic numbers. Hence*

$$\sum_{i=1}^{n} \beta_i e^{\alpha_i} \neq 0$$

for any algebraic numbers $\beta_1, \beta_2, \ldots \beta_n$, *unless* $\beta_1 = \beta_2 = \cdots = \beta_n = 0$.

The Lindemann-Weierstrass Theorem immediately yields the classical results that $e, \pi, \ln 2$ are transcendental.

- Assume towards a contradiction that Euler's number e is algebraic. We put $\alpha_1 = 0, \alpha_2 = 1, \beta_1 = -e, \beta_2 = 1$ and get the contradiction $-e \cdot e^0 + 1 \cdot e^1 \neq 0$. (Hermite 1873)

- Assume towards a contradiction that π is algebraic. We put $\alpha_1 = 0, \alpha_2 = i\pi, \beta_1 = \beta_2 = 1$ and get the contradiction $1 \cdot e^0 + 1 \cdot e^{i\pi} \neq 0$. (Lindemann 1882)

- Assume towards a contradiction that $\ln 2$ is algebraic. We put $\alpha_1 = 0, \alpha_2 = \ln 2, \beta_1 = -2, \beta_2 = 1$ and get the contradiction $-2 \cdot e^0 + 1 \cdot e^{\ln 2} \neq 0$.

Corollary 52. *Let the <u>distinct</u> algebraic numbers $\alpha_1 \neq 0, \alpha_2 \neq 0, \ldots \alpha_n \neq 0$ be all nonzero and the algebraic numbers $\beta_1 \neq 0, \beta_2 \neq 0, \ldots \beta_n \neq 0$ be nonzero. Then the sum*

$$\sum_{i=1}^{n} \beta_i e^{\alpha_i} \quad \text{is transcendental.}$$

Theorem 43 (Baker 1967). *Let the numbers $e^{\alpha_1}, e^{\alpha_2}, \ldots e^{\alpha_n}$ be algebraic and the numbers $\alpha_1, \alpha_2, \ldots \alpha_n$ be $\mathbf{Z}$-linearly independent. Then*

$$\sum_{i=1}^{n} \beta_i \alpha_i \quad \text{is transcendental}$$

for any algebraic numbers $\beta_1, \beta_2 \ldots \beta_n$, unless $\beta_1 = \beta_2 = \cdots = \beta_n = 0$.

Baker's Theorem immediately yields the classical results that e^π and $2^{\sqrt{2}}$ are transcendental.

- Assume towards a contradiction that e^π is algebraic. We put $\alpha_1 = \pi, \alpha_2 = i\pi, \beta_1 = -i, \beta_2 = 1$ and get the $\beta_1 \alpha_1 + \beta_2 \alpha_2 = (-i) \cdot \pi + 1 \cdot i\pi = 0$ is transcendental.

- Assume towards a contradiction that $2^{\sqrt{2}}$ is algebraic. We put $\alpha_1 = \sqrt{2} \ln 2, \alpha_2 = \ln 2, \beta_1 = \sqrt{2}, \beta_2 = -2$ and get the contradiction that $\beta_1 \alpha_1 + \beta_2 \alpha_2 = 0$ is transcendental.

Theorem 44 (Gelfond-Schneider Theorem). *Let t be irrational and $a \neq 0, 1$ be any two algebraic numbers. Then a^t is transcendental. Indeed <u>all</u> values of the power a^t are transcendental.*

Reason. The numbers $\alpha_1 := \ln a$ and $\alpha_2 = t \ln a$ are by assumption $\mathbf{Z}$-linearly independent. Assume towards a contradiction that a^t is algebraic. We put $\beta_1 := -t, \beta_2 := 1$ and get the contradiction that $\beta_1 \alpha_1 + \beta_2 \alpha_2 = 0$ is transcendental. $\square$

Theorem 45 (M. Waldschmidt). *Given are two complex numbers t and z. Suppose that t is irrational and $z \neq 0$. Then at least one of the numbers t, e^z, e^{tz} is transcendental. The exponentials e^z and e^{tz} are defined by the power series of the exponential function.*

Reason. The numbers $\alpha_1 := z$ and $\alpha_2 = tz$ are by assumption $\mathbf{Z}$-linearly independent. Assume towards a contradiction that all three numbers t, e^z, e^{tz} are algebraic. We put $\beta_1 := -t, \beta_2 := 1$ and get the contradiction that $\beta_1\alpha_1 + \beta_2\alpha_2 = (-t) \cdot z + 1 \cdot tz = 0$ is transcendental. $\qquad\qquad\qquad\qquad\qquad\qquad\qquad\qquad\qquad\qquad\qquad\qquad\square$

Remark. The Theorem 45 of M. Waldschmidt is just a clarified version of the famous Theorem 44 of Gelfond-Schneider.

Indeed, for any two algebraic numbers a and t, where t is irrational and $a \neq 0, 1$, the Theorem of M. Waldschmidt implies that <u>all</u> values of the power a^t are transcendental:

To see this, we put $z := \ln a + 2\pi i k$ with arbitrarily chosen values of $\ln a$ and the integer k. One gets $a = e^z$, and e^{tz} is <u>one of many</u> values of a^t,—it still depends on the integer k since $z \neq 1$. From Waldschmidt's Theorem one concludes that at least one of the three numbers t, e^z, e^{tz} is transcendental. Since the first two numbers t and e^z are algebraic by assumption, we see that the third number $e^{tz} \in a^t$ is transcendental. Hence <u>all</u> values of a^t are transcendental.

Remark. A thorough account of the results just reviewed is contained in Alan Baker's book [3] on *"Transcendental Number Theory"*. The book *"Making Transcendence Transparent"* of Burger and Tubbs [5] is readable and informative, too. Finally one can recommend the internet source

`http://en.wikipedia.org/wiki/Baker%27s_theorem`

Corollary 53 (Alan Baker, (1967b) inhomogeneous version). *Under the assumption that the numbers $\alpha_1, \alpha_2, \ldots \alpha_n$ and $\beta_0, \beta_1, \ldots \beta_n$ are algebraic and all are nonzero, the sum*

$$\beta_0 + \sum_{i=1}^{n} \beta_i \log \alpha_i \neq 0$$

is nonzero. The above source gives the idea and Baker gives an effective lower bound for this sum.

IX.2.2   Which Lunes are Algebraic, Which Constructible?

We look once more at the generic figure IX.1.1. I call a lune *algebraic* for which the radius a and b, the angle functions $\sin\alpha, \cos\alpha, \sin\beta, \cos\beta$ of the central angles, and the area of the lune are algebraic numbers.

Theorem 46. *All algebraic lunes satisfy Hippocrates' two basic assumptions*

(a) *The two circular sectors corresponding to the lune's arcs have the same area.*

(b) *The central angles of the two circular arcs are commensurable.*

Proof. Since the area of the kite $\square ACDB$ is algebraic, and the area of the lune $\mathcal{L}$ is assumed to be algebraic, their difference is algebraic. Since

$$\text{area}(\mathcal{L}) = \text{area}(\mathcal{S}_A) + \text{area}(\square ACDB) - \text{area}(\mathcal{S}_B)$$

we conclude that the difference of the areas of the circular sectors $\mathcal{S}_A$ and $\mathcal{S}_B$:

$$\text{area}(\mathcal{S}_A) - \text{area}(\mathcal{S}_B) = \alpha a^2 - \beta b^2$$

is algebraic, as a consequence of the assumptions of the theorem. The angles $\alpha \neq 0$ and $\beta \neq 0$ at the centers A and B, corresponding to the two arcs of the lune have to be measured in radian measure.

The assumptions imply that $e^{i\alpha}, e^{i\beta}$ and a, b and are all algebraic and nonzero and that the area $\alpha a^2 - \beta b^2$ is algebraic. We use Alan Baker's Corollary 53 with $\alpha_1 := i\alpha, \alpha_2 := i\beta$ and $\beta_0 := -i(a^2\alpha - b^2\beta), \beta_1 := a^2, \beta_2 := -b^2$. If $\beta_0 \neq 0$, we could conclude that the the sum

$$\beta_0 + \beta_1 \log \alpha_1 + \beta_2 \log \alpha_2$$

would be non zero, a contradiction. The only way out is that $\beta_0 = 0$. Hence $a^2\alpha = b^2\beta$ and Hippocrates' assumption (a) holds.

We now use the Theorem 45 of Gelfond Schneider and M. Waldschmidt with $x = \beta/\alpha$ and $z = i\alpha$. Assume towards a contradiction that x is irrational. We would conclude that at least one of the numbers $\beta/\alpha, e^{i\alpha}, e^{i\beta}$ is transcendental. The latter two are algebraic by assumption and

$$\frac{\beta}{\alpha} = \frac{a^2}{b^2}$$

is algebraic, too, by the first part of the proof. Hence β/α is rational, confirming Hippocrates' assumption (b). $\qquad\square$

Main Theorem 4 (N.G. Tschebatorev and A.W. Dorodnov 1947). *There exist only five squarable and constructible lunes, corresponding to the cases in which $n : m$ is $2 : 1$, $3 : 1$, $3 : 2$, $5 : 1$, $5 : 3$.*

Remark. The final result was obtained by A. W. Dorodnow, after many earlier partial results.

- E. Landau showed in 1903 that the $p : 1$ lune is not constructible if p is a prime but not a Fermat prime.

- L. Tschakaloff showed in 1926 that the $17 : 1$ lune is not constructible, neither any $p : m$ lune where p is a prime and $p > m$.

- N. Tschebotarov [22] showed in 1934 that no $n : m$ lune with n, m both odd and $m > 5$ is constructible.

Bibliography

[1] Emil Artin, *Algebra with Galois Theory*, Courant Institute of Mathematical Sciences, AMS, New York University, 2007.

[2] Michael Artin, *Algebra*, second ed., Prentice Hall, 2011.

[3] Alan Baker, *Transcendental Number Theory*, Cambridge University Press, 1990.

[4] Arthur Baragar, *Constructions using a compass and twice-notched straightedge*, The American Mathematical Monthly **109** (2002), 151–164.

[5] Edward B. Burger and Robert Tubbs, *Making Transcendence Transparent*, Springer, 2004.

[6] David M. Burton, *Elementary Number Theory*, McGraw-Hill, 1997.

[7] Th. Clausen, *Vier neue mondförmige Flächen, deren Inhalt quadrirbar ist*, Crelle **21** (1840), 375–376.

[8] Gerd Fischer, *Lehrbuch der Algebra*, forth ed., Springer Spektrum, 2017.

[9] Karl F. Gauss, *Disquisitiones Arithmeticae*, Leipzig, 1801.

[10] Marvin J. Greenberg, *Old and new results in the foundations of elementary plane Euclidean and Non-Euclidean geometries*, The American Mathematical Monthly **117** (2010), 198–219.

[11] Pierre Antoine Grillet, *Abstract Algebra*, second ed., Springer, 2007.

[12] Robin Hartshorne, *Geometry: Euclid and Beyond*, second printing, Springer, 2002.

[13] Johann Gustav Hermes, *Über die Teilung des Kreises in 65537 gleiche Teile*, Nachrichten von der Gesellschaft der Wissenschaften zu Göttingen, Mathematisch-Physikalische Klasse **3** (1894), 170–186.

[14] David Hilbert, *Foundations of Geometry*, 2nd English ed., Open Court, La Salle, 1971.

[15] Ming-Chang Kang, *A note on cyclotomic polynomials*, Rocky Mountain J. of Math. **29** (1999), 893–907.

[16] Oystein Ore, *Number theory and its history*, McGraw-Hill, 1948.

[17] Magnus Georg Paucker, *Geometrische Verzeichnung des regelmäßigen Siebzehn-Ecks und Zweyhundersiebenundfünfzig-Ecks in den Kreis*, Jahresverhandlungen der Kurländischen Gesellschaft für Literatur und Kunst (1822), 160–219.

[18] M. M. Postnikov, *The problem of squarable lunes*, The American Mathematical Monthly **107** (2000), 645–651.

[19] Friedrich Julius Richelot, *De resolutione algebraica aequationis $x^{257} = 1$, sive de divisione circuli per bisectionem anguli septies repetitam in partes 257 inter se aequales commentatio coronata*, Journal für die reine und angewandte Mathematik **9** (1832).

[20] F. Rothe, *Old and New Topics in Geometry, Volume I: Projective, Neutral and Basic Euclidean Geometry*, Liteprime, 2022.

[21] ______ , *Old and New Topics in Geometry, Volume II: Advanced Euclidean and Hyperbolic Geometry*, Liteprime, 2022.

[22] N. Tschebotarow, *Über quadrierbare Kreisbogenzweiecke*, Math. Zeitschrift **39** (1934), 161–175.

[23] B.L. van der Waerden, *Science Awakening I*, Kluwer Academic Publishers, 1954.

[24] Stan Wagon, *Editor's corner: The Euclidean algorithm strikes again*, The American Mathematical Monthly **97** (1990), 125–129.

Index